Fats in Food Technology

Sheffield Food Technology

Series Editors: P.R. Ashurst and B.A. Law

A series which presents the current state of the art of chosen sectors of the food and beverage industry. Written at professional and reference level, it is directed at food scientists and technologists, ingredients suppliers, packaging technologists, quality assurance personnel, analytical chemists and microbiologists. Each volume in the series provides an accessible source of information on the science and technology of a particular area.

Titles in the series:

Chemistry and Technology of Soft Drinks and Fruit Juices
Edited by P.R. Ashurst

Natural Toxicants in Food
Edited by D.H. Watson

Technology of Bottled Water
Edited by D.A.G. Senior and P.R. Ashurst

Environmental Contaminants in Food
Edited by C.F. Moffat and K.J. Whittle

Handbook of Beverage Packaging
Edited by G.A. Giles

Technology of Cheesemaking
Edited by B.A. Law

Mechanisation and Automation in Dairy Technology
Edited by A.Y. Tamime and B.A. Law

Enzymes in Food Technology
Edited by R.J. Whitehurst and B.A. Law

Food Flavour Technology
Edited by A.J. Taylor

Fats in Food Technology
Edited by K.K. Rajah

Fats in Food Technology

Edited by

KANES K. RAJAH
Consultant in Oils, Fats and Dairy Produce
Director, Centre for Entrepreneurship
University of Greenwich
London, UK

Sheffield
Academic Press

CRC Press

1002670825

First published 2002
Copyright © 2002 Sheffield Academic Press

Published by
Sheffield Academic Press Ltd
Mansion House, 19 Kingfield Road
Sheffield S11 9AS, UK

ISBN 1-84127-225-6

Published in the U.S.A. and Canada (only) by
CRC Press LLC
2000 Corporate Blvd., N.W.
Boca Raton, FL 33431, U.S.A.
Orders from the U.S.A. and Canada (only) to CRC Press LLC

U.S.A. and Canada only:
ISBN 0-8493-9784-7

This book contains information obtained from authentic and highly regarded sources. Reprinted
material is quoted with permission, and sources are indicated. Reasonable efforts have been
made to publish reliable data and information, but the author and the publisher cannot assume
responsibility for the validity of all materials or for the consequences of their use.

Trademark Notice: Product or corporate names may be trademarks or registered trademarks,
and are used only for identification and explanation, without intent to infringe.

Printed on acid-free paper in Great Britain by
Antony Rowe Ltd, Chippenham, Wiltshire

British Library Cataloguing-in-Publication Data:
A catalogue record for this book is available from the British Library

Library of Congress Cataloging-in-Publication Data:
A catalog record for this book is available from the Library of Congress

Preface

This book is exciting in a number of ways.

First, it is a book about fats in food technology—their roles, behaviour and the benefits that they impart to consumers. It is about fats that are 'naturally present' in foods (e.g. milk fat in cheese) or fats that have been added to help with physical and chemical properties (e.g. cocoa butter in chocolate). It is a book which has useful information about market issues that have driven change and disciplines that have helped to regulate the trade and use of fats and oils in foods.

My initial challenge was to find authors who could write to such exacting and wide-ranging criteria. In this, I have been privileged to be able to gather together an internationally respected team of authors from the United States, Europe and Japan, to contribute, either independently or in joint initiatives, a total of nine chapters. All have senior-level commercial experience of R&D in oils and fats technology and direct exposure to technical developments in world markets.

Consequently, all established products are reviewed systematically and new ideas are presented, not only from the recent literature but often from the personal R&D experiences of the authors. Where efficiencies in processing or economy in the costs of raw materials can be achieved, these have—either implicitly or explicitly, by the choice of appropriate examples or formulations—been discussed within the relevant chapters.

Authors have attempted to provide relevant market information in respect of regulation, legislation or directives governing the major markets, especially within the United States and Europe. Market trends and changes which facilitate a better understanding of the scope and potential for fat technology are also presented. In an integrated approach, the issues concerning greater consumer awareness of health, diet and lifestyle are interwoven into some of the relevant chapters—such as, typically, lower fat products and high moisture emulsions. The technology of non-aqueous fat systems has been brought up-to-date. Equally, that of water-in-oil and oil-in-water emulsions is discussed far more extensively in this book than elsewhere.

The book begins with a presentation on the physical properties of fats and emulsions. The chapter on bakery fats deals with solid, fluid and powdered fats. The chapter on confectionery products has been widened to cover both chocolate and sugar confectionery fats. New developments in dairy cream technology are explained within the chapter on water-continuous emulsions. Cream liqueur and ice cream production processes are also included. Hydrogenation and fractionation, which are the most widely used techniques of fat modification, are covered as separate subjects in the same chapter. Products from these processes often replace, complement or supplement each other. This is evident from the discussion and also in the examples seen in those chapters dealing with end-use, e.g. bakery, spreads and confectionery. The chapter on spreadable products includes the results of some secondary market research on important developments in butter, margarine and low fat spreads technology and packaging. The significant growth in fat-based sweet and savoury spreads is also acknowledged within this important chapter. Culinary fats appear as a separate chapter because they focus on some of the unique features and benefits of fats, frying oil, ghee and vanaspati and speciality fats in the kitchen: flavour, eating quality, texture, aroma and benefits to health. The treatment of emulsifiers is comprehensive and it guides the reader through the technology of current products used in recipes and formulations to ensure that shelf-life, emulsion stability and anti-spattering properties, for example, are optimised. The chapter on hard and semi-hard cheeses is unique. This is the first comprehensive discussion of the technology of the fats used or found in fermented products.

The book should be helpful to anyone who has an interest in the technology of fat-containing products. Food technologists in either the dairy industry or the edible oils industry (and indeed the food industry generally) should find that this volume provides important ideas for product and process development. For scientists in academic research establishments, the book offers important insights into some of the most significant scientific developments that have been commercialised. The book will also serve as a useful source of reference for traders and marketing personnel in the oils, fats and butter industries.

This has been a major challenge and a creative experience for all of us. It has been possible only because those who participated gave of their best, unflinchingly. My warmest thanks, therefore, to my fellow authors for their hard work and for generously sharing their knowledge, insight and experience in producing such excellent chapters. Any perceived shortcomings in this book are entirely my own responsibility. I am equally grateful to the Series Editor, Professor Barry Law, and the Publisher, Dr Graeme Mac-Kintosh, for their unstinting support and counsel as I traversed some tricky terrain at various stages of the project. To Graeme in particular, thank you for that most endearing marksman-like quality—timing, aim and the most effective execution at every stage, which ensured that I delivered to plan.

Kanes K. Rajah

Contributors

Dr Albert J. Dijkstra Carbougnères, F-47210 St. Eutrope-de-Born, France

Dr Timothy P. Guinee Dairy Products Research Centre, Teagasc, Moorepark, County Cork, Republic of Ireland

Dr Tetsuo Koyano Meiji Seika Kaisha Ltd, 5-3-1 Chiyoda, Sakado-Shi, Saitama 350-0289, Japan

Professor Barry A. Law The Australian Cheese Technology Programme, Food Science Australia, Sneydes Road, Private Bag 16, Werribee, Victoria 3030, Australia

Mr John Podmore 66 Shrewsbury Avenue, Aintree Village, Liverpool L10 2LF, UK

Dr Kanes K. Rajah Centre for Entrepreneurship, Maritime Greenwich Campus, Old Royal Naval College, Park Row, Greenwich SE10 9LS, UK

Dr H.M. Premlal Ranjith Diotte Consulting & Technology, The Conifers, 36 Bishops Wood, Nantwich, Cheshire CW5 7QD, UK

Dr David J. Robinson Van den Bergh Foods Ltd, London Road, Purfleet, Essex RM19 1SD, UK

Professor Kiyotaka Sato Physical Chemistry Laboratory, Faculty of Applied Biological Science, Hiroshima University, Higashi-Hiroshima 739-8528, Japan

Dr Clyde Stauffer Technical Foods Consultants, 750 Southmeadow Circle, Cincinnati, OH 45231, USA

Mr Ian Stewart Britannia Food Ingredients, Britannia Way, Goole DN14 6ES, UK

Dr Ralph Timms Britannia Food Ingredients, Britannia Way, Goole DN14 6ES, UK

Contents

1   Physical properties of fats in food   **1**
    TETSUO KOYANO and KIYOTAKA SATO

    1.1   Introduction   1
    1.2   Basic physical properties of fat crystals   2
          1.2.1   Polymorphism of fats   2
          1.2.2   Phase behaviour of fat mixtures   6
          1.2.3   Microstructure, texture and rheological properties   8
    1.3   Structure–function relations in food fats   9
          1.3.1   Fats in bulk phase   9
          1.3.2   Fats in oil-in-water emulsions   11
          1.3.3   Fats in water-in-oil emulsions   13
    1.4   Fat blending in confectionery fats   15
          1.4.1   Cocoa butter equivalents and cocoa butter replacers   15
          1.4.2   Cocoa butter and milk fat   16
    1.5   Fat crystallisation in oil-in-water emulsions   20
          1.5.1   Crystallisation in emulsions of palm oil in water   21
          1.5.2   Crystallisation in emulsions of palm mid-fraction in water   22
    1.6   Conclusions   25
    Acknowledgements   25
    References   25

2   Bakery Fats   **30**
    JOHN PODMORE

    2.1   Introduction   30
    2.2   Production of margarine and shortening   31
    2.3   Crystallisation behaviour   33
    2.4   Processing   36
    2.5   Plastic bakery fats   37
          2.5.1   Short pastry   39
          2.5.2   Cake   41
          2.5.3   Puff pastry   46
    2.6   The influence of emulsifiers in baking   48
    2.7   Control of quality in margarine and shortening manufacture   51
    2.8   Liquid shortenings   54
    2.9   Fluid shortenings   55
    2.10 Powdered fats, flaked fats and fat powders   56
          2.10.1 Methods of manufacture   56
          2.10.2 Applications of fat powders and powdered fats   60
    2.11 Fat in biscuit baking   61
          2.11.1 The function of fats in biscuits   62
          2.11.2 Biscuit filling creams   64

        2.11.3 Spray fats   64
        2.11.4 Fat bloom   65
    2.12 Conclusions   66
    References   67

3   Water continuous emulsions   **69**
    H. M. PREMLAL RANJITH

    3.1 Introduction   69
        3.1.1 The structure of water continuous emulsions   69
        3.1.2 Milk fat globule structure   71
    3.2 Preparation of water continuous emulsions   73
        3.2.1 Dairy creams   73
        3.2.2 Recombined creams   79
        3.2.3 Ice-cream mix   82
        3.2.4 Heat treatment of emulsions   89
        3.2.5 Preparation of dressings   105
    3.3 Factors affecting water continuous emulsions   109
        3.3.1 Emulsion stability of high-fat creams   109
        3.3.2 Defects in ice cream   117
        3.3.3 Defects in mayonnaise and salad dressing   119
    References   120

4   Hydrogenation and fractionation   **123**
    ALBERT J. DIJKSTRA

    4.1 Introduction   123
    4.2 Hydrogenation   124
        4.2.1 Kinetics and mechanism   125
        4.2.2 Industrial hydrogenation processes   136
    4.3 Fractionation   141
        4.3.1 Theoretical aspects   142
        4.3.2 Wax removal by winterisation   143
        4.3.3 Industrial fractionation processes   144
    4.4 Discussion   153
    References   154

5   Fats for chocolate and sugar confectionery   **159**
    IAN M. STEWART and RALPH E. TIMMS

    5.1 Introduction   159
    5.2 Production and properties   160
        5.2.1 Cocoa butter and milk fat   160
        5.2.2 Symmetrical/SOS-type alternative fats   166
        5.2.3 High-trans-type alternative fats   168
        5.2.4 Lauric-type alternative fats   170
        5.2.5 Comparison and compatibility   171
    5.3 Legislation and regulatory aspects   174
        5.3.1 Legislation   174
        5.3.2 Adulteration and its detection   178
    5.4 Applications   179

5.4.1	Real chocolate	179
5.4.2	Compound chocolate	181
5.4.3	Toffees and other sugar confectionery	184
5.4.4	Truffles	185
5.4.5	Bloom and rancidity	186
5.5	Conclusions	188
	References	189

6 Spreadable products   **192**
DAVID J. ROBINSON and KANES K. RAJAH

6.1	Introduction	192
6.1.1	Definition of spreads: margarine, low(er) fat spreads and butter	192
6.1.2	Summary of product development	194
6.1.3	Summary of process development	196
6.1.4	Summary of ingredient development	199
6.1.5	Summary of packaging developments	202
6.2	Legislation	204
6.2.1	EU regulations	205
6.2.2	US regulations	205
6.2.3	Codex standards	207
6.3	Emulsion technology	207
6.3.1	Properties of emulsions	207
6.3.2	Emulsifiers and hydrophilic–lipophilic balance values	211
6.3.3	Stabilisers	212
6.3.4	Preservatives and microbiological stability	212
6.3.5	Emulsion preparation	213
6.4	Process technology	216
6.4.1	Current yellow fat range	216
6.4.2	Scraped-surface cooling	216
6.4.3	Churning technology	220
6.4.4	Storage conditions	222
6.5	Yellow fat blends	222
6.5.1	Trans-fatty-acid-free oil blends	222
6.5.2	Some properties of butter	223
6.5.3	Oils high in lauric and palmitic fatty acids	223
6.5.4	Long-chain fatty acids	225
6.6	Nonyellow fat range	225
	References	226

7 Emulsifiers and stabilizers   **228**
CLYDE E. STAUFFER

7.1	Introduction	228
7.2	Surface activity	229
7.2.1	Amphiphiles	229
7.2.2	Surface and interfacial tension	230
7.3	Interface formation	232
7.3.1	Division of internal phase	232
7.3.2	Emulsions	233
7.3.3	Foams	234

	7.3.4	Wetting	234
7.4	Stabilization		235
	7.4.1	Creaming	235
	7.4.2	Flocculation	236
	7.4.3	Ionic stabilization	237
	7.4.4	Steric hindrance	238
	7.4.5	Foam drainage and film breakage	238
	7.4.6	Viscosity and gelation	240
7.5	Food emulsifiers		240
	7.5.1	Monoglycerides	241
	7.5.2	Monoglyceride derivatives	242
	7.5.3	Sorbitan derivatives	245
	7.5.4	Polyhydric emulsifiers	247
	7.5.5	Anionic emulsifiers	249
	7.5.6	Lecithin	250
7.6	Hydrophilic–lipophilic balance		252
	7.6.1	Basic principal of the concept	252
	7.6.2	Experimental determination of hydrophilic–lipophilic balance	254
7.7	Polysaccharide stabilizers and thickeners		256
	7.7.1	Gums	256
	7.7.2	Modified starch	258
	7.7.3	Cellulose derivatives	259
7.8	Applications		259
	7.8.1	Margarine and dairy products	259
	7.8.2	Baking	264
	7.8.3	Coatings	267
	7.8.4	Dressings and sauces	268
7.9	Regulatory aspects		272
	References		273

8 Role of milk fat in hard and semihard cheeses **275**
    TIMOTHY P. GUINEE and BARRY A. LAW

8.1	Introduction		275
8.2	Effect of fat on cheese composition		277
	8.2.1	Fat level	277
	8.2.2	Effect of degree of fat emulsification	281
8.3	Proteolysis		283
	8.3.1	Primary proteolysis (where moisture in nonfat solids differs)	284
	8.3.2	Secondary proteolysis (where moisture in nonfat substance differs)	286
8.4	Effect of fat on cheese microstructure		287
	8.4.1	Microstructure of rennet-curd cheese	287
	8.4.2	Microstructure of pasteurized processed cheese products and analogue cheese products	291
	8.4.3	Effect of fat level	292
	8.4.4	Effect of fat emulsification	295
	8.4.5	Effect of fat on heat-induced changes in microstrucutre	295
8.5	Effect of fat on cheese microbiology		297
8.6	Effect of fat on cheese flavour		299
8.7	Effect of fat on cheese yield		301
8.8	Effect of fat on cheese texture and rheology		304

	8.8.1 Contribution of fat to cheese elasticity and fluidity	304
	8.8.2 Effect of fat level on fracture-related properties, as measured by using large strain deformation	306
	8.8.3 Effect of degree of fat emulsification on fracture-related properties, as measured by using large strain deformation	307
8.9	Effect of fat on the functional properties of heated cheese	310
	8.9.1 Effect of fat level	315
	8.9.2 Effect of milk homogenization and degree of fat emulsification	319
8.10	Conclusions	322
	References	323

9 Culinary fats: solid and liquid frying oils and speciality oils   332
JOHN PODMORE

9.1	Introduction	332
9.2	Salad and cooking oils	333
9.3	Frying fats	337
	9.3.1 Shallow (pan) frying	338
	9.3.2 Deep frying	338
	9.3.3 Selection of frying media	343
9.4	Oils for roasting nuts	347
9.5	Ghee	347
	9.5.1 Ghee attributes and quality	348
	9.5.2 Uses of ghee	350
9.6	Vanaspati	350
9.7	Speciality oils	352
	9.7.1 Almond oil	353
	9.7.2 Groundnut oil	353
	9.7.3 Hazelnut oil	353
	9.7.4 Sesame seed oil	355
	9.7.5 Safflower oil	355
	9.7.6 Grapeseed oil	356
	9.7.7 Walnut oil	357
	9.7.8 Rice bran oil	357
9.8	Concluding remarks	358
	References	358

Appendix   Nomenclature for fatty acids and triglycerides   360

Index   361

1 Physical properties of fats in food

Tetsuo Koyano and Kiyotaka Sato

1.1 Introduction

Oils and fats are important ingredients in a wide variety of manufactured foods, and constitute a significant part of food recipes. The major foods in which they are used are all discussed in detail in this volume. However, it is important to note that the forms in which oils and fats are made available to food manufacturers have changed significantly over the years, particularly since the 1960s, largely because of the major shifts that have taken place in consumer lifestyles and the increasing concerns with health, food safety and a balanced diet. Many of the food products that are now available to consumers reflect this new direction. Important examples arising out of the lipid research that has followed are *trans*-fatty acids, free high-melting fats, very-low-yellow fat emulsions, spreadable butter, molecularly designed structured fats with new nutritional advantages and so on. All of these initiatives have required an in-depth understanding of the behaviour of the fats concerned so that they can be used effectively as ingredients in food. Consequently, the study of their physical properties is of major interest and is covered in this chapter.

In general, fats form networks of crystal particles, maintaining specific poly-morphic forms, crystal morphology and particle–particle interactions. The control of the physical properties of food fats has therefore been of importance in research efforts and can be considered under four headings:

- clarification of molecular and crystal structures of triacylglycerols (TAGs) with different fatty acid moieties (Kaneko, 2001; Kaneko *et al.*, 1998)
- solidification and transformation mechanisms of TAG crystals (Sato, 1996, 1999; Sato and Koyano, 2001; Sato *et al.*, 2001)
- rheological and texture properties that are dominated mainly by fat crystal networks (Boode *et al.*, 1991; Marangoni and Hartel, 1998; Walstra *et al.*, 2001)
- influences of minor lipids on fat crystallisation kinetics (Wright *et al.*, 2000).

The first topic is of an introductory nature and so will not be elaborated in this chapter (for more details, see the cited references). The remaining three topics are related to observed systems of food fats, with which this chapter is mainly concerned.

The chapter begins with a brief review of the three basic physical properties of fats by collecting together recent work on the solidification and transformation of the fats in bulk and in emulsion states. We will then focus on fundamental aspects of the crystallisation and transformation of fats employed in real food systems, with use of important examples, such as cocoa butter, palm oil and palm mid-fractions. Since these natural fats are multi-TAG systems, knowledge about the fundamental properties of pure TAGs composing the natural fats may be necessary, as will be argued. Those who wish to compare real fats with pure fats are directed to the literature (Sato, 1996; Sato and Koyano, 2001; Sato and Ueno, 2001; Sato *et al.*, 1999).

1.2   Basic physical properties of fat crystals

The physical properties of the food fats are influenced primarily by three factors: polymorphism (structural, solidification and transformation behaviour); the phase behaviour of fat mixtures; and the rheological and textural properties exhibited by fat crystal networks. In this section we cover the fundamentals and look at recent research work on these three properties.

1.2.1   Polymorphism of fats

Polymorphism is defined as the ability of a chemical compound to form different crystalline or liquid crystalline structures. The melting and crystallisation behaviour will differ from one polymorph to another.

In fat crystals, three main polymorphs, α, β' and β, are defined in accordance with subcell structure: α polymorphs have a hexagonal subcell (H); β' polymorphs have an orthorhombic–perpendicular subcell ($O_\perp$); and β polymorphs have a triclinic–parallel subcell ($T_{//}$) [Larsson, 1966; see figure 1.1(a)]. Table 1.1 summarises basic physical properties of the three typical polymorphic modifications. Polymorph α is least stable, easily transforming to either the β' form or the β form, depending on the thermal treatment. Polymorph β', the metastable form, is used in margarine and shortening because of its optimal crystal morphology and fat crystal networks, which give rise to optimal rheological and texture properties. The most stable β form tends to form large and plate-like crystal shapes, resulting in poor macroscopic properties in shortening and margarine. Figure 1.1(b) shows chain-length structure, illustrating the repetitive sequence of the acyl chains involved in a unit cell lamella along the long-chain axis (Larsson, 1972). A double chain-length structure (DCL) is formed when the chemical properties of the three acid moieties are the same or very similar. In contrast, when the chemical properties of one or two of the three chain moieties are largely different from those of the moieties, a triple chain-length (TCL) structure is formed because of chain sorting. The relevance of

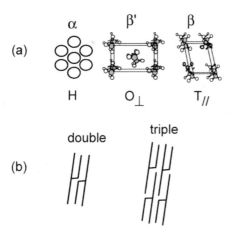

Figure 1.1 Polymorphic structures of fats: (a) subcell and (b) chain length.

Table 1.1 Three typical polymorphic forms of fats and their main physical properties

Form	Stability	Density	Melting point	Morphology
α	Least stable	Lowest	Lowest	Amorphous-like
β'	Metastable	Intermediate	Intermediate	Rectangular
β	Most stable	Highest	Highest	Needle shaped

chain-length structure is revealed in the mixing phase behaviour of different types of the TAGs in solid phase: when the DCL fats are mixed with TCL fats, phase separation readily occurs.

In food fats, transformation from polymorph β' to polymorph β often causes deterioration of the end product, mostly because of changes in crystal morphology and network, as indicated in table 1.1. The β-type polymorph is found in confectionery fats made of cocoa butter. There are two β-type crystals: a metastable β_2 form is more useful than the more stable β_1 form (Sato and Koyano, 2001). Atomic-level structure analyses of the TAGs have been attempted to resolve the microscopic mechanism of the polymorphic $\beta'-\beta$ transformation. Results were reported first for the β forms (as reviewed for the β forms in Kaneko 2001), and have been reported quite recently for the β' form (van Langevelde et al., 2000; Sato et al., 2001).

The macroscopic aspects of fat polymorphism concern behaviour at melting and crystallisation and the subsequent transformation. With respect to the melting temperature, T_m, for the three forms of any particular TAG, a general tendency is that T_m is lowest for the α form, intermediate for the β' form and highest for the β form. In contrast, the crystallisation behaviour is more complicated and is determined primarily by the type of crystallising medium, the crystallisation temperature and the rate of cooling. In general, the β form is usually crystallised from solution phase, but all three forms may be crystallised from neat liquid. When crystallisation is from the neat liquid, the relative rates and extents of crystallisation of the three polymorphs are determined by the rate of nucleation, which is highest for the α form, intermediate for the β′ form and lowest for the β form. This behaviour was measured with precision for tripalmitoyl glycerol (PPP; Sato and Kuroda, 1987) and for the other fat crystals (Blaurock, 1999). Once the less-stable forms are crystallised, they transform to more stable forms in a postcrystallisation process in solid phase or through liquid mediation (Sato et al., 1999). As a consequence, the morphology of fat crystals is determined by the polymorphic modification, by the thermal processes of crystallisation and by subsequent transformation. It is worthy of note that various morphologies of β′-form PPP crystal have been found for different temperature treatments. In particular, the β′ form showed needle-like crystals after slow crystallisation, similar to β-form crystals, which usually exhibit a long needle shape (Kellens et al., 1992).

As the physical properties of food fats are greatly influenced by fat polymorphism, it is a prerequisite for those who are engaged in the material production of oils and fats to know how the fatty acid composition influences the fat polymorphism. Two categories of fatty acid composition may be considered: mono-acid TAGs in which the three fatty-acid moieties of the TAG are of the same type, and mixed-acid TAGs in which different fatty-acid components are connected to three different glycerol carbons on the TAG. The following diversity in fatty-acid composition of TAGs can be found.

- Mono-acid TAGs:
  - the acids may be saturated
    - the number of carbon atoms in the fatty-acid chain, N_c, may be odd or even
  - the acids may be unsaturated
    - the number of carbon atoms in the fatty-acid chain, N_c, may be odd or even
    - there may be a *cis* or a *trans* conformation around the double bond
    - the number of double bonds may vary
    - the position of the double bonds may vary
- Mixed-acid TAGs:
  - there may be three saturated acids with different chemical species
  - there may be three unsaturated acids with different chemical species

- there may be three acids containing saturated and unsaturated species
- the different fatty acids may be connected to carbon atoms of different stereo-specific number (*sn*)

In 1988 Hagemann summarised the melting behaviour of TAGs with different combinations of fatty-acid moieties with different chemical species. Hagemann showed a general tendency in the melting behaviour of mono-acid TAGs to be as follows:

- In saturated mono-acid TAGs, the melting points of the α, β' and β forms increase when N_c is increased from 8 to 30. With respect to quantitative dependence of the melting point of the polymorphs on N_c, the melting points of the α form increase smoothly with N_c, whereas the melting points of the β' and β forms increase in a 'zig-zag' manner with N_c odd or evens.
- In the monounsaturated mono-acid TAGs, the melting points of the β forms are available, showing specific dependence on double-bond conformation and on position of the double bond. For example, *trans* unsaturated TAGs showed higher melting points than those of *cis* unsaturated TAGs at every double-bond position.

Since 1988 much work has been done on the polymorphic behaviour of mixed-acid TAGs. It is important to understand such behaviour as natural oils and fats contain these mixed-acid TAGs (for reviews, see Sato and Ueno, 2001; Sato *et al.*, 1999), the polymorphic behaviour of mixed-acid TAGs differing greatly from that of mono-acid TAGs. For example, table 1.2 shows variations in polymorphic occurrence and melting behaviour of a series of TAGs in which the two fatty-acid chains at the *sn*-1 and *sn*-3 positions are stearic acid, and in which the fatty acid at the *sn*-2 position may be stearic (to give SSS), elaidic (SES), oleic (SOS) (Sato *et al.*, 1989), ricinoleic (SRS) (Boubekri *et al.*, 1999) or linoleic (SLS) (Takeuchi *et al.*, 2000). Three typical polymorphs of α, β' and β polymorphs are revealed in SSS, all being stacked in a DCL structure. Substitution of the *sn*-2 acid with elaidic acid (SES) caused a decrease in melting point for the three polymorphs, which exhibit basically the same properties as those of SSS. However, large differences are produced when the *sn*-2 acid is substituted with oleic, ricinoleic or linoleic acid, revealing a new polymorph, γ, and variation in chain-length structure from double (α form) to triple (the other, more stable, forms). In addition, the β form is absent in SRS and SLS, and the β' form does not occur in SLS. These peculiar properties of mixed-acid TAGs involving *cis*-unsaturated fatty-acid moieties may be partly understood in terms of chain–chain interactions between the saturated and the *cis*-unsaturated fatty-acid moieties (Kaneko *et al.*, 1998; Sato and Ueno, 2001).

Table 1.2 Polymorphic occurrence of SSS, SES, SOS, SRS and SLS

Polymorph[a]	Melting point (°C)				
	SSS	SES	SOS	SRS	SLS
α-2	55.0	46.0	23.5	25.8	21.6
β'-2	61.6	58.0	–	–	–
γ-3	–	–	35.4	40.6	34.5
β-2	73.0	61.0	–	–	–
β'-3	–	–	36.5	–	–
β'$_2$-3	–	–	–	44.3	–
β$_2$-3	–	–	41.0	–	–
β'$_1$-3	–	–	–	48.0	–
β$_1$-3	–	–	43.0	–	–

Note: SSS, tristearoylglycerol; SES, 1,3-distearoyl-2-elaidoyl-*sn*-glycerol; SOS, 1,3-distearoyl-2-oleoyl-*sn*-glycerol; SRS, 1,3-distearoyl-2-ricinoleyl-*sn*-glycerol; SLS, 1,3-distearoyl-2-linoleoyl-*sn*-glycerol.
[a]Suffixes 2 and 3 refer to double and triple chain-length structures, respectively.

The variation in polymorphic properties of saturated–unsaturated mixed-acid TAGs (see table 1.2) may have critical significance for our understanding of the polymorphism of natural oils and fats such as milk fat and palm oil, which contain large amounts of mixed-acid TAGs (Gunstone, 1997).

1.2.2 Phase behaviour of fat mixtures

Naturally occurring fats and lipids are mixtures of different types of TAG. The complicated behaviour they exhibit with regard to melting, crystallisation and transformation, crystal morphology and aggregation are partly a result of the physical properties of the component TAGs (discussed above) and, more importantly, partly a result of the phase behaviour of the mixture. To resolve this complexity in mixed-fat systems, a fundamental study of binary and ternary mixtures of specific TAG components is necessary (Rossel, 1967).

Three typical mixture phases occur in binary solid mixtures of fats in the case where the two components are miscible, for all concentration ratios, in the liquid state. These are: solid solution phase, eutectic phase and compound formation. For TAG mixtures, two factors affect the mixing phase behaviour simultaneously: chain–chain interactions and polymorphism. Chain–chain interactions are influenced by the chemical nature of the component TAGs varying with chain length (N_c), the saturation or unsaturation and the isomeric conformation (*cis* or *trans*) of the unsaturated chains. The effect of polymorphism is revealed in the formation of miscible and eutectic phases, miscible mixtures being formed with less stable polymorphs (the α and β' forms) and eutectic phases tending to occur with the stable polymorph (the β form). In addition, differences in

Table 1.3 Typical binary mixing behaviour of triacylglycerols (TAGs)

TAG Polymorphic form[a]	Mixing phase and polymorphic form
PPP α-2, β′-2, β-2	Miscible phase: α-2 and β′-2
SSS α-2, β′-2, β-2	Eutectic phase: β-2
SOS α-2, γ-3, β′-3, $β_2$-3, $β_1$-3	Compound formation: α-2, β′-2, β-2
OSO α-2, β′-2, β-3	
POP α-2, γ-3, β′-2, $β_2$-3, $β_1$-3	Compound formation: α-2, β′-2, β-2
PPO α-3, β′-3	

Note: PPP, tripalmitoylglycerol; SSS, tristearoylglycerol; SOS, 1,3-distearoyl-2-oleoyl-*sn*-glycerol; OSO, 1,3-dioleoy-2-stearoyl-*sn*-glycerol; POP, 1,3-dipalmitoyl-2-oleoyl-*sn*-glycerol; PPO, 1,2-dipalmitoyl-3-oleoyl-*rac*-glycerol.
[a] Suffixes 2 and 3 refer to double and triple chain-length structures, respectively.

chain-length structure (DCL or TCL) affect mixing systems, the formation of the miscible phase for fats with different chain-length structure being prohibited. Three examples illustrate the effects of polymorphism and structure on the binary mixture behaviour of TAGs (table 1.3).

In the mixture of saturated mono-acid TAGs, a eutectic phase with a limited region of miscible phase was formed for the stable polymorph when the difference in N_c is not larger than two, as shown for the mixture of PPP and SSS. However, a miscible mixture was formed for the α and β′ polymorphs in the mixture of PPP and SSS. This means that, when the mixed PPP and SSS liquids is chilled to form the α or β′ form, and when further polymorphic transformation into the β form is induced, the mixing changes from being miscible to separated. This was clearly demonstrated by an in-situ X-ray diffraction measurement by using a synchrotron radiation X-ray beam (Kellens *et al.*, 1991). This result indicates the importance of the effect of polymorphism on the mixture system and must be kept in mind when one is searching for an optimal blend of TAGs for use in food where the phase separation of fats is not to be preferred.

The formation of molecular compound crystals has been observed among mixtures of saturated/unsaturated mixed-acid TAGs. The formation of a molecular compound may be viewed as a special case of eutectic mixing systems that occurs as a result of specific molecular interactions between the component TAGs. Two examples are shown in table 1.3; for an SOS/OSO mixture (Koyano *et al.*, 1992) and for a POP/PPO mixture (Minato *et al.*, 1997). In both mixture systems, a common result was obtained in that the polymorphic behaviour was largely different for the component TAGs, yet the molecular compound was formed for all three polymorphic forms. It is worthy of note that all of the molecular compounds were packed in the DCL structure, although DCL and TCL structures are revealed in the component TAGs. A mechanistic treatment has been carried out on the formation of molecular compound systems of SOS/OSO and POP/PPO, taking into account chain–chain interactions

(Kaneko et al., 1998; Sato et al., 1999). The significance of the formation of a molecular compound in food applications is that the rate and extent of the $\beta'-\beta$ transformation in molecular compound systems are remarkably higher than those occurring in the component TAGs.

1.2.3 Microstructure, texture and rheological properties

One of the most important macroscopic physical properties of food fats is the rheology, affecting the spreadability of margarine and other spreads, the 'snap' of chocolate and the smoothness, mouth feel and stability of bulk fats and emulsion products (de Man, 1999; van den Tempel, 1961). In addition, control of rheological properties is necessary in production processes in a factory.

The rheological properties of food fats are determined by many factors that can be grouped into two categories (Marangoni and Hartel, 1998):

- internal factors, involving the molecular compositions of fats (TAGs, ingredients and additives), the polymorphism of crystals of the constituent TAGs and the microstructure of fat crystals (morphology, crystal size distribution and crystal network formation)
- external processing conditions, involving temperature variation, shear, flow velocity, and so on.

Among the internal factors, the microstructure of the fat crystals greatly influences the rheological properties.

Much progress has been made recently by Marangoni and his colleagues on the analysis of fat crystal microstructure. These workers discussed the macroscopic physical properties of food fats in terms of formation and internal assembly of fat microstructures, the rheological properties of final products being assessed by conventional analytical methods such as thermal measurements, and determination of solid fat content (SFC) and turbidity. One aspect of their research was focused on the hardness of chocolate made with Salatrim® (a trademark of Pfizer), which is much softer than chocolate made with cocoa butter (Narine and Marangoni, 1999). This phenomenon of hardness is caused by differences in the microstructure of chocolate fats. The TAGs involved in Salatrim® consist of asymmetric molecules ($C_3-C_3-C_{18}$) that result in strong repulsive interactions between TAG molecules in the crystal and induce platelet structures. In contrast, cocoa butter consists of symmetric TAGs such as POP and SOS that pack together densely to make tightly packed crystals. Hence, macroscopic structure is a consequence of the interaction of microstructures and is thus affected by microstructure characteristics. In the case of Salatrim®, a random macrostructure is formed because there are few attractive interactions between the constituent microstructures, thus the macroscopic structure is weak. In the case of cocoa butter, a strong three-dimensional crystal network is formed.

The effects of chemical interesterification on crystallisation, SFC, fat microstructure and rheological properties of various fat blends have also been examined systematically by Marangoni and co-workers (Marangoni and Rousseau, 1998a, 1998b; Rousseau *et al.*, 1998).

1.3   Structure–function relations in food fats

In this section we discuss how the fat structures discussed above influence the macroscopic functions of foods in bulk, in oil-in-water (O/W) emulsion states and in water-in-oil (W/O) emulsion states (figure 1.2), looking at specific example materials for each case.

1.3.1   Fats in bulk phase

In this section we discuss the physical properties of fats in a bulk phase, taking cocoa butter in chocolate as a model. Cocoa butter forms a continuous fat phase in chocolate, with small particles of sugar, cocoa mass, milk powder (in milk chocolates) and other ingredients, including food emulsifiers, dispersed within it. In the preparation of most commercial-grade chocolates, temperature

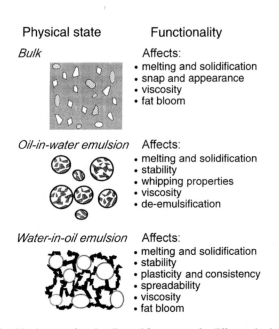

Figure 1.2 Relationships between functionality and fat structure for different physical states. The bulk phase of chocolate is shown, consisting of a continuous phase of cocoa butter crystals, with sugar and other solid powders shown as small features. Oil is represented by open 'circular' shapes; water is shown as dark fragments.

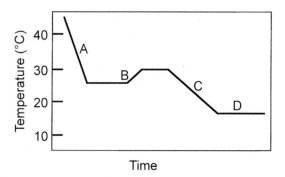

Figure 1.3 Temperature variation during the tempering process of chocolate: A, first cooling; B, reheating; C, recooling; D, storage.

treatment of the sample, called tempering, is varied in the manner shown in figure 1.3, after the mixing, blending, grinding and conching processes. In the following we will describe the events occurring at each stage of tempering for which the extent and polymorphism of cocoa butter crystals are most critical. There are six polymorphs of cocoa butter: forms I to VI (Wille and Lutton, 1966). The polymorphism of the cocoa butter crystals described above is of form IV (β′ type) and form V (β type); fat bloom is caused by the transformation from form V to form VI. Note that both are of a β-type polymorph (Bricknell and Hartel, 1998; Sato and Koyano, 2001). The snapping, appearance and demolding properties of chocolate are best when the cocoa butter crystals are of form V (β type), but the formation of fat bloom is a serious problem and is a result of the growth of needle-like crystals after the transformation to and crystal growth of form VI.

- Stage A (cooling period): This consists of the nucleation and crystal growth of metastable forms of cocoa butter as a result of heterogeneous interactions between cocoa butter molecules and preexisting high melting fats. The metastable forms such as form IV cause fat bloom if no further tempering is applied.
- Stage B (reheating): This comprises transformation from the metastable forms to the stable form (form V) of cocoa butter crystals by raising the temperature. The form-V crystals dominantly formed at this stage serve as seed crystals, leading the rest of cocoa butter liquid to crystallise in form V. Transformation and further crystal growth of cocoa butter in form V are accelerated by shear force operated in the tempering machine (MacMillan et al., 1998; Ziegleder, 1985).
- Stage C (cooling): crystal growth of form V develops, with the consequence that shrinkage of the cocoa butter occurs so that demolding is enabled. Rheological properties at this stage are extremely important for

producing chocolate bars or indeed any other enrobed chocolates,[1] the critical step being delicately determined by the solid fat content of the fat.

- Stage D (storage): storage of chilled chocolate enables stabilisation of the fat crystal network and size distribution.

Even after the whole production process has been completed, the fat bloom phenomena may arise, causing further problems. Fat bloom is caused by exposing chocolate to temperature fluctuations up to 30°C during the storage period (Padley, 1997). This can become a serious problem if, during the winter, the chocolate is stored in warehouses, sales outlets or in the home of the consumer, where temperatures can drop to extremely low levels. Similarly, during the summer, temperatures may rise considerably.

Various techniques have been applied to control the crystallisation and transformation processes of cocoa butter, shown in figure 1.3. Blending of different fats with cocoa butter, in accordance with regulations for blending (which differ from one country to another) is one of the main ways to modify the melting and crystallisation properties of chocolate fats (Ali and Dimick, 1994; Faulkner, 1981; Hogenbirk, 1984; Koyano et al., 1993; Narine and Marangoni, 1999; Sabariah et al., 1998; Timms, 1980; Uragami et al., 1986). In addition, stability against fat bloom can be improved by fat blending (Lohman and Hartel, 1994; Tietz and Hartel, 2000). Emulsifiers are added to chocolate fats to modify rheological properties so that interactions between fat crystals, hydrophilic particles such as sugar and milk powders and water phases are mediated through the emulsifiers (Katsuragi, 1999). The emulsifiers also affect fat bloom stability, acting at growing crystal surfaces of cocoa butter (Aronhime et al., 1988).

The addition of high-melting fats having the same crystal structure as form V, namely form β_2 of 1,3-dibehenoyl-sn-2-oleoyl-glycerol (BOB), is quite effective in crystallising cocoa butter in form V without the need for tempering (crystal seeding; Hachiya et al., 1989a, 1989b; Koyano et al., 1990). This property was employed to produce enrobing chocolate at temperatures as high as 38°C at which the viscosity of the chocolate was so decreased that enrobing conditions became very stable. External forces, such as shear stress (see above) and ultrasound irradiation (Baxter et al., 1995), are also effective in modifying crystallisation rate and the polymorphic transformation of cocoa butter.

1.3.2 Fats in oil-in-water emulsions

O/W (i.e. water-continuous) emulsions in food are observed in the main body of whip cream, ice cream, coffee cream, and so on, and their production processes

---

[1]Enrobing is a process involving covering nuts or baked snacks, for instance, with chocolate. In this process, stabilisation of the chocolate viscosity is critical in ensuring the chocolate coating is of uniform thickness.

involve pre-emulsification, homogenisation, pasteurisation, rehomogenisation and cooling. Although some details differ from one product to another, the most preferable physical properties of O/W emulsions containing a fat phase are optimal melting and solidification properties, emulsion stability when chilled during storage and during the final usage stage, and a good crystal morphology and network exhibiting optimal rheological and whipping properties (figure 1.2). The interactions of emulsifiers and proteins in the oil and water phases are of immense significance but are outside the scope of this chapter.

The crystallisation of the oil phase in the O/W emulsion is an important process, affecting the coagulation of the emulsion when chilled, the de-emulsification process of whipping creams, the freezing of ice creams and so on (Boekel and Walstra, 1981; Dickinson and McClements, 1996; Skoda and van den Tempel, 1963). It has been widely recognised that, when the O/W emulsion is chilled, emulsion stability is influenced by the size and orientation of the fat crystals present in the fat phase as manifested by the following external conditions: particle size and distribution, extent of creaming, rate of cooling (determining rate of nucleation and crystal growth), nature of emulsification when using emulsifiers and/or proteins, fatty acid composition, polymorphism of fat crystals, and so on. It has been reported that fat crystallisation is initiated at the O/W interface (Krog and Larsson, 1992). However, in general, the factors which most influence the fat crystallisation and crystallisation mechanisms, such as rate and extent of nucleation and crystal growth, polymorphism, crystal orientation and emulsion stability, have been poorly understood (see next section).

As with regard to the fatty acid composition of TAGs in the oil phase, high-melting TAGs are adsorbed at the oil–water interface, and these TAGs are first crystallised on chilling and dominate the successive crystallisation of the whole oil phase. Therefore, the adsorption property of the TAGs at the air–water interface may be of primary influence. Fahey and Small (1986) examined the surface properties of the air–water interface with monolayers of TAGs of 1,2-dipalmitoyl-3-acylglycerols (PPn), in which n, the number of carbon atoms of the sn-3 fatty acids, varied from 2 to 10 (figure 1.4). They found that, in the monolayers of PPn with shorter sn-3 chains ($n = 2$), the sn-3 chains are completely submerged regardless of surface pressure [figure 1.4(a)]. With increasing chain length of sn-3 acid, the adsorption behaviour changed as follows: at $n = 3$–6, the sn-3 chains lie flat on the surface at moderate surface pressure [figure 1.4(b)] but are completely submerged at high pressure [figure 1.4(a)]. However, at $n = 8$–10, the sn-3 chains are not submerged at monolayer pressures and thereby prevent monolayer crystallisation [figure 1.4(c)].

From this, one can see that some TAG fractions having short-chain-length fatty-acid moieties present in the oil phase of the O/W emulsion may be adsorbed at the oil–water interface in such forms as illustrated in figures 1.4(a) and 1.4(b). Upon chilling, the TAGs thus adsorbed first crystallise and play a role as

(a)   (b)   (c)

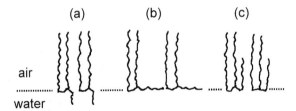

air

water

Figure 1.4 Adsorbed states of triacylglycerols at the air–water interface under (a) high surface pressure, (b) medium surface pressure and (c) low surface pressure.

precursors for further crystallisation, in which the orientation and aggregation of the finally crystallised fats are modified by the first-crystallising TAGs initiated at the oil–water interface. This mechanism may be used to interpret the results reported by Arishima *et al.* (1993), who found that the emulsion stability of a palm-oil–water emulsion after crystallisation was improved by adding 10% PP4.

1.3.3 Fats in water-in-oil emulsions

Margarine and spreads are typical food fats found in the form of W/O emulsions and consist of vegetable fats and oils. The optimal functional, and by implication, physical properties required of margarine and spreadable fats are listed in figure 1.2, where spreadability, plasticity and consistency are most important (de Man, 1983; de Man *et al.*, 1992). For this purpose, semisolid fats are to be preferred, with optimal SFC values of 50%–60% at around 5°C, which gradually decrease with increasing temperature, until complete melt at about 38°C. Furthermore, special texture and spreadability are needed for margarine for industrial uses, such as the roll-in type used in bakeries. Fat blending of high-melting-point (40–55°C), medium-melting-point (20–40°C) and low-melting-point (< 20°C) fats has been tried, and it has been found that the medium-melting-point fats play an important role, exhibiting optimal spreadability for margarine.

The preferable polymorphic form of margarine or spreadable fats is β′, since β′ crystals exhibit a very fine crystal network which comprises two types of particle–particle interactions: primary and secondary (de Man, 1982; Naguib-Mostafa *et al.*, 1985). Primary interactions with strong binding forces form a three-dimensional crystal network throughout the continuous fat phase. Secondary interactions result in small crystals with weaker binding forces. The transformation from the β′ to the β form causes serious deterioration, giving a sandy texture, hardening, reduced spreadability and oil–fat separation and coalescence of water droplets (emulsion instability) in extreme cases. These properties result from granular crystals of the β form, which tend to grow with a rectangular needle morphology (Bennema *et al.*, 1992).

The formation of granular crystals in fat blends similar to those of margarine fats containing palm oil and other vegetable oils was analysed at the polymorphic level (Watanabe *et al.*, 1992). Although palm oil is categorised as a β-tending fat, fats containing palm-oil fractions show formation of granular crystals in long-term storage. Chemical and physical analyses for TAG compositions, polymorphism and melting points of the granular crystals led to the conclusion that the granular crystals are of β polymorph of POP (1,3-dipalmitoyl-*sn*-2-oleoyl-glycerol). Figure 1.5 shows X-ray diffraction (XRD) spectra of the granular crystal grown in two fat blends: blend A (palm oil+canola oil) and blend B (POP+canola oil), which were kept at around 15–20°C in long-term storage. The XRD wide-angle (short-spacing) spectra [figure 1.5(b)] show that the granular crystals are of β form. The small-angle (long-spacing) spectra for wavelength 6.30 nm (001 reflection) and 3.10 nm (002 reflection) mean that the β form of POP is a TCL structure (Sato *et al.*, 1989). The XRD small-angle spectrum at wavelength 4.26 nm of blend B corresponds to normal fats exhibiting evidence of the β′ form with a DCL structure. The differences in chain-length structure between POP β and the other fats may accelerate fat separation and granular crystal formation. By contrast, the addition of a small amount of PPO in POP retarded the β′–β transformation because of the formation of molecular compound crystals of PPO/POP with a DCL structure (Minato *et al.*, 1997).

Therefore, various ways of prohibiting crystallisation and transformation into the β form of food fats have been developed, with use of many techniques, as reviewed by Chrysam (1996). For example, one can choose β′-tending fat resources (such as cottonseed oil) and blend them with β-tending fats (such as soybean, safflower, etc.) (Wiedermann, 1978), one can add food emulsifiers that retard transformations from the β′ to the β form (Aronhime *et al.*, 1988; Sato and Kuroda, 1987), one can add diacylglycerols (Mohamed and Larsson, 1992), or

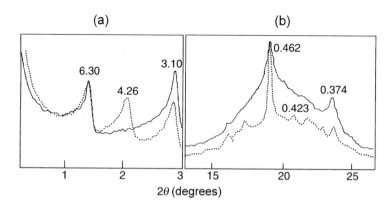

Figure 1.5 X-ray diffraction spectra of two fat blends in (a) small-angle and (b) wide-angle areas. Blend A, palm oil+canola oil; blend B, 1,3-dipalmitoyl-*sn*-2-oleoyl-glycerol (POP)+canola oil; θ, angle. .... Blend A; ⎯⎯ Blend B.

one can use interesterification techniques with different fats and oils (Chrysam, 1996).

The fatty-acid compositions of TAGs are closely related to β'-tending properties. TAGs containing different types of fatty acids are more stable in the β' form, as exemplified in pure TAGs (Hagemann, 1988). In natural fats, milk fats having long-chain and short-chain saturated and unsaturated fatty acids are β'-tending; palm oil is also categorised as a β'-tending fat because of the presence of asymmetric mixed-acid TAGs such as POO (1-palmitoyl-2,3-dioeloyl-rac-glycerol) and PPO. Recent work on single-crystal structure analyses of the β' form of 1,3-dilauroyl-sn-2-caproyl-glycerol (CLC; van Langevelde et al., 2000) and 1,2-dipalmitoyl-sn-3-myristoryl-glycerol (PPM; Sato et al., 2001) has indicated that chain–chain interactions through methyl end stacking combined with glycerol group conformations may stabilise the β' structures (Hernqvist and Larsson, 1982).

1.4   Fat blending in confectionery fats

As shown in figure 1.2, the continuous phase in chocolate consists of cocoa butter. Milk fat and vegetable fats are blended with cocoa butter in accordance with the chocolate recipe primarily to improve the physical properties of the chocolate such as melting, solidification, rheology, bloom stability and so on. The vegetable fats thus blended are called cocoa butter equivalent (CBE) or cocoa butter replacer (CBR) (Padley, 1997). Precise understanding of the interactions between cocoa butter and CBE or CBR is a prerequisite for controlling the desired quality of finished products and avoiding unexpected deterioration caused by inadequate fat blending. This will be covered here briefly, and is covered more comprehensively in chapter 5, on confectionery fats.

1.4.1   Cocoa butter equivalents and cocoa butter replacers

The TAGs involved in cocoa butter exhibit unique chemical compositions; POP 2-oleoyl-palmitoyl-stearoyl-glycerol (POS) and 1,3-distearoyl-oleoyl-glycerol (SOS) constitute over 80% of cocoa butter TAGs. The rest is rich in liquid TAGs, such as POO and SOO (1-stearoyl-2,3-dioeloyl-rac-glycerol). The three main TAGs combine oleic acid (C18:1) at the sn-2 positions of glycerol groups and saturated acids (C16:0 or C18:0) at the sn-1,3 positions (Sat–O–Sat type TAGs). Therefore, the three TAGs have a very similar chemical structure, except the length of the hydrocarbon chain differs by two carbon atoms. As a result, the three main TAGs in cocoa butter can form mixed crystals and exhibit a sharp melting profile (Sato and Koyano, 2001). The mixed crystals formed by POP, POS and SOS are TCL structures because of the difference in fatty-acid structure that combines the oleic acid moiety at the sn-2 position. Differences in chemical structure of milk-fat or vegetable-fat TAGs from those of cocoa butter TAGs result in specific interactions with cocoa butter.

It follows that the CBEs must consist mainly of Sat–O–Sat type TAGs, so that the CBEs are miscible with cocoa butter at all concentration ratios. Even slight differences in chain length of the fatty acids among the TAGs in the CBE and cocoa butter can lead to changes in the physical properties of specific cocoa butter/CBE mixed systems (Koyano *et al.*, 1993). The details have been reviewed (Padley, 1997) and are further elaborated in Chapter 5.

CBR is categorised as a fat that can be mixed with cocoa butter in only limited ratios, because CBR is incompatible with cocoa butter. Lauric fats and hydrogenated vegetable fats are typical examples of CBRs. The CBR crystallises in a DCL structure because of the high content of saturated fatty acids or *trans* fatty acids which are formed during the hydrogenation process. The DCL structure is very different from the TCL structure which is formed in the stable cocoa butter crystals. Therefore, it is very difficult to form a miscible crystalline phase, and eutectic phases are formed between CBR and cocoa butter. Figure 1.6 shows iso-solid diagrams obtained from the study of cocoa butter and lauric fat mixtures (Hogenbirk, 1984), indicating an apparent eutectic point formed by the two fats. Expert recipe formulation is required to find an optimal formula for blending these fats, otherwise serious deterioration in the final chocolate products will occur, such as lowering of the melting point, graining, due to individual crystallisation caused by TAG separation, or fat-bloom formation.

1.4.2   Cocoa butter and milk fat

The first marriage of cocoa butter and milk fat began in 1876 when Swiss chocolate-makers put milk into chocolate. However, the chemistry between there two fats continues to be of major research interest, as huge volumes of

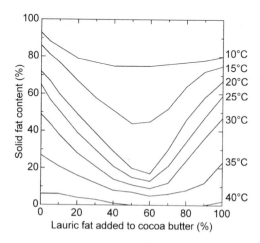

Figure 1.6 Solid fat content of cocoa butter/lauric fat mixtures at various temperatures.

milk chocolate are produced and consumed in confectionery markets worldwide (Hartel, 1996).

Milk fat consists of a wide variety of fatty acids: short-chain and long-chain fatty acids and saturated and unsaturated fatty acids (Hartel and Kaylegian, 2001). Therefore, fractionation of milk fat requires tremendous scientific and technological investigation (Deffense, 1993; Hartel and Kaleygian, 2001; ten Grotenhuis *et al.*, 1999; van Aken *et al.*, 1999). Furthermore, it has been reported that minor components of milk fats exert a significant influence on the crystallisation behaviour when it is mixed with cocoa butter and other confectionery fats, having, for instance, a softening effect and antibloom properties (Barna *et al.*, 1992; Jeffery, 1991; Ransom-Painter *et al.*, 1997; Tietz and Hartel, 2000). To clarify the mechanisms of these effects and to enable the use of milk-fat fractions as a cocoa butter improver, many researchers have worked on the problem of confectionery fat blending by studying intact milk fat and milk-fat fractions. In the next section the effects of milk-fat fractions on the crystallisation behaviour of cocoa butter and finished chocolate will be discussed based on preliminary experimental results.

1.4.2.1 *Phase behaviour of cocoa-butter/milk-fat fractions*

Typically, milk fat exhibits three major differential scanning calorimetry (DSC) melting peaks, as shown in figure 1.7. Each fraction of milk fat has a different effect on the cocoa butter fraction: the high-melting fraction inhibits bloom formation, the middle-melting fraction shows a eutectic effect and the low-melting fraction lowers the melting point of the cocoa butter.

We have studied the crystallisation behaviour of cocoa butter mixed with intact milk fat. As shown in figure 1.7, melting enthalpy is decreased with increasing amount of added milk fat. This result shows good agreement with results of other, previous, studies (Sabariah *et al.*, 1998). Barna *et al.* (1992) indicated that the peak top temperature of DSC melting curves decreased with increasing concentration of milk fat. They further considered that the lowering of the peak top temperatures are also caused by a decrease in fusion of enthalpy, which is essentially a result of the decrease in the amount of cocoa butter crystals present in the fat mixtures.

The endothermic DSC peaks in figure 1.7 show the melting of stable crystals of cocoa butter as well as additional peaks when the added milk-fat concentration is increased. The total area of the additional peaks increases with increasing amount of milk-fat fraction, indicating eutectic effects in the binary mixture phase of cocoa butter and milk fat.

X-ray analysis was carried out on the samples described in figure 1.7, and the observed small-angle (long-spacing) data are shown in figure 1.8. When the concentration of milk fat was below 40%, single long spacing (6.3 nm) was detected, corresponding to the TCL structure of the stable polymorph of cocoa

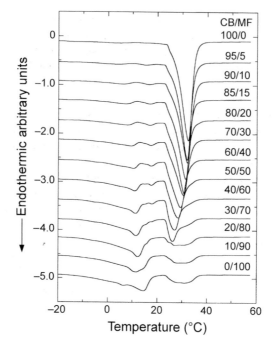

Figure 1.7 Differential scanning calorimeter (DSC) heating thermopeaks for cocoa butter (CB) and milk fat (MF) mixtures (composition is expressed in percentages).

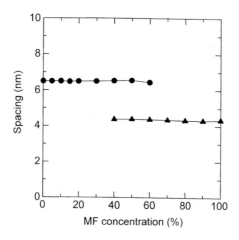

Figure 1.8 Small-angle X-ray diffraction spectra of cocoa butter/milk fat (MF) mixtures.

butter (Loisel *et al.*, 1998). When the concentration of milk fat was between 40% and 60%, two small-angle values are observed. At a milk-fat concentration of 60% and above the spectrum at spacing 6.3 nm disappears and a new spectrum at 4.2 nm appears. The spectrum at 4.2 nm corresponds to the DCL structure.

These results indicate the formation of a eutectic phase, because different crystal structures are revealed to exist in the same mixture system at specific mixture ratios of cocoa butter and milk fat. From these observations, we conclude that milk fats do not easily form miscible crystals when blended with cocoa butter.

1.4.2.2 Softening effects

We have examined the softening effects of milk-fat fractions on the finished chocolate for various mixture levels. The hardness of the chocolate was measured by a penetration method (figure 1.9). The hardness of chocolate decreased (the chocolate softened) as the amount of milk fat increased. Timms (1980) and Kaylegian (1997) suggested that the softening effect of milk fat is due to dissolution of cocoa butter crystals in liquid fat combined with the formation of a eutectic phase between the cocoa butter and the middle-melting milk-fat fraction. It has been suggested (Giovanni, 1988; Hogenbirk, 1984) that the softening effect is related to the presence of liquid oil in the milk fat. We assume that the decrease in solid fat content is the main reason for the softening effect, which is caused by the dilution of cocoa butter by the addition of milk-fat fractions. A decrease in solid fat content makes the structure of the chocolate more fragile.

1.4.2.3 Bloom inhibition

The effects of addition of milk fat on the retardation of fat-bloom formation have been well known for a long time, but the exact mechanism is still open to question. The true mechanism that underlies the effect of bloom inhibition may involve interactions between the various TAGs involved in milk fats and the polymorphic nature of cocoa butter, and some other components such as minor lipids. It has been reported that the high-melting fraction in milk fat exhibits

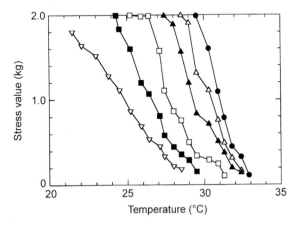

Figure 1.9 Stress values for cocoa butter (control) and cocoa butter/milk fat (MF) mixtures. —●— Control; —△— MF fraction = 5%; —▲—MF fraction = 10%; —□—MF fraction = 20%; —■— MF fraction = 30%; —▽— MF fraction = 40%.

an antibloom effect (Kaylegian, 1997), and that the lipid content, in particular polar lipids such as diacylglycerols and phospholipids, involved in milk fat have a significant effect on the retardation of fat bloom (Tietz and Hartel, 2000). Bricknell and Hartel (1998) also studied the effects of chocolate microstructures with respect to sugar particles on fat-bloom formation. Whatever components are found to improve antibloom properties, a microscopic understanding of the fat-bloom processes is needed (Sato and Koyano, 2001).

1.5   Fat crystallisation in oil-in-water emulsions

The crystallisation of solid fats in O/W emulsion droplets influences stability, rheology and appearance of emulsions (Boekel and Walstra, 1981; Boode et al., 1991; Dickinson and McClements, 1996). Therefore, it is important to analyse fat crystallisation processes in O/W emulsions. The rate and extent of crystalli-sation, the effect of polymorphism and emulsifiers, the influence of emulsion droplet size and droplet–droplet interactions and the effect of rate of cooling and subsequent temperature history on fat crystallisation behaviour must be clarified, as summarised by Povey (2001). The elucidation of these complicated crystallisation processes in O/W emulsions can be achieved through two-step studies: (a) by monitoring in situ the crystallisation process under a well-defined simple model; (b) by extending this model to more complicated systems containing polymorphic fats under varying temperature treatments and so on.

For this purpose, ultrasonic velocity measurement has been employed to monitor *in situ* crystallisation processes (Dickinson et al., 1991) based on the principle that the event of crystallisation of a liquid oil phase dispersed in a water phase can be monitored by means of the ultrasonic sound velocity, which increases as the transformation from the liquid to the solid phase progresses. In addition, DSC is also effective for monitoring fat crystallisation in O/W emulsions (Katsuragi et al., 2000).

Our recent studies on the kinetic properties of the nucleation processes of palm oil and palm mid-fractions (PMFs) in O/W emulsions show a remark-able acceleration when highly hydrophobic food emulsifiers are added. The emulsions were formed by using Tween® 20 as the main emulsifier (sorbitan monolaurate polyglycol ether; Tween is a trademark of ICI/Uniqema), dis-tilled water and oils (water:oil, 80:20). The concentration of Tween® 20 was 2wt% with respect to the oil and water phases. The average droplet size was 0.8–1.0 µm. Highly hydrophobic sucrose fatty-acid oligoesters (SOEs) (1) were employed as additives in the oil phase: for example, palmitic acid (P-170), stearic acid (S-170), oleic acid (O-170) and lauric acid (L-195) moieties were used, the melting and crystallisation temperatures of which are listed in table 1.4. The average degree of esterification of the SOEs emplyed in the study was approximately 5.

CH$_2$COOR

O

H   CH$_2$OR   H

OR   H

OR   H   OR   CH$_2$OH

OR

H   OR   OR   H

R = fatty acid moiety

1

Table 1.4 Crystallisation (T_c) and melting (T_m) temperatures of sucrose fatty-acid oligoesters (SOEs) measured by differential scanning calorimetry

SOE	T_m (°C)	T_c (°C)
L-195	17.9	10.7
P-170	52.6	52.1
S-170	58.2	56.2
O-170	n.d.	n.d.

n.d., not detected.

1.5.1 Crystallisation in emulsions of palm oil in water

It is known that the rate of crystallisation of palm oil is low compared with that of other fats, probably because of the presence of POP and other mixed-acid TAGs. In the O/W emulsion involving palm oil as the oil phase, the crystallisation rate is reduced even further, as observed by ultrasonic velocity measurements (figure 1.10). In figure 1.10(a) the linearly increasing ultrasonic velocity values (V values) showed on cooling a sudden increase from the linear line at around 30°C, because of the crystallisation of the palm stearin fraction. Further cooling also increased the V value at around 14°C because of the crystallisation of the palm olein fraction. The addition of S-170 and P-170 at 1wt% with respect to the oil phase did not change the V values, meaning that these additives did not affect the crystallisation of palm oil in the bulk phase.

In the O/W emulsion [figure 1.10(b)], the V values started to decrease at around 40°C because of the structuring of water molecules and they continued to decrease on further cooling. The sudden increase in the V value at 11°C is caused by the crystallisation of palm oil. On heating from 0°C, the V value still increased because of a very slow rate of crystal growth of palm oil in the emulsion, but melting is revealed in the decrease of the V value at around 18°C (palm olein) and at around 35°C (palm stearin). Figure 1.10 shows that, owing to emulsification, the crystallisation temperature (T_c) decreased from 30°C to 11°C but that the melting behaviour did not change.

The addition of SOEs with long saturated acid moieties increased the nucleation rate and, as a whole, increased the amount of crystal fraction of palm oil. This is primarily attributable to accelerated nucleation, although the growth

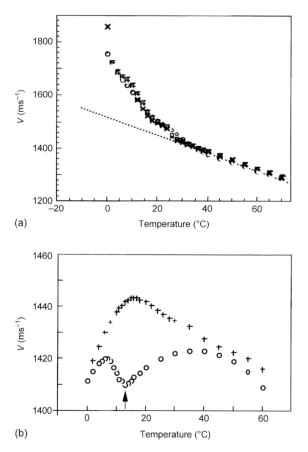

Figure 1.10 Ultrasonic velocity values (V) of (a) bulk palm (cooling) and (b) a palm-oil/water emulsion. S-170, stearic acid; P-170, palmitic acid. ⊙ Pure; ▢ S-170 1%; ✕ P-170 1%; ○ cooling; + heating.

rate of palm oil was decreased (Hodate *et al.*, 1997). Figure 1.11(a) shows T_c increases up to 20°C and that the V values of fat-crystallised emulsion also increased by $10 \, \text{m s}^{-1}$ by the addition of S-170 (1 wt% with respect to palm oil). This was not ascribed to the rate of crystal growth, since the crystal growth rate was reduced by the addition of S-170 and P-170 [figure 1.11(b)]. This means that heterogeneous nucleation of palm oil in the emulsion was accelerated by the addition of the SOEs.

1.5.2   Crystallisation in emulsions of palm mid-fraction in water

PMF, a fraction of palm oil having melting points in-between those of palm stearin and palm olein, has been employed for vegetable-fat-based creams. The effect of the addition of SOEs on the crystallisation of PMF in the emulsion were monitored by ultrasound and X-ray diffraction methods. Crystallisation

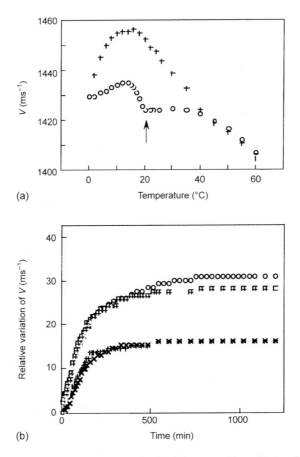

Figure 1.11 (a) Ultrasonic velocity values (V) of palm-oil/water emulsion with the addition of S-170, and (b) relative variation of V values during isothermal crystallisation with the four additives. The concentrations of the additives were all 1 wt% with respect to palm oil. O-170, oleic acid; P-170, palmitic acid; S-170, stearic acid. (a) ○, cooling; +, heating (b) ○, pure palm oil; +, P-170 added; ×, S-170 added; 卅, O-170 added.

was found to be accelerated and the polymorphic nature modified from α-tending to β′-tending. figure 1.12(a) shows the increase in the V value of PMF/water emulsions with P-170 additive at different concentrations. Since the crystal growth of PMF was also retarded by the addition of SOEs, heterogeneous nucleation was promoted by the additives. The variation in T_c as a function of P-170 concentration showed a two-stage process [figure 1.12(b)]. It is worth noting that the acceleration of PMF crystallisation resulted in the conversion of polymorphic crystallisation from α-tending, without SOE additives, to β′-tending, because the increase in T_c up to 17°C made it possible to crystallise out the β′ form, which otherwise was not predominant because of the rapid crystallisation of the α form.

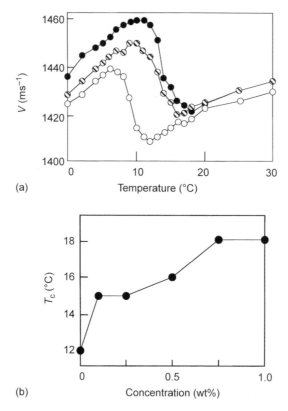

(a)

(b)

Figure 1.12 (a) Ultrasonic velocity values (*V*) of palm mid-fraction (PMF)/water emulsion with the addition of P-170 (palmitic acid) during the cooling process, and (b) crystallisation temperature (*T*_c) of PMF at various concentrations of P-170. ○ Pure PMF; ◐ + 0.50 wt%; ● + 1.00 wt%.

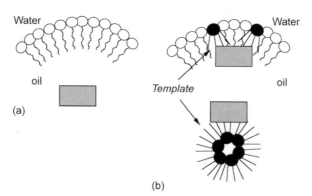

Figure 1.13 A model of heterogeneous nucleation of fat at the interface of an oil-in-water emulsion, as accelerated by sucrose fatty-acid oligoester (SOE): (a) pure emulsion, and (b) the influence of SOE additives. ⋔ SOE; ⌇ Tween® 20; ▬ Crystal.

From the experimental results on the natural fats described here and *n*-hexadecane as a model system (Kaneko *et al.*, 1999; Katsuragi *et al.*, 2001), and using other experimental techniques such as DSC and synchrotron radiation X-ray beams, enabling us to observe *in situ* the crystallisation events in the emulsified state, we have reached a conclusion about the mechanisms of hetero-geneous nucleation, as illustrated in figure 1.13. The acceleration of nucleation occurs at the oil–water interface at lower concentrations of additive molecules adsorbed at the interface, and occurs around reversed hexagonal micelles formed by the SOE molecules at high concentrations.

The acceleration effects of nucleation of fats in the O/W emulsions discussed in this section were observed by using hydrophobic polyglycerin esters and diacylglycerols having high-melting fatty-acid moieties. With regard to the fats, the crystallisation of palm kernel oil and cocoa butter in O/W emulsions was also accelerated by the additives described above. Therefore, a general conclusion may be drawn that the use of hydrophobic emulsifiers is remarkably efficient for the acceleration of fat crystallisation, which is usually reduced in emulsion phases because of the size effects of the crystallising media in the emulsions. This may indicate fat crystal network formation in the various dispersed phases, where the food-fat crystals play an important role.

1.6   Conclusions

The solidification and transformation properties of fat materials deserve further study regarding fundamental aspects, as discussed in this chapter. It is worthy of note again that the interrelations between polymorphism, solidification kinet-ics and crystal particle networks underly the apparently complicated physical behaviour of various food fats. Synchrotron radiation, X-ray diffraction and ultrasonic velocity techniques have great potential and applicability in such studies. Further research is to be carried out in this area.

Acknowledgements

The authors express their thanks to Dr S. Ueno and Mr Y. Hamada of Hiroshima University and to Dr T. Katsuragi of Mitsubishi Chemical Co. for cooperative work and valuable discussion.

References

Ali, A.R.M. and Dimick, P.S. (1994) Melting and solidification characteristics of confectionery fats: anhydrous milk fat, cocoa butter and palm kernel stearin blends. *J. Am. Oil Chem.*, **7**, 803-806.
Arishima, T., Miyabe, T., Matsumura, Y. and Matsumoto, M. (1993) Fat crystals and whipped cream, in *Collected Abstracts of the 32nd National Conference of the Japan Oil Chemists' Society*, Tokyo, p. 101.

Aronhime, J.S., Sarig, S. and Garti, N. (1988) Dynamic control of polymorphic transformation in triglycerides by surfactants: the button syndrome, *J. Am. Oil Chem. Soc.*, **65**, 1144-1150.

Barna, C.M., Hartel, R.W. and Metin, S. (1992) Incorporation of milk fat fractions into milk chocolate, *The Manufacturing Confectioner* (June), 107-116.

Baxter, J.F., Morris, G.J. and Gaim-Marsoner, G. (1995) Process for retarding fat bloom in fat-based confectionery masses, EU patent application 95306833.5.

Bennema, P., Vogels, L.J.P. and de Jong, S. (1992) Morphology of β phase monoacid triacylglycerol crystals: theory and observations, *J. Cryst. Growth*, **123**, 141-162.

Blaurock, A. (1999) Fundamental understanding of the crystallisation of oils and fats, in *Physical Properties of Fats, Oils and Emulsifiers* (ed. N. Widlak), AOCS Press, Champaign, IL, pp. 1-32.

Boekel, V. and Walstra, P. (1981) Stability of oil-in-water emulsions with crystals in the dispersed phase. *Coll. & Surf.*, **3**, 109-118.

Boode, K., Bisperink, C. and Walstra, P. (1991) Destabilization of O/W emulsions containing fat crystals by temperature cycling. *Colloid Surf.*, **61**, 55-74.

Boubekri, K., Yano, J., Ueno, S. and Sato, K. (1999) Polymorphic transformations in *sn*-1,3-distearoyl-2-ricinoleyl-glycerol. *J. Am. Oil Chem. Soc.*, **76**, 949-955.

Bricknell, J. and Hartel, R.W. (1998) Relation of fat bloom in chocolate to polymorphic transition of cocoa butter. *J. Am. Oil Chem. Soc.*, **75**, 1609-1615.

Chrysam, M.M. (1996) Margarines and spreads, in *Bailey's Industrial Oil and Fat Products, 5th edn Vol. 3. Edible Oil and Fat Products: Products and Application Technology* (ed. Y.H. Hui), John Wiley, New York, pp. 65-114.

Deffense, E. (1993) Milk fat fractionation today: a review. *J. Am. Oil Chem. Soc.*, **70**, 1193-1201.

de Man, J.M. (1982) Microscopy in the study of fats and emulsions. *Food Microstructure*, **1**, 209-222.

de Man, J.M. (1983) Consistency of fats: a review. *J. Am. Oil Chem. Soc.*, **60**, 6-11.

de Man, J.M. (1999) Relationship among chemical, physical, and textural properties of fats, in *Physical Properties of Fats, Oils and Emulsifiers* (ed. N. Widlak), AOCS Press, Champaign, IL, pp. 79-95.

de Man, L., D'Souza, V. and de Man, J.M. (1992) Polymorphic stability of some shortenings as influenced by the fatty acid and glyceride composition of the solid phase. *J. Am. Oil Chem. Soc.*, **69**, 246-250.

Dickinson, E. and McClements, D.J. (1996) *Advances in Food Colloids*, Blackie Academic and Professional, London.

Dickinson, E., McClements, D.J. and Povey, M.W. (1991) Ultrasonic investigation of the particle size dependence of crystallization in *n*-hexadecane-in-water emulsions. *J. Colloid Interface Sci.*, **142**, 103-110.

Fahey, D.A. and Small, D.M. (1986) Surface properties of 1,2-dipalmitoyl-3-acylglycerols. *Biochemistry*, **25**, 4468-4472.

Faulkner, R.W. (1981) Cocoa butter equivalents are truly specialty vegetable fats. Paper presented at the 35th PMCA Production Conference, June, pp. 67-73; copy available from the Pennsylvania Manufacturing Confectioners' Association Production Conference Committee, PO Box 68, Perkiomenville, PA 18074.

Giovanni, L.B. (1988) Practical aspects of the eutectic effect on confectionery fats and their mixtures. *The Manufacturing Confectioner* (May), 65-80.

Gunstone, F.D. (1997) Major sources of lipids, in *Lipid Technologies and Applications* (eds. F.D. Gunstone and F.B. Padley), Marcel Dekker, New York, pp. 19-50.

Hachiya, I., Koyano, T. and Sato, K. (1989a) Seeding effects on solidification behavior of cocoa butter and dark chocolate: I. Kinetics of solidification. *J. Am. Oil Chem. Soc.*, **66**, 1757-1762.

Hachiya, I., Koyano, T. and Sato, K. (1989b) Seeding effects on solidification behavior of cocoa butter and dark chocolate: II. Physical properties of dark chocolate. *J. Am. Oil Chem. Soc.*, **66**, 1763-1770.

Hagemann, J.W. (1988) Thermal behavior and polymorphism of acylglycerides, in *Crystallization and Polymorphism of Fats and Fatty Acids* (eds. N. Garti and K. Sato), Marcel Dekker, New York, pp. 9-95.

Hartel, R.W. (1996) Application of milk fat fractions in confections. *J. Am. Oil Chem. Soc.*, **73**, 945-954.

Hartel, R.W. and Kaleygian, E.K. (2001) Advances in milk fat fractionation: technology and applications, in *Crystallization Processes in Fats and Lipid Systems* (eds. N. Garti and K. Sato), Marcel Dekker, New York, pp. 381-427.

Hernqvist, L. and Larsson, K. (1982) On the crystal structure of the β'-form of triglycerides and structural changes at the phase transitions liq-α-β'-β. *Fette Seifen Anstrchim.*, **84**, 349-354.

Hodate, Y., Ueno, S., Yano, J., Katsuragi, T., Tezuka, Y., Tagawa, T., Yoshimoto, N. and Sato, K. (1997) Ultrasonic velocity measurement of crystallization rates of palm oil in oil-water emulsion. *Colloid Surf.*, **128**, 217-224.

Hogenbirk, G. (1984) Compatibility of specialty fats with cocoa butter. *The Manufacturing Confectioner* (June), 59-64.

Jeffery, M.S. (1991) The effect of cocoa butter origin, milk fat and lecithin levels on the temperability of cocoa butter systems. Paper presented at the 45th PMCA Production Conference, 30-36; copy available from the Pennsylvania Manufacturing Confectioners' Association Production Conference Committee, PO Box 68, Perkiomenville, PA 18074.

Kaneko, F. (2001) Polymorphism and phase transitions of fatty acids and triacylglycerols, in *Crystallisation Processes in Fats and Lipid Systems* (eds. N. Garti and K. Sato), Marcel Dekker, New York, pp. 53-97.

Kaneko, N., Horie, T., Ueno, S., Yano, J., Katsuragi, T. and Sato, K. (1999) Impurity effects on crystallization rates of *n*-hexadecane in oil-in-water emulsions. *J. Cryst. Growth*, **197**, 263-270.

Kaneko, F., Yano, J. and Sato, K. (1998) Diversity in the fatty-acid conformation and chain packing of *cis*-unsaturated lipids. *Curr. Opin. Struct. Biol.*, **8**, 417-425.

Katsuragi, T. (1999) Interactions between surfactants and fats, in *Physical Properties of Fats, Oils, and Emulsifiers* (ed. N. Widlak), AOCS Press, Champaign, IL, pp. 211-219.

Katsuragi, T., Kaneko, N. and Sato, K. (2000) DSC study of effects of addition of sucrose fatty acid esters on oil phase crystallisation of oil in water in oil emulsion. *J. Jpn. Oil Chem. Soc. (Yukagaku)*, **49**, 255-262.

Katsuragi, T., Kaneko, N. and Sato, K. (2001) Effects of addition of hydrophobic sucrose fatty acid oligoesters on crystallization rates of *n*-hexadecane in oil-in-water emulsions. *Coll. Surf. B*, **20**, 229-237.

Kaylegian, E.K. (1997) Milk fat fractions in chocolate. *The Manufacturing Confectioner* (May), 79-84.

Kellens, M., Meeussen, W., Gehrke, R. and Reynears, H. (1991) Synchrotron radiation investigations of the polymorphic transitions in saturated monoacid triglycerides. Part 2: polymorphism study of a 50:50 mixture of tripalmitin and tristearin during crystallization and melting. *Chem. Phys. Lipids*, **58**, 145-158.

Kellens, M., Meeussen, W. and Reynaers, H. (1992) Study on the polymorphism and the crystallization kinetics of tripalmitin: a microscopic approach. *J. Am. Oil Chem. Soc.*, **69**, 906-911.

Koyano, T., Hachiya, I. and Sato, K. (1990) Fat polymorphism and crystal seeding effects on fat bloom stability of dark chocolate. *Food Structure*, **9**, 231-240.

Koyano, T., Hachiya, I. and Sato, K. (1992) Phase behavior of mixed systems of SOS and OSO. *J. Phys. Chem.*, **96**, 10514-10520.

Koyano, T., Kato, Y., Hachiya, I., Umemura, R., Tamura, K. and Taguchi, N. (1993) Crystallization behavior of ternary mixture of POP/POS/SOS. *J. Jpn. Oil Chem. Soc.*, **42**, 453-457.

Krog, N. and Larsson, K. (1992) Crystallization at interfaces in food emulsions—a general phenomenon. *Fat Sci. Technol.*, **94**, 55-57.

Larsson, K. (1966) Classification of glyceride crystal forms. *Acta Chem. Scand.*, **20**, 2255-2260.

Larsson, K. (1972) Molecular arrangement in glycerides. *Fette Seifen Anstrichm.*, **74**, 136-142.

Lohman, M.H. and Hartel, R.W. (1994) Effects of milk fat fractions on fat bloom in dark chocolate. *J. Am. Oil Chem. Soc.*, **71**, 267-276.

Loisel, C., Keller, G., Lecq, G., Bourgaux, C. and Ollivon, M. (1998) Phase transitions and polymorphism of cocoa butter. *J. Am. Oil Chem. Soc.*, **75**, 425-439.

MacMillan, S.D., Roberts, K.J., Rossi, A., Wells, M., Polgreen, M. and Smith, I. (1998) Quantifying the effect of shear on the crystallization of confectionery fats using on-line synchrotron radiation SAXS/WAXS techniques, in *The Proceedings of the World Congress on Particle Technology, Brighton*, pp. 96-103.

Marangoni, A.G. and Hartel, R.H. (1998) Visualisation and structural analysis of fat crystal networks. *Food Technol.*, **52**, 46-51.

Marangoni, A.G. and Rousseau, D. (1998a) The influence of chemical interesterification on the physicochemical properties of complex far systems: 1. Melting and solidification. *J. Am. Oil Chem. Soc.*, **75**, 1265-1271.

Marangoni, A.G. and Rousseau, D. (1998b) The influence of chemical interesterification on the physicochemical properties of complex far systems: 3. Rheological and fractality of the crystal network. *J. Am. Oil Chem. Soc.*, **75**, 1633-1636.

Minato, A., Ueno, S., Smith, K., Amemiya, Y. and Sato, K. (1997) Thermodynamic and kinetic study on phase behaviour of binary mixtures of POP and PPO forming molecular compound systems. *J. Phys. Chem. B*, **101**, 3498-3505.

Mohamed, H.M.A. and Larsson, K. (1992) Effects on phase transitions in tripalmitin due to the presence of dipalmitin, sorbitan-monopalmitate or sorbitan-tripalmitate. *Fat Sci. Technol.*, **94**, 338-341.

Naguib-Mostafa, A., Smith, A.K. and de Man, J.M. (1985) Crystal structure of hydrogenated canola oil. *J. Am. Oil Chem. Soc.*, **62**, 760-762.

Narine, S.H. and Marangoni, A.G. (1999) The difference between cocoa butter and Salatrim lies in the microstructure of the fat crystal network. *J. Am. Oil Chem. Soc.*, **76**, 7-13.

Padley, F.B. (1997) Chocolate and confectionery fats, in *Lipid Technologies and Applications* (eds. F.D. Gunstone and F.B. Padley), Marcel Dekker, New York, pp. 391-432.

Povey, M.J.W. (2001) Crystallisation of oil-in-water emulsions, in *Crystallisation Processes in Fats and Lipid Systems* (eds. N. Garti and K. Sato), Marcel Dekker, New York, pp. 251-288.

Ransom-Painter, K.L., Williams, S.D. and Hartel, R.W. (1997) Incorporation of milk fat and milk fat fractions into compound coatings made from palm kernal oil, *J. Dairy Sci.*, **80**, 2237-2248.

Rossel, J.B. (1967) Phase diagrams of triglyceride system, in *Advances in Lipid Research, Vol. 5* (eds. R. Paoletti and D. Kritchevsky), Academic Press, New York, pp. 53-408.

Rousseau, D., Marangoni, A.G. and Jeffrey, K.R. (1998) The influence of chemical interesterification on the physicochemical properties of complex fat systems: 2. Morphology and polymorphism. *J. Am. Oil Chem. Soc.*, **75**, 1265-1271.

Sabariah, V., Ali, A.R.M. and Chong, C.L. (1998) Chemical and physical characteristics of cocoa butter substitutes, milk fat and Malaysian cocoa butter. *J. Am. Oil Chem. Soc.*, **75**, 905-910.

Sato, K. (1996) Polymorphism of pure triacylglycerols and natural fats, in *Advances in Applied Lipid Research* (ed. F.B. Padley), JAI Press, London, pp. 213-268.

Sato, K. (1999) Solidification and phase transformation behaviour of food fats—a review. *Fett/Lipid*, **101**, 467-474.

Sato, K. and Koyano, T. (2001) Crystallization properties of cocoa butter, in *Crystallization Processes in Fats and Lipid Systems* (eds. N. Gari and K. Sato), Marcel Dekker, New York, pp. 429-455.

Sato, K. and Kuroda, T. (1987) Kinetics of melt crystallization and transformation of tripalmitin polymorphs. *J. Am. Oil Chem. Soc.*, **64**, 124-127.

Sato, K. and Ueno, S. (2001) Molecular interactions and phase behaviour of polymorphic fats, in *Crystallization Processes in Fats and Lipid Systems* (eds. N. Garti and K. Sato), Marcel Dekker, New York, pp. 177-209.

Sato, K., Arishima, T., Wang, Z.H., Ojima, K., Sagi, N. and Mori, H. (1989) Polymorphism of POP and SOS: I. Occurrence and polymorphic transformation. *J. Am. Oil Chem. Soc.*, **66**, 664-674.

Sato, K., Goto, G., Yano, J., Honda, K., Kodali, D.R. and Small, D.M. (2001) Atomic resolution structure analysis of β′ polymorph crystal of a triacylglycerol: 1,2-dipalmitoyl-3-myristoyl-*sn*-glycerol. *J. Lipid Res.*, **42**, 338-345.

Sato, K., Ueno, S. and Yano, J. (1999) Molecular interactions and kinetic properties of fats. *Prog. Lipid Res.*, **38**, 91-116.

Skoda, W. and van den Tempel, M. (1963) Crystallisation of emulsified triglycerides. *J. Colloid. Sci.*, **18**, 568-564.

Takeuchi, M., Ueno, S., Yano, J., Floter, E. and Sato, K. (2000) Polymorphic transformation of 1,3-distearoyl-*sn*-2-linoleoyl-glycerol. *J. Am. Oil Chem. Soc.*, **77**, 1243-1249.

ten Grotenhuis, E., van Aken, G.A., van Malssen, K.F. and Schenk, H. (1999) Polymorphism of milk fat studied by differential scanning calorimetry and real-time X-ray powder diffraction. *J. Am. Oil Chem. Soc.*, **76**, 1031-1039.

Tietz, R.A. and Hartel, R.W. (2000) Effects of minor lipids on crystallization of milk fat—cocoa butter blends and bloom formation in chocolate. *J. Am. Oil Chem. Soc.*, **77**, 763-771.

Timms, R.E. (1980) The phase behaviour of mixtures of cocoa butter and milk fat. *Lebensm. Wiss. Technol.*, **48**, 61-65.

Uragami, A., Tateishi, T., Murase, K., Kubota, H., Iwanaga, Y. and Mori, H. (1986) The development of hard butter by solvent fractionation system. *J. Jpn. Oil Chem. Soc. (Yukagaku)*, **36**, 995-1000.

van Aken, G.A., ten Grotenhuis, E., van Langevelde, A.J. and Schenk, H. (1999) Composition and crystallization of milk fat fractions. *J. Am. Oil Chem. Soc.*, **76**, 1323-1331.

van den Tempel, M. (1961) Mechanical properties of plastic disperse systems at very small deformations. *J. Colloid Sci.*, **16**, 284-296.

van Langevelde, A., van Malssen, K., Driessen, R., Goubitz, K., Hollander, F., Peschar, R., Zwart, P. and Schenk, H. (2000) Structure of $CnCn + 2Cn$-type (n = even) β′-triacylglycerols. *Acta Cryst. B*, **56**, 1103-1111.

Walstra, P., Kloek, W. and van Vliet, T. (2001) Fat crystal network, in *Crystallization Processes in Fats and Lipid Systems* (eds. N. Garti and K. Sato), Marcel Dekker, New York, pp. 289-328.

Watanabe, A., Tashima, I., Matsuzaki, N., Kurashige, J. and Sato, K. (1992) On the formation of granular crystals in fat blends containing palm oil. *J. Am. Oil Chem. Soc.*, **69**, 1077-1080.

Wiedermann, L.H. (1978) Margarine and margarine oil, formation and control. *J. Am. Oil Chem. Soc.*, **55**, 823-829.

Wille, R.L. and Lutton, E.S. (1966) Polymorphism of cocoa butter. *J. Am. Oil Chem. Soc.*, **43**, 491-496.

Wright, A.J., Hartel, R.W., Narine, S.S. and Marangoni, A.G. (2000) The effects of minor components on milk fat crystallization. *J. Am. Oil Chem. Soc.*, **77**, 463-475.

Ziegleder, G. (1985) Improved crystallization behaviour of cocoa butter under shearing. *Intern. Z. Lebensm. Techn. Verf.*, **36**, 412-416.

2 Bakery fats

John Podmore

2.1 Introduction

Fats have been used for many years in food preparation to provide structure, flavour and nutritive value. Climate and agricultural practices have influenced the fat used in food preparation. For example, people in northern climates use plastic fats such as butter and suet, whereas in more southerly climates liquid oils such as olive oil are more popular.

Economic conditions and population growth led to the invention of margarine as a substitute for butter, and developments in refining technology and fat modification techniques allowed the use of a widening range of fats in margarines and as alternatives for lard and suet.

Progress in the technology of butter production, oil refining and modification and in margarine and shortening manufacture has provided the food processor with a wide variety of fats and oils with differing functional properties to meet product and process needs.

The structural and crystalline properties of fats determine their functionality in food. This is readily illustrated in the manufacture of baked products such as short pastry, cake and puff or flaky pastry. Advances in emulsion technology and emulsifier systems have been applied to bakery products, giving improvements in bread volume and shelf-life as well as leading to recipe balance in other baked products and altering the requirement for plastic fats so that fluid and liquid shortenings can be used. The use of powdered fat and fat powders can add convenience in a number of food sectors (e.g. in prepared cake mixes, toppings and bread improvers).

In the more developed countries, nutritional demands, combined with rapid changes in lifestyles and eating habits which require quicker and easier food preparation at the point of use, have challenged suppliers in several important ways. These demands require the manufacturer of oils and fats to deliver products with the desired functionality but with improved nutritional and health characteristics—such as lower fat content, high in polyunsaturated and monounsaturated fatty acids and with a lower component of 'trans' fatty acids. These developments will lead to a search for novel oils and altered approaches to blending and modification.

In less-developed countries, consumers are not so demanding, and change is slower. The requirement for high levels of fat in the diet still exists. However, as these economics strengthen, they will develop the technology to exploit the oils that are readily available to them. For instance, Malaysia uses the fractionation process and some of the most up-to-date technology to produce a range of palm-oil-based bakery fats and margarines.

Fats in their natural state have been used as a bakery ingredient for almost · countless years improving the palatability and nutritive value of foods. Climate and agricultural practices have influenced the fats used in food preparation. Traditionally, in Europe butter, lard and suet have been extensively used whereas in the warmer Mediterranean area liquid vegetable oils, particularly olive oil, are the traditional culinary oils.

Increasing industrialisation and population expansion in the later part of the nineteenth century led to a shortage of traditional fats. These conditions stimulated the invention of the first substitute food—margarine. Margarine was invented in 1869 by a French chemist, Mégè Mouries (French patent 86480), and the invention was exploited by Dutch butter exporters. Starting from modest beginnings the margarine industry has developed into an important and sophisticated food-processing industry. Additionally, it has had important repercussions on the agricultural industry, for as margarine production has expanded it has stimulated an expansion in the production and export of tropical oils and oilseeds, and these now represent a substantial proportion of world trade in agricultural products.

Since the late nineteenth century butter production has seen improvements in agricultural techniques and processing methods combined with a more scientific appreciation of milk and milkfats which has led to a consistent and highly · characteristic product which endows cooked products with distinctive flavours and characteristic textures.

Progress in the understanding of the function of the ingredients in food now means that the fat processor and food manufacturer can work together to improve the food products available to the consumer. Crystalline form and product consistency have a profound influence on the performance of fats in foods, particularly in baked products. Thus an understanding of physical properties such as crystallisation behaviour, polymorphism and crystal structure in fats is necessary to control production processes such that they can be 'tailor made' to suit particular applications.

2.2   Production of margarine and shortening

The modern processor has available bland oxidatively stable and low coloured edible oils of vegetable and animal origin, achieved by the processes shown in figure 2.1. The quality standards for edible oils continue to be raised and so handling and refining practices are being continually improved which, combined

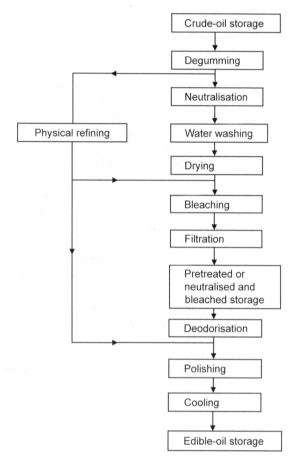

Figure 2.1 The stages of oil and fat refining.

with a clearer understanding of the influence of the minor components on shelf-life and flavour stability, leads to modified refining methods.

The refining process must be carried out to remove those impurities that have an adverse effect on oil quality but to avoid damaging the triacylglycerols. There is also a requirement that beneficial minor components be retained. Important minor components to be retained are tocopherols and phytosterols, which are biologically active and show antioxidant activity. Lower-temperature methods of refining and deodorisation are being applied to minimise this loss.

The refiner has the option of chemical or physical refining. The choice between the two methods depends on the economics of the processes, the quality of the crude oil and the ability to handle soapstock successfully. Most types of vegetable oil can be physically refined, a major exception being cottonseed oil because of the presence of gossypol (de Greyt and Kellens, 2000). These natural oils can be modified by hydrogenation, interesterification and fractionation, used either singly or in combination, to produce fats that bear no relation to the original material.

In the 1990s there was a change in emphasis away from hydrogenation as the way of providing the hard stock for the formulating of shortening and margarine oil blends. There were two main reasons for this change. The first reason is the ready availability of relatively inexpensive palm fractions and increasing confidence in their performance in bakery fat formulations. The second reason is the finding that 'trans' fatty acids are implicated in the development of coronary heart disease (Willet et al., 1992). Since the hydrogenation reaction can generate high levels of 'trans' fatty acids there has been a trend toward reducing reliance on hydrogenated oils in formulations. Instead palm stearins from the fractionation of palm oil have been found to be a valuable alternative to hydrogenated oils. However, the use of palm stearins, with their flat melting profile, giving rise to higher solid fat contents in the 30–40°C temperature range than seen with hydrogenated oils, has led to an increase in use of interesterification in order to mitigate this effect.

Blending oils and fats to achieve the required solid-to-liquid ratio is a major part of the processor's skill as it is critical to the firmness and texture of the finished product. Added to this is the influence of the crystal habit of the oils and fats selected and their polymorphism. Thus the processor requires an understanding of these characteristics when preparing blends for margarines and shortenings.

Irrespective of how the hardstock is obtained the basic requirements for blending are unchanged in that the desired solid-to-liquid ratios and crystallising characteristics be achieved in order to provide a stable finished product of the correct firmness, texture and crystal form.

2.3 Crystallisation behaviour

In common with all other long-chain molecules, fats and fatty acids exhibit polymorphism—that is, the ability to exist in more than one crystalline form and so possess multiple melting points. Triglycerides occur in any one of three basic polymorphs, designated α, β' and β (Bailey, 1950):

- the α form is the most loosely packed arrangement and hence is the least stable and has the lowest melting point
- the β' form is more stable than the α form but transforms irreversibly to the β form
- the β form is the most closely packed and is the polymorph with the highest melting point

Work by Timms (1984) describes the behaviour of a monoacid triglyceride, showing that with rapid cooling the α form is obtained which, on slow heating, melts to resolidify and give the β' form. After further slow heating it melts and resolidifies in the β form. Most fats possess an α form that is so unstable that it can be ignored; some also possess both β' and β forms; others possess only a stable β' or a stable β form. Some examples are shown in table 2.1.

X-ray studies on tristearin have shown that the triglycerides pack side by side in separate layers. The triglyceride molecules form the shape of a chair, and the molecules are arranged in pairs, head to tail. Figure 2.2 shows the packing arrangements possible in pairs of two or three fatty acids. Figure 2.3 shows the main features of the molecular packing of the three polymorphs of tristearin. It can be seen that for the α form the fatty-acid chains are perpendicular to a basal plane (that plane containing the methyl end-groups). In the β' form the fatty-acid chains are tilted at an angle to the basal plane. Each fatty acid has its hydrocarbon chain in regular zig-zag formation in a plane perpendicular to its neighbour. The β form, which also has the fatty acids tilted to the basal

Table 2.1 Crystallisation performance of some natural edible oils

β' form	β form
Cottonseed oil	Soybean oil
Palm oil	Sunflower oil
Tallow	Groundnut oil
Butter fat	Coconut oil
High-erucic-acid rapeseed oil	Palm kernel oil
	Lard
	Low-erucic-acid rapeseed oil

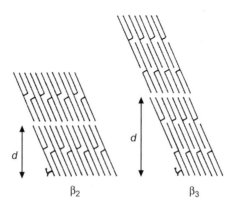

Figure 2.2 Triglycerides in the β_2 and β_3 polymorphic forms (de Jong, 1980, as cited in Timms, 1984).

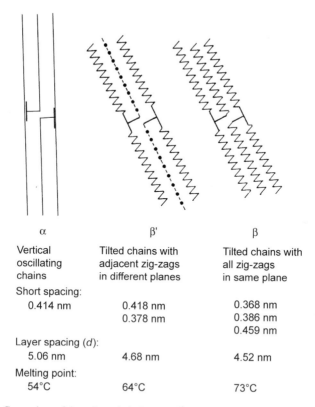

	α	β'	β
	Vertical oscillating chains	Tilted chains with adjacent zig-zags in different planes	Tilted chains with all zig-zags in same plane

Short spacing:
0.414 nm   0.418 nm   0.368 nm
0.378 nm   0.386 nm
0.459 nm

Layer spacing (*d*):
5.06 nm   4.68 nm   4.52 nm

Melting point:
54°C   64°C   73°C

Figure 2.3 Comparison of the polymorhpic forms α, β' and β, as exemplified by tristearin (StStSt).

plane, has the zig-zag planes of the hydrocarbon chains parallel to the same plane. These descriptions show an increasing closeness in packing and hence increasing melting point and stability.

Where there is a wide variety of molecular size and type of triglyceride, (e.g. cottonseed oil and beef tallow) the β' form rather than the β form predominates because it is more able to accommodate the distortion of the chain packing necessary for a solid solution.

From the foregoing comments it can be seen that each triglyceride has it own polymorphic and melting behaviour. However, in a mixture of triglycerides the individual triglycerides do not behave independently but take on a totally new character in terms of crystallisation behaviour. The systems are so complex that it is easier in the case of natural fats to describe them in terms of their different phases; thus the physical properties of a fat can be discussed in terms of its phase behaviour. In a fat or fat blend at a given temperature there will always be a liquid phase and a solid phase, and the solid phase can have several components, which can change with temperature and composition.

A phase diagram (Birker and Padley, 1987) can be used to show how blended fats interact for example to produce minima points (eutectic behaviour) or maxima points (solid solutions). Where the fats are compatible the isosolids line is horizontal for all compositions.

Thus, to summarise, the major features defining the firmness, texture and performance of a blended margarine and shortening are (Opfer, 1975):

- the proportion by weight of crystals, which is governed by the solid-to-liquid ratio
- the melting point of the crystals
- the crystal geometry; that is their size, shape and alignment
- the degree of formation of mixed crystals
- the ability of the crystals to flocculate into a network which increases firmness

Normally, the greater the quantity of solid triglyceride the greater the rigidity of the network because of the increased number of crystals and thus cohesive forces between them. These last prevent flow at stresses below those appropriate for the desired consistency. Changes in temperature will obviously change the product firmness and plastic behaviour by altering the quantity of crystals present, the hardness and the viscosity of the liquid triglycerides.

The crystal modification present will also influence the firmness and texture of the finished product in that the smaller and finer β' crystals can stabilise more liquid component than can the larger and coarser β crystals.

Fat crystallisation is initiated by nucleation in a supercooled system. In the manufacture of margarine and shortening the cooling rate, agitation and degree of supercooling control the rate of crystal growth and thus crystal size and crystal agglomeration, which affect the textural and melting properties of the fat product.

2.4 Processing

The vast majority of margarine and shortening is now manufactured on scraped surface heat exchangers (Joyner, 1953), and though they vary in design the basic principles apply to them all. The heat exchanger is made of two concentric tubes and in the annular space thus created a compressible refrigerant is circulated. The inner tube has a heated shaft which runs the length of the tube on which are mounted floating scraper blades. As the shaft rotates these blades scrape the internal surface of the tube.

In the process the liquid emulsion or fat blend is pumped along the tube at a fixed speed and the rotating blades remove the chilled product from the walls. This constant renewal of the cooling surface and the turbulence created leads to supercooling and the initiation of crystal nuclei and hence crystallisation. The supercooled and partly crystallised product can then be pumped to a worker unit where crystallisation is completed and the heat of crystallisation released.

It can be seen that it is important that the flow rate of product does not vary, irrespective of the fact that as it is chilled there is a viscosity increase. There is also a temperature differential across the product flow leading to a range of crystalline compositions being created; so, in order to minimise product variability tight and continuous control must be maintained at all stages on the evaporation rate of the refrigerant.

The worker unit (or units) is (are) also tubular and can contain a system of beaters to ensure that the crystal structure is developed in a dynamic environment, hence controlling the size of the crystal aggregates and giving a smooth plastic texture. It is also possible to make no provision for mechanical agitation to induce growth of large crystals from the mass in order to provide a product firm enough to pack into wrapped units.

There are now available systems of much greater complexity to improve the texture and plasticity of a widening range of blend types and that meet greater specificity in the requirements of the user. Factors such as shaft rotation speed, scraper blade design, size of the annular space and size and location of worker units all can affect the final texture of the product. These are described in detail in chapter 6.

2.5 Plastic bakery fats

Fats and oils in their natural state have been used as a bakery ingredient for many years to improve the mouth feel and palatability of the finished foods. The growing sophistication of the bakery industry, in terms of both product range and automation techniques, requires greater control of the ingredients used, including fats. The bakery industry is also now becoming more concentrated, with bakery plants becoming larger, more automated and more specialised. These changes in turn mean that process control can be substantially improved and fats specified to meet precise performance criteria.

In designing shortenings and margarine for bakery use it is important to understand the application so that the functional properties required can be designed into the product by way of the oil blend used.

The availability of relatively inexpensive palm-oil fractions and concerns about overfishing in some of the world's seas has meant there has been a major shift in European countries to the use exclusively of vegetable oil products in bakery fats. However, once the blend is selected the quality and the process control techniques necessary to maintain the desired properties must be applied.

Comparison of the rheological properties of butter with those of margarine shows there are major differences. Butter is a considerably more complex system than is margarine. The fat system of butter is less homogeneous than that of margarine as butter is made up of liquid and crystalline fat, fat globules and globule membrane fragments interspersed with moisture droplets (Mulder and Walstra, 1974).

The globular fat influences the texture and consistency in that the solid fat inside each globule causes it to go rigid with increase in firmness. However, these globules cannot form solid networks with the crystals outside. Thus, when butter and margarine consistency is compared a margarine with less solid fat than butter is equal in firmness to the butter.

The functions of fat in bakery applications are (Hodge, 1986):

- shortening power and lubricity
- batter aeration
- emulsifying properties
- provision of an impervious layer
- improvements in keeping properties
- provision of flavour.

The functionality of fat will be discussed in terms of short pastry, cake and puff or flaky pastry in order to demonstrate how fat contributes to the structure and eating quality of the product and hence how the fat can be blended and processed to maximise these functions (Pyler, 1973). Some typical fat-blend recipes are shown in table 2.2. The role of fat in biscuit manufacture will also be discussed.

Table 2.2 Typical oil blends used in pastry fats

Product	All-vegetable blend	%	Not all-vegetable blend	%
Sweet paste	Hyd. palm oil, mpt 50°C	10	Hyd. marine oil, mpt 35°C	50
	Hyd. rapeseed oil, mpt 35°C	35	Tallow	25
	Palm oil	20	Rapeseed oil	25
	Rapeseed oil	35		
Boiled pie paste	Hyd. palm oil, mpt 44°C	15	Tallow	90
	Hyd. soybean oil, mpt 36°C	55	Rapeseed oil	10
	Rapeseed oil	30		
Puff pastry	Hyd. palm oil, mpt 45°C	65	Hyd. fish oil, mpt 46°C	40
	Palm oil	25	Palm stearin	35
	Rapeseed oil	10	Rapeseed oil	25
	Plastic blend		Fluid blend	
Bread dough	Palm stearin	45	Hyd. rapeseed stearin	8
	Hyd. rapeseed oil mpt 35°C	35	Rapeseed oil	92
	Rapeseed oil	20		

Note: Hyd., hydrogenated; mpt, melting point.

2.5.1 Short pastry

Short pastry is used in a wide range of savoury and fruit products. The major ingredients are simply flour, fat and water. When mixing flour and water the wheat proteins are hydrated to form 'gluten' during the preparation of the dough.

Wheat contains four classes of protein, based on solubility in certain solvents: albumins, which are water soluble; globulins, which are also water soluble; glutenins, which are acid and alcohol soluble; and gliadins, which are soluble in aqueous alcohol. It is the glutenins and gliadins that provide the gluten of the wheat that gives rise to the tough and extensible network in flour–water doughs.

The development of an elastic flour–water dough requires access to water of the wheat protein, sufficient water to hydrate the protein in the flour and sufficient energy in mixing to cause the glutenins to aggregate and develop into an elastic mass.

When a flour–water dough is baked it develops into a hard brittle texture, often described as being 'flinty'. The function of the fat in a flour–water system is to coat flour particles and so limit the extent of hydration by minimising moisture ingress. The interruption in development of the gluten results in planes of weakness and so the product becomes 'shorter' and more inclined to melt in the mouth. In simplistic terms, it can be seen that too little fat will result in a tough and harsh eating pastry, and too much will so interrupt the gluten development that the dough will be loose and soft to handle and too fragile when baked.

The same comments apply when the shortening or margarine is too firm or too soft. A firm fat with a very high solid triglyceride content will not smear easily and so will not distribute itself successfully in the dough to interrupt gluten development. Consequently, use of such fats lead to a flinty product exhibiting shrinkage. Liquid or fluid shortening, at the other extreme, leads to sloppy and soft and unworkable doughs. Thus a fat for short pastry should be of firm consistency so that when being mixed into the dough it retains sufficient body under shear conditions to be distributed as protective thin films and droplets throughout the dough.

There are three types of short pastry:

- sweet paste, for use in fruit pies, jam tarts and so on
- savoury paste, for use in meat pies, pasties, quiche Lorraine and so on
- rich paste, containing little or no added water, typified by Viennese and confectioners' biscuits such as Shrewsbury biscuits (basic recipes are given in table 2.3)

In sweetened pastes sugar reduces the water availability, thereby reducing gluten development. Sweet pastes are made by either the 'rubbing-in' method or the 'creaming' method. In the former fat and flour are mixed together before the addition of other ingredients, with the sugar in solution, after which the

Table 2.3 Basic recipes for short pastry doughs

	Ingredient (%)[a]						
Pastry	flour	baking powder	salt	fats	sugars	water	milk
Sweet	100	0	0.78	50	18.75	0	16.63
Unsweetened	100	3.13	3.13	50	0	27.5	0
Rotary-moulded short biscuit	100	1–2	1	32	30	12	1
Shortbread	100	0	0	50	25	0	0
Sweet paste	100	0	0	50	18.75	0	0
Wine biscuits	100	0	0	62.5	50	0	0
Shrewsbury biscuit	100	0	0	31.25	25	0	0
Viennese biscuit	100	0	0	65.6	25	0	12.5
Choux paste	100	0	0	50	0	125	0

[a]Amount as a percentage of the weight of flour.

ingredients are mixed to a paste in the shortest possible time. When creaming is used equal parts of fat and flour are creamed together, then the rest of the ingredients are added followed by the balance of the flour.

Savoury pastries may be subdivided according to two principle methods of production: boiled paste, used for items such as pork pies; or cold-water paste, used for cornish pasties or quiche Lorraine. Boiled pie paste is made commercially by rubbing the fat into the flour and then adding boiling water, which contains salt. A relatively stiff paste results as a result of the gelatinisation of flour starch, which can be either 'blocked-out' into tins or hand raised. The pastry when baked has a crisp and slightly greasy feel. The fat used generally has a higher melting point than that used for sweetened paste, as it is necessary for the fat to solidify as the temperature falls to avoid fat loss, to give an oily paste. The fat must also show a good plastic character and not become hard or brittle. Animal fats such as lard and tallow are favoured for this application, although vegetable-oil blends are also used. Animal fats have the additional advantage of contributing to the flavour of the meat filling.

In cold-water paste cold water is substituted for hot or boiling water. Basic recipes are shown in table 2.4. In some cases the fat is creamed with some flour before the water is added. These pastes have a high potential for gluten development, which affects the handling characteristics and influences the texture of the baked pastry to make it firm and slightly brittle, as the final eating quality will be influenced by the filling.

The fat used for a savoury paste is the same as that used in sweetened paste; that is, one that can give good distribution during the minimum mixing time. The traditional fat used for short pastry for savoury products is lard, which, because of its particular triglyceride structure (Carlin, 1944), crystallises in the β polymorph, which has led to the belief that the β form is preferred for short

Table 2.4 Recipes for savoury pastries

Ingredient[a]	Pork pies[b]	Cornish pasties[c]	Quiche Lorraine[c]
Flour (%)	100	100	100
Shortening (%)	44	45	48
Water (%)	31	20	38
Salt (%)	1.5	1	1
Baking powder (%)	0	0	1.2

[a]Amounts are given as a percentage of the weight of flour.
[b]Boiled-water method.
[c]Cold-water method.

pastry manufacture. However, compounded shortenings containing β and β' polymorphs have been found to perform well in short pastry recipes to give a good 'short' texture and good mouth feel.

A feature of considerable importance is the ability of the fat to retain its plastic characteristics over a wide temperature range: realistically, 15–30°C. This is a function of the solid-to-liquid ratio of the fat blend and a relatively high proportion of triglycerides with three saturated fatty acids so that a significant proportion of solid crystalline material is retained at higher temperatures. There is a balance to be achieved in formulating shortening blends in ensuring that undue firmness is not achieved but that adequate solids are present at higher temperature. One must also bear in mind that the fat will contribute to the short pastry flavour, and thus a high content of residual solid material could detract from the flavour.

Products such as margarine and butter are not extensively used alone in short pastry as, on a strict weight-for-weight basis, more has to be used because these products are only 80% fat (the functional ingredient); thus they are often combined with lard or shortening in order to enrich the flavour of the pastry. Butter has a place in high-quality sweet paste because of the superb flavour and the textural 'bloom' it can impart to the pastry.

2.5.2 Cake

The mechanism by which a fat functions in a cake has been the subject of a considerable amount of research work. The process has been described by Shepherd and Yoell (1976) in terms of batter preparation and of the changes taking place throughout the period of baking. It has been shown that cakes are highly dependent on fat for proper aeration; as well as ensuring successful aeration, fat also contributes to crumb texture and mouth feel.

The first step in making, for example, a maderia cake is to blend the ingredients; the method of blending the ingredients has some influence on the fat particle size in the batter. The traditional methods of batter preparation are the sugar batter method, in which the sugar and fat are creamed together first, or the flour batter method, in which the fat and flour are blended first. The all-in method, where the batter preparation is completed in one stage, has become more popular with the introduction of high-speed mixers. Popular in large commercial bakeries are continuous mixers, where a loose slurry of ingredients is fed to a mixing head, where air is injected into the batter.

Examination of batters prepared by these various methods has shown that the air is held initially in the fat phase (when plastic fats are used), the method of batter preparation having an influence on the distribution and fineness of the fat particle size in the batter. The finer the distribution of the fat and air the better the final cake volume and crumb structure. It has been suggested (Stauffer, 1996) that, in the case of single-stage mixing, when the air is trapped in the water phase rather than in the fat phase the protein present stabilises the foam. The risk of foaming by the fats and oils present can be prevented by the inclusion of α-tending emulsifiers such as propylene glycol monoesters of acetylated monoglyceride which in sufficient concentration form a film at the oil–water interface, which protects the protein foam.

Figure 2.4 shows data from a simple 'creaming test', where fat and sugar are beaten together, and demonstrates that the plastic properties of the fat are important in its ability to incorporate and retain air. By definition, a plastic material contains solid and liquid portions, and this is the case with a plastic shortening or margarine. In the creaming process there must be enough liquid oil available to envelop the air bubble, and sufficient crystalline fat to stabilise the system. The small β′ crystals are the most effective in stabilising air bubbles, as they can readily locate at the air–oil interface and so stabilise the air bubble.

Crystal aggregates that break up during the process can also stabilise the system. The proportion of crystalline triglyceride at the working temperature

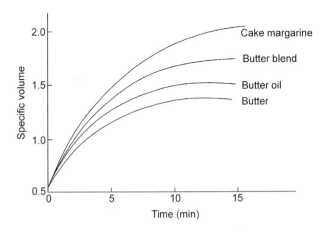

Figure 2.4 Rate of incorporation of air into a fat–sugar cream.

must be above a certain minimum, which practice has shown to be 5%. However, traditionally, most commercial bakery fats contain about 20% solid triglyceride at the working temperature. The increasing use of all vegetable-oil blends combined with the demand for more nutritionally acceptable formulations has led to the use by major bakeries of shortenings and margarine with crystalline contents nearer the 5% level, particularly as working environments now have better temperature control.

The test demonstrates the plastic behaviour of the fat and its resistance to work softening, which, if it happens to a significant degree, results in the coalescence of the air cells and a loss of volume in the cream. The creaming curves in figure 2.4 show the expected behaviour of the texturised butter oil as it initially creams very quickly and then fails. Butter also demonstrates the limitations of a low solid-to-liquid ratio, leading to a short plastic range. The incorporation of butter oil as a margarine oil blend component significantly improves the performance, though it still does not match the 'tougher' cake margarine.

Following the incorporation of the aqueous ingredients and flour it can be seen that the final batter is a multiphase system where flour particles are suspended in the aqueous phase, but the water continuous phase still has parts that are a water-in-oil (W/O) emulsion. The application of heat to this system at the start of the baking process has little effect; however, at about 37°C the irregularly shaped fat particles begin to melt and become droplets of oil; at this point the w/o emulsion parts of the batter invert to being an oil-in-water (O/W) emulsion. As the temperature continues to rise the fat withdraws from the air bubbles, which are left in the more viscous aqueous phase, to produce a foam, probably stabilised by the egg protein preventing coalescence of the air bubbles. The flour particles and fat droplets are now distributed through the continuous aqueous phase.

Convection currents in the still fluid batter cause bulk flow such that the air bubbles act as nuclei, for the increase in volume of the total batter, as the carbon dioxide and water vapour moves into them. Studies by Carlin (1944) have shown that no new air cells are created during baking; thus all the air cells that create the cake texture are introduced during batter preparation.

A continued rise in temperature to 65–70°C results in the start of gelatinisation of the flour and coagulation of the egg protein. The expansion of the air bubbles is very rapid and at 95–100°C the structure becomes fixed.

Plastic fats by design must be easily whipped into a batter and yet retain sufficient structure in order to retain the incorporated air. These characteristics are achieved by using an oil blend with a flat melting curve to ensure the solid fat content at low temperatures is not so great as to cause hardness or brittleness and with sufficient solid fat content at higher temperature to ensure there is sufficient crystalline material present to stabilise the incorporated air. The oil blend when processed must also preferentially crystallise in the β' modification.

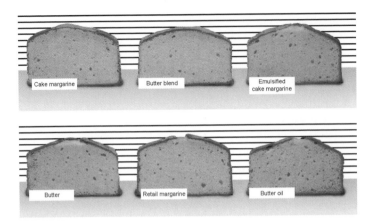

Figure 2.5 Commercial Madeira cake made with a range of margarines and fats. Top row, from left to right: cake margarine, butter blend, emulsified cake margarine. Bottom row, from left to right: butter, retail margarine, butter oil.

Figure 2.5 demonstrates the effects of toughness and wide plastic range on a finished cake. The volumes of cake made with butter and a range of commercial margarines show that use of butter or butter oil gives a lower volume and lighter texture compared with a retail packet margarine. Special cake margarine shows a finer distribution of air cells in the finished cake and this can be achieved when milkfat is included in the formulation. In this case again, butter confers a richness of flavour not achieved by margarine.

Shortenings can be designed to be general purpose in that they can be used in cake and short pastry applications. However, margarine is preferred in cake manufacture since it not only has the functionality described above but also can contribute a richness and flavour to a cake not normally found with shortening. Butter is of particular value in giving a highly characteristic flavour. However, margarines are an excellent vehicle for emulsifier systems, which can significantly improve performance.

Emulsifiers now have an important position in the manufacture of bakery products. This is best illustrated by a consideration of high-ratio cakes. These cakes were developed from a better understanding of the function of all the basic ingredients in a cake recipe—flour, sugar, egg and fat. The 1930s saw the development in the USA of superglycerinated, or high-ratio, shortenings, which brought about a significant change in the baking industry. Emulsifiers, mainly monoglycerides and diglycerides, were introduced into shortenings. The monoglycerides and diglycerides contributed to a finer dispersion of the fat particles and so a greater number of smaller sized fat globules. As a result, the emulsifier strengthened the batter, which allowed the introduction of additional liquids, which in turn allowed increased sugar to be dissolved in the system.

Table 2.5 Yellow layer cake

Ingredient	'Old' formula (%)[a]	'New' formula (%)[a,b]
Flour	100	100
Sugar	100	140
Shortening	50	55
Eggs	50	65
Milk	50	110
Baking powder	2	6
Salt	2	3

[a]Amounts are given as a percentage of the weight of flour.
[b]High ratio of sugar to flour.

Recipe amendments now became possible. These are illustrated in table 2.5 (Hartnett, 1977). The more typical recipe balance is for the weight of sugar to be equal to that of the flour, for the total weight of liquids to equal that of the flour and for the proportions of fat and egg to be the same. These proportions were important, as too much sugar affected starch gelatinisation, whereas too much liquid weakened the structure. The introduction of emulsifier allowed the use of greater proportions of sugar and liquids relative to flour weight.

As well as the addition of emulsifiers, the flour used to make a high-ratio cake must be milled to a finer powder and treated with chemicals in order to allow it to absorb greater quantities of liquid, with the gluten-forming proteins largely broken down.

This type of cake altered the rules of recipe balance and the methods of mixing the ingredients, allowing simplification and automation and achieving cakes that were generally moister and more tender.

Sponge goods, that is, those based on egg, sugar and flour, were traditionally considered as a separate product type from cake because the presence of small amounts of fat could dramatically interfere with the aeration and stability of the sugar and egg foam. In traditional sponge manufacture the egg and sugar are whisked together to form a stable foam and the flour is carefully folded in to avoid loss of whipped-in air. There are variations on this method to improve aeration and industrial efficiency (Bent, 1997).

The development of enriched sponges has now begun to blur the differences between cake and sponge goods in that many large manufacturers now include fat in sponge goods for products such as gateaux bases, the fat content varying widely from 5–25% relative to the egg content. The function of the fat in an enriched sponge is to improve flavour, mouth feel and shelf-life. In making an enriched sponge the structure is first formed during the whisking of the egg and sugar; the fat or oil is then added very quickly at a high temperature (93°C) on slow mixing just prior to folding in the flour. Butter is a preferred fat in enriched sponges because of the distinctive flavour contribution it makes, although margarines are used. Liquid vegetable oils are used extensively as they add succulence and tenderness.

2.5.3 *Puff pastry*

Puff pastry is an important example of another basic and unique type of bakery product in that it has a light flaky and layered structure which during baking increases in volume up to eight times compared with the original dough. In the preparation of puff pastry, layers of dough, with a well-developed gluten network, are arranged so that two layers of dough enclose a layer of fat; then, by a complex system of folding and rolling, a structure of alternate layers of dough and fat are built up. The layers of fat behave as impervious barriers to the moisture vapour and gases generated during baking. The retained gases expand and so stretch the gluten network to give the well-known puffed or flaky texture.

There are essentially three basic methods for the manufacture of puff pastry, namely, the English, French and Scotch methods. They differ mainly in how the laminating fat is used. The following descriptions of the manual methods of manufacture of puff pastry illustrate the methods.

- English: a rectangular sheet of dough is covered to two thirds of its area with a layer of fat or margarine. The uncovered part of the dough sheet is folded over half of the fat layer and then the remaining half of the fat layer and adhering dough is folded over the first fold to create a unit with two fat layers and three dough layers. The piece is turned through 90° and then sheeted out to its original length. This process of 'half turns' is repeated until the desired number of layers is achieved.
- French: this method starts like the English method, but in this case half the dough sheet is covered with a layer of fat and the uncovered dough is folded over the fat layer. Formation of the alternate fat and dough layers then proceeds as described for the English method.
- Scotch: this is an all-in mixing method. The fat is cut into cubes of roughly 2–3 cm, and these are added to the mixer with all the other ingredients. After a short mixing time, to ensure the laminating fat is retained as distinct lumps, the dough is sheeted out and then folded and turned as described for the English method.

The fat blend for a puff pastry margarine must have a tough and plastic texture as it is required to be rolled and stretched and sheeted out to as thin a layer as possible and yet remain continuous. The mechanical stress of the rolling process must not cause the margarine to soften unduly as this would lead to the loss of its property as a layering fat. Further, any brittleness in the texture may cause penetration into the dough during manufacture. The melting point of the fat blend must be such that it keeps the dough layers apart in the initial stages of baking without giving the final baked product a 'waxy' mouth feel.

Butter is often used in the manufacture of puff pastry but it requires handling in such a way that the applied stress is kept below the yield value of the butter to ensure that its plastic behaviour is good and that it layers well. The low solid-to-liquid ratio of butter at ambient temperatures means its layering capability is poor. Figure 2.6 shows examples of puff pastry made with butter and with refrigerated milkfat. Additionally, a processed sample of fractionated butter improved the performance as did the chilling of the butter at the point of usage. Butter and butter oil fractions confer the benefits of a characteristic flavour and of being easily digestible as they melt below body temperature.

The laminating of the dough and fat into the paste involves a series of folding and reduction steps. As the number of laminations increases the baked specific height increases (height per unit weight of paste). There is an optimum of 162 theoretical fat layers, above which there is a fall in specific height.

Table 2.6 shows a basic puff pastry recipe. Within a particular recipe it is possible to vary the relative amounts of dough fat and layering fat, which influences the structure and volume of the baked pastry. For example, by increasing the relative amount of dough fat the paste becomes softer to handle and shorter to eat; however, the final volume will be reduced.

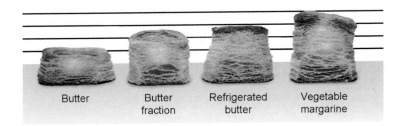

Butter   Butter fraction   Refrigerated butter   Vegetable margarine

Figure 2.6 Comparison of layering fats in puff pastry. From left to right: butter, butter fraction, refrigerated butter, vegetable margarine.

Table 2.6 The main basic puff-pastry recipes (quantities are in grammes)

Ingredient	Half paste	Three-quarter paste	Full paste
Flour	3.178	3.178	3.178
Salt	57	57	57
Cake margarine	227	284	398
Water	1.532	1.475	1.419
Layering fat	1.362	2.090	2.781
Total fat	1.589	2.374	3.179

Danish pastry is another form of laminated product, having one type of fat in the dough and a second type of fat for layering. A good Danish pastry will have 25% layering fat (based on flour). Danish pastry dough is softer than that found in puff pastry. The dough in Danish pastry is enriched and contains yeast (rather like a rich bun dough) and is given a short fermentation time. The layering fat is added by the English method, though the number of half turns are reduced in order to restrict the volume increase. The product is cut into the desired shape, gently proved and then baked. Butter is finding a widening application in the manufacture of Danish pastry and croissants as it enriches the product but maintains a light eating quality. The volume increase expected is much less than in puff pastry; even so, the butter is often refrigerated before use.

Puff-pastry margarines are made from oil blends that have a high solid-to-liquid ratio, often at the expense of the final melting point. Slip melting points of 42°C and higher are not uncommon. In manufacture, the margarine emulsion is shock chilled from a high temperature in order to give quickly a very stable crystal network which is then subjected to a heavy working and kneading routine to prevent the establishment of larger-crystal networks and to obtain a proper balance of reversible and irreversible bonds to prevent the finished margarine becoming too rigid, causing brittleness and flintiness in use.

Puff-pastry margarine is made in specially designed tubular chillers to give the shock chilling and plasticising necessary at the high pressures experienced. However, it is still often manufactured on the chilling drum and complector system, which is claimed to give better plasticity. In this case the oil blend or emulsion is spread across the surface of a rotating horizontal drum chilled internally with liquid ammonia. As the drum rotates the layer of fat or emulsion is rapidly chilled and crystallises. The flakes are then stored usually in a hopper to allow crystallisation to be completed as shown by a significant rise in temperature. The flakes are then plasticised by forcing them through a tube with one or two rotating screws to be extruded in blocks ready for wrapping. The post-crystallisation working can be adjusted to the final hardness desired.

2.6   The influence of emulsifiers in baking

Emulsifying agents, particularly those esters formed from fatty acids and poly-valent alcohols such as glycerol, propylene glycol, sorbitol and sucrose, and their modifications made by esterification with organic acids such as acetic acid, citric acid, lactic acid and diacetyl tartaric acid, have been used extensively in bakery products and other foodstuffs. The function of emulsifiers (Krog and Laurisden, 1976) in food systems falls into three broad categories:

- stabilisation of emulsions and aerated systems
- improvements to texture and shelf-life of starch-based products
- dough conditioning by interaction with wheat gluten

In the stabilisation and aeration of cake batters emulsifiers can play a very important role, and the physical state of the emulsifier has a marked influence on the batter. The effect of an emulsifier incorporated in the fat or margarine has been mentioned in section 2.5.2 regarding high-ratio shortenings, and in more conventional recipes the emulsifier improves the distribution of the fat and so promotes the distribution of the air. In addition, hydrates of emulsifiers have been used for many years to improve aeration in cake batter, in particular in fat-free sponge cakes. It has been demonstrated (Krog, 1975) that with distilled monoglyceride dispersions in water at varying concentrations and temperatures a series of liquid crystalline mesophases can be formed. These mesophases are a result of hydration, where water penetrates through layers of the polar groups of the crystalline monoglyceride above the Krafft point (the critical temperature at which micelles are formed). On cooling, the hydrocarbon chains crystallise again and the water between the lipid bilayers forms an α crystalline gel structure. As well as the lamellar type of mesophase, cubic and hexagonal structures have been identified. Monoglycerides based on saturated fatty acids form viscous gels where the lamellar structure dominates, whereas unsaturated fatty-acid monoglycerides predominate in a cubic structure. Figure 2.7 is the phase diagram of a distilled monoglyceride in water. The diagram shows that above 50°C the monoglyceride absorbs water to form a dispersion, at 60–65°C. This is a lamellar structure with the water fixed between the polar groups of the monoglyceride. In this form the monoglyceride is at its most effective in forming complexes with starch.

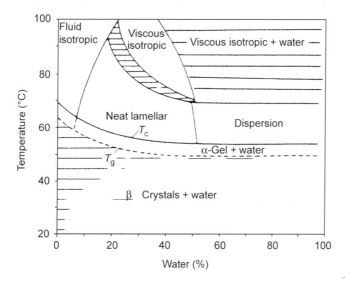

Figure 2.7 Phase diagram of a monoglyceride in water.

The use of these gels in cake batter has been shown to give a much more uniform air distribution than the shortening containing monoglyceride or mono-diglyceride. The finer air distribution increases the viscosity, which leads to a better cake volume and texture. Table 2.7 shows improvement in the cake and batter volumes in high ratio cake recipes when the emulsifier is used as a hydrate compared with when it is incorporated in the shortening. The reasoning for the better performance of the hydrated form of the monoglyceride compared with the simple solution in a shortening oil blend is that the monoglyceride has a very fine crystal size in the aqueous dispersion, which ensures optimum distribution in the batter.

There are several other emulsifiers that show crystalline properties in water dispersions and that have been found to improve batter aeration properties, such as lactic acid esters of monoglycerides and propylene glycol esters of fatty acids.

Work on starch-based products (Krog and Nybo Jensen, 1970) has shown that certain emulsifiers, particularly distilled monoglycerides, have crumb-softening and antistaling properties in wheat bread. The process of bread staling has been shown to be a result of the amylose fraction of wheat starch. During baking, some amylose leaks out of the starch granule and dissolves in the water available to form a gel between the swollen granules of the fresh bread. On cooling of the baked bread the gel contributes to the initial firmness of the breadcrumb; however, with time, the amylose recrystallises (retrogrades) to its insoluble form and so the bread becomes hard and brittle.

It is now generally accepted that amylose in its helical form has a lipophilic core; hence in this form the amylose can be stabilised by straight-chain hydrocarbon molecules, such as those found in fatty acids. The saturated dis-tilled monoglycerides have a steric configuration that can easily be enclosed in the amylose helix. The insoluble helical complex raises the gelatinisation temperature of the starch and thus reduces the total gelatinised starch in the

Table 2.7 Comparison of the performance monoglyceride gels with high ratio shortenings in some cake recipes

	High ratio shortening			Hydrate 2–4% on flour (typical use levels)		
Cake	batter gravity	vol.	crumb and grain	batter gravity	vol.	crumb and grain
115% White cake	0.96	1878	Sl. open Sl. irreg.	0.86	2085	Sl. irreg. good
130% White cake	1.04	1580	Compact Sl. irreg.	0.93	1865	Sl. closed good
100% Yellow cake	1.02	1680	Sl. irreg.	0.90	1980	Sl. irreg.

Note: Sl., Slight.

bread crumb. The monoglyceride-complexed amylose will not retrograde as does the unreacted amylose, thus leading to less amylose being available to be part of the starch gel, to give a softer crumb. Additionally, the amylose–monoglyceride complex does not take part in transporting moisture from the surrounding protein network, with the result that this network becomes less rigid and hence gives a softer crumb.

Anionic emulsifiers such as the sodium or calcium salts of stearoyl 2-lactylates, diacetyl tartaric esters of monoglyceride and succinylated mono-glycerides have been found to impart dough strengthening characteristics in fermented doughs. The effect of dough strengtheners or conditioners is to improve dough processing characteristics and also to give increased volume and a finer texture to the baked product.

During the processing of a dough a gluten network is developed which traps the carbon dioxide produced by the yeast to give the final volume and texture to the finished bread or cake. Any weakening of the gluten during processing will lead to a loss of gas and will result in poor volume. The emulsifiers used as dough conditioners interact with the gluten to improve gas retention and dough elasticity to provide tolerance to variations in fermentation time and temperature and to mechanical shock.

There is still some doubt about the mechanism of the interaction of gluten and emulsifiers; however, it has been shown (Larsson, 1980) that the emulsifier can replace some of the flour lipid in association with the gluten, suggesting there are lamellar emulsifier structures in the aqueous films at the interface between the gluten strands and starch.

2.7 Control of quality in margarine and shortening manufacture

Since the fat blend is the major component in the manufacture of margarine and shortening most analytical control effort is directed towards ensuring not only that the oil blends used are of good edibility and oxidative stability but also that they have the specified solid-to-liquid ratio and the correct crystal habit.

The tests used to judge the quality and edibility of fats and oils are well documented (Cocks and Van Rede, 1966). Classical tests for free fatty-acid content, peroxide value and colour are well known and have been supplemented by tests such as the anisidine value, used to assess secondary oxidation products, and accelerated stability tests, such as the active oxygen method (Swift's test) and the Rancimat test, which are both based on bubbling air through the oil or fat at an elevated temperature. Finally, the flavour of the oil must be judged by an expert panel to ensure it is bland or near bland. The processor will receive these oils from the refinery with the assurance that qualitative standards have been achieved. It is then important to ensure the blend and solid characteristics of the oils are correct (Zurcher and Hadorn, 1979).

The margarine and shortening manufacturer can either receive refined, deodorised oils which must then be blended, or, by consultation with the refiner, the manufacturer can receive complete blends. There are arguments for and against both types of operation in terms of quality, efficiency and process control. The system selected usually depends on the way the processor's production organisation has been built up.

Establishment of the correct blend for the duty the margarine and shortening are to perform requires close consultation with the user and an understanding of the user's process by the margarine manufacturer. A knowledge of the fatty-acid composition and the triglyceride structure of the fats available, using gas chromatographic techniques (Christie, 1973), ensure the crystal habit of the fat is correct.

It is important to know the extent to which a fat or fat mixture crystallises at the temperatures of practical interest and the extent of crystallisation at various temperatures. The modern method for measuring the solids content of fats is based on the difference in molecular mobility in liquid and solid triglycerides (Waddington, 1986) and, as the solid-to-liquid ratio varies with temperature, then a temperature profile of the solids content of the blend can be obtained. The technique in question is wide-line nuclear magnetic resonance (NMR). Pulsed NMR has replaced the much more time-consuming technique of dilatometry, which measures the solid fat content by volume contraction during crystallisation.

The phase behaviour of oil blends can be evaluated by plotting pulsed NMR solids content data as a function of temperature and composition. The interaction of two oils or triglyceride types can be shown with these diagrams. These so-called isosolid diagrams show components are compatible by giving horizontal isosolid lines. However, where eutectics or compounds are formed the isosolid lines are not horizontal and so product defects can be forecast; for example a margarine may rapidly develop a grainy or brittle texture as a result of compound formation (Birker and Padley, 1987).

The analyses discussed above, that is gas chromatography and solids content determination, are used to establish the oil blend in terms of the user's require-ments for performance and eating qualities. Once the parameters are established then blend control can be effected by routine analyses, such as of the solids content of the individual and blended oils and the iodine value, supplemented with gas chromatography of the fatty-acid methyl esters.

In the production of margarine the manufacturer is faced with additional control problems in that a product is being made that has two phases which when processed must be completely stable; also, the fat and water levels must reach the statutory levels, and any added salt must attain the level specified by the consumer (Andersen and Williams, 1965).

The production of stable w/o emulsions is facilitated by the addition of emul-sifying agents; traditionally monodiglycerides have been used. The influence of emulsifiers on such features as air incorporation, batter stability, etc., has led to a greater sophistication. The emulsifiers also ensure that the water droplets in the emulsion are small (about 5 μ), which leads to a good bacteriological standard and prevents their coalescence. The presence of large droplets would provide a medium with sufficient nutrients where bacteria could grow.

The phases in margarine manufacture can be mixed either by means of a batch process or continuously. The batch process is the more traditional method, the oil phase and aqueous phase, with their soluble ingredients, being premixed in the form of a suspension at a temperature sufficiently high to ensure that the crystallisation of the highest melting component does not take place. The 'premix' is then fed to a chiller by way of a buffer tank. In many factories, continuous metering systems have been installed where proportioning pumps blend the fat and aqueous ingredients immediately prior to the chiller, and the system relies on the agitation and shear characteristics in the chilling tube to give a correctly distributed water globule size.

Both systems can be found in modern factories and can be substantially automated. Most modern systems now include in-place cleaning facilities, and emulsions can receive a high-temperature treatment similar to pasteurisation prior to chilling. Automation of the process relies principally on monitoring the product temperature, refrigerant demand and product back pressures. The desired targets are fixed experimentally and then automatically monitored within fixed ranges to ensure a consistent final product. The parameters are fixed on the basis of the solids profile of the oil blend and its rate of crystallisation.

Moisture content can be automatically monitored with in-line equipment as well as by evaporation loss. Fat and salt contents are measured by conventional laboratory tests. Bakery margarines can include milk solids, either in the form of whey solids or spray-dried skimmed milk powder. These products are added to improve the flavour, and the presence of the lactose can improve crust browning. There has been a move to simplify bakery margarines by removing the milk solids and making them into simple fat–water emulsions. This step has been supported by the improved quality of the flavours used and a desire not only to lower the cost but also to ensure microbiological standards are more easily achieved.

In modern margarine factories the possibility of microbial contamination has been almost completely eliminated. However, the finished product must be regularly examined for the presence of spoilage organisms, yeasts, moulds, lipolytic bacteria and food-poisoning organisms, and the surface and the atmosphere should also be monitored. Additionally, close control of the pH value of the aqueous phase of the margarine, combined with maintaining a small water globule size, inhibits the proliferation of spoilage organisms.

As discussed earlier, the processing conditions have some influence on the texture of the finished fatty product; hence the chilling and working conditions need to be very closely defined; for example attention must be paid to parameters such as emulsion or oil-feed temperature, throughput speed, refrigerant evaporation temperature and product temperature. These controls will lead to a consistent product, and the firmness can be confirmed by estimation of the yield value. Final quality testing is done by user tests; for example, one can measure the air incorporation achieved in a standard bakery mixing-machine, one can manufacture a basic cake that is sensitive to fat performance and, in the case of puff-pastry fats, one can make test vol-au-vent cases in order to measure the volume increase. These user tests can be supplemented by the use of objective tests such as penetrometer tests to show hardness and texture, profile analysis to indicate initial firmness, plasticity, brittleness, gumminess and so on, which affects the functionality of the product.

Bakery fats can be processed into products other than the plastic fat described so far. These other forms will be discussed later.

2.8   Liquid shortenings

Liquid shortenings by definition are clear and fully liquid at ambient temperature. As discussed in section 2.5.2, batter aeration is dependent on the ability of a plastic fat to retain air; liquid oils do not have this property. However, the advent of high-ratio cakes showed that dependency on the fat's plasticity for aeration was reduced by the inclusion of emulsifiers. Thus the developments in continuous methods for cake mixing and the need for bulk storage of ingredients led to the introduction of fully liquid shortening. In early applications it was shown that shortenings made by dissolving various emulsifiers in liquid vegetable oils provided an effective alternative to plastic fats (Hartnett and Thalheimer, 1979). The use of liquid oils containing emulsifiers or emulsifier combinations in place of plastic shortening in cake manufacture appears to maintain the tenderness and moistness of the cake for longer. This may be, in part, due to the high level of liquid oil being used in the cake recipe making the cake initially more tender. A further observed advantage is that there can be up one-third reduction in total fat in the recipe, compared with a plastic fat, without loss of volume or eating qualities (Hegenbart, 1993).

Liquid shortenings have been exploited principally in the USA but have failed to gain popularity in the UK, in part because of the high cost of the emulsifier systems and because the systems have a much greater temperature sensitivity than claimed. For example at cooler temperatures (18–20°C) the emulsifiers are often precipitated from solution, with the consequence that the product becomes highly variable.

Liquid shortenings have now largely replaced solid fats in bread manufacture. Liquid shortenings when used in place of plastic fats give lower loaf volumes and a more open crumb structure. The use of liquid shortenings in bread doughs requires the use of dough conditioners to overcome the shortcomings mentioned above. Typical dough conditioners are sodium steroyl lactylate and diacetyl tartaric esters of monoglycerides. Their function in the dough system arises from their ability to form a hydrogen bond complex with both the protein and the starch fractions of flour, with the effect that the starch–water– protein matrix is strengthened during the critical rising and setting stages of baking.

Dough strengtheners, because of their ability to complex with the starch in wheat, also behave as dough softeners in the same way that the monoglycerides of fatty acids do, as described in section 2.6.

2.9 Fluid shortenings

These are pourable shortenings but are distinguished from liquid shortenings in that they contain suspended solid particles. Fluid shortenings were principally developed as frying media to provide a stable but pourable frying oil. Oils such as soybean oil and rapeseed oil, with a relatively high content of linolenic acid, were 'brush' hydrogenated to reduce the linolenic acid level and hence improve the oxidative stability. Instead of winterising the product, to give a clear oil, the addition of a small quantity of a fully hydrogenated fat and a technique for maintaining the solid material in suspension led to a pourable 'slurry'.

A variety of techniques have been patented (Schroeder and Wynne, 1968; Rossen, 1970) for creating a slurry-like fluid over a wide range of temperatures. The critical feature of the suspended particles is their size; if too large they will settle quickly, and if too small, though settling slowly, they will pack closely. It has been found that β' crystals are too small, and β crystals are better as they are too large to pack closely and treatment to prevent aggregation of the crystals ensures a stable slurry.

The systems described (Haighton and Mijinders, 1968) usually require slow crystallisation of the shortening following the addition of a high melting fat component to a liquid vegetable oil and a final homogenisation of the flocculent precipitated crystalline mass. These products can also be made on scraped-surface heat exchangers where the control on the rate of cooling is critical to ensure the correct crystal modification is created.

Slurry shortenings still rely heavily on added emulsifiers for their functionality. A number of emulsifier systems have been shown to work well, and experience with pilot-scale trials has shown that acceptable performance can be achieved with blends containing α monoglycerides, polyglycerol esters, propylene glycol monostearate, lactic acid monoglycerides and sodium steroyl lactylate. Used in high-ratio cake recipes these give finished cakes approaching the volume and texture of a cake made with conventional high-ratio shortenings. Acceptable quality cakes have been obtained with a reduction of fat in the cake recipe.

It has been found that by the introduction of sorbitan tristearate into a fluid shortening formulation a higher solids content can be achieved and fluidity maintained. This then allows the shortening to function adequately in a range of cake and short pastry applications without addition of other emulsifiers.

Fluid shortenings have temperature limitations in the same way as liquid shortenings. Storage at temperatures below 12–14°C can lead to the product setting, and temperatures in excess of 35°C cause some melting of the crystals formed, and subsequent cooling will cause the formation of large crystals which will settle.

2.10 Powdered fats, flaked fats and fat powders

The claimed advantages for fat powders, powdered fat and flaked fat are ease of handling in transport, dosing and simplified storage. Blending with the growing number of other dry ingredients is eased.

Before considering the manufacture and application of these forms of fat it is necessary to define the differences between them. Powdered fats and flaked fats are both similar in that they are entirely made of fat or of fat and emulsifier. However, they are manufactured by different methods, as their names imply, though there is some overlap in their application. Fat powders, though they contain substantial amounts of fat also contain nonfatty material which acts as a carrier. This applies a restriction in their use in that the nonfatty component must be compatible with the final recipe of the user.

2.10.1 Methods of manufacture

2.10.1.1 Powdered and flaked fat
In the manufacture of powdered, granulated and flaked fat there are certain common features to be considered. The fat in its final form must be a solid at ambient temperatures, and the flake or particle must be such that the crystallisation must go as quickly as possible to completion so that late crystallisation of the liquid core does not release sufficient heat to cause lumping or caking in the product.

The technique of cooling and crystallising on a cooling drum to create a flaked product has long been used in the margarine industry. The method is similar in creating flaked fat though generally the flakes are thicker so that they can be handled easily in conveying and packaging. Additionally, so that they can crystallise quickly, the fat is usually of a high melting point with high solid fat content at ambient temperature (e.g. a pulsed NMR measurement of 70% at 20°C and of 30% at 30°C). Any fat hydrogenated to a high enough melting point can be used. However, a fat with a wide variety of triglyceride types exhibiting little polymorphism is preferred. The inclusion of coconut or palm kernel oils in their hydrogentated form in the fat formulation can assist in achieving the required percentage solids while slightly improving mouth feel.

Flaked fats can be further pulverised to make them more granular in texture, to improve the flow properties. However, this requires the application of low temperatures.

Powdered fats (Lamb, 1987) are manufactured by the technique of spray chilling—that is, by dosing the fully liquid fat or fat blend into a tower through which cold air is being circulated. The fat (Munch, 1986) must be injected into the upper portion of the tower as a fine spray. Since the globule size is a critical factor in the success of the operation a range of systems have been developed such as rotary atomisers, both disc and centrifugal, and high-pressure nozzles to control the globule. The system is selected on the requirements of the processor and of the products—that is, the requirements for flexibility, throughput, product viscosity and formulation complexity.

The major parameter for the successful spray cooling of fat is that the globule size should be such that the total particle is fully solidified before leaving the tower. It can be such that the holding time in the cold air stream is also important. The difference between the air temperature and the melting point of the fat also plays a part. Thus in principle the smaller the droplet radius and the greater the difference between the melting and the air temperatures the more rapid the solidification. Other features that have an influence are product temperature, viscosity and the injection pressure. These influence the droplet size and the amount of energy to be removed in the cooling.

Liquid nitrogen has been tested as the cooling medium in the manufacture of powdered fats. The great differential between the melting point of the fat and the liquid nitrogen at $-70°C$ means that the rate of crystallisation is enhanced, ensuring crystallisation is complete before the fat leaves the cooling chamber, and there is also the possibility that the particle size can be varied without risk of a molten core, thereby widening the range of applications.

The design of the tower is of great importance in that the powder take-off system must not allow outside air or moisture into the tower, and the chilled air must be filtered and cooled for reuse.

It can been seen again that a relatively high melting fat is required in this system so that rapid hardening is achieved and lumping in store is prevented. The advantage of this type of fat over flaked fat is easier control on dispensing and easier dispersion with other ingredients. This is particularly important when emulsifiers or fats containing high levels of emulsifiers are being introduced as an ingredient as the quantity is likely to be very small and so its successful distribution is more difficult.

2.10.1.2 Fat powders

Fat powders are popularly made by the technique of spray drying (Blenford, 1987). The fat is first made into an o/w emulsion with an aqueous solution of carrier powder (e.g. milk power, starch, dextrin). The emulsion is then supplied to the spray tower atomiser by a high-pressure homogeniser to ensure the feed is homogeneous. The fine spray of emulsion droplets is projected into a hot air stream to evaporate the moisture. The moist air and fine particles are collected in a cyclone, and the dry fat powder is collected at the base of the drying chamber.

The design of spray driers has advanced significantly from the single-stage spray-drying system, in which the drier discharge tended to be at a relatively high temperature, to the double-stage and triple-stage systems, which reduce the energy consumption so that the product can be produced at lower temperatures. Thus more temperature-sensitive and high-fat products could be handled. Figure 2.8 shows the layout for a simple single-stage system.

As with powdered fats the design of the atomiser is critical to the success of the plant. The geometry of the spray chambers is also very important, where the air flow can be counter or concurrent, and it is important that the spray must not strike the tower wall until dehydrated, otherwise it will stick and burn onto the wall and ultimately interrupt the air stream. The volume of air, its velocity and temperature must be controlled to be consistent with the heat sensitivity of the product.

Microencapsulation. Microencapsulation of fats and oils has now been developed to a considerable extent. The method gives a product where the fat is at the core of a nonfatty substance; thus liquid oils can be used without risk of leakage and the temperature sensitivity of the product is reduced.

Microencapsulation utilises a spray-drying technique. The fat and oil are thoroughly emulsified with a water solution of the water-soluble coating material,

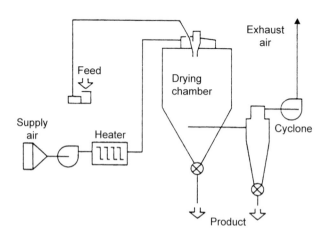

Figure 2.8 A single-stage spray-drier system.

such as gelatine, gum arabic, starch or dextrin. The water is evaporated off as described above to leave dry particles in a shell or capsule of dry colloid in which the fatty material is embedded or encapsulated in the form of a minute droplet.

Various other processes. There are several other methods for manufacturing fat powders; for example systems have been developed where molten fat can be sprayed into a stream of dry particles in a spray chiller. This simple expedient of feeding a 'carrier'-like flour into the area of atomisation in a spray chiller gives the opportunity for lower melting fats to be used. The system can be used for the manufacture of dry food mixes. Simple mixing of a powdered fat with a dry component can be used to manufacture fat powders. In this system heating takes place as a result of the mechanical and shear forces applied; thus a cooling phase prior to packaging of the product is required.

There are also a number of recently patented systems for the manufacture of fat powders. For example European Patent Application 0289069 (Hamaguchi, 1988) describes a system of blending a fat or oil with a hydrophilic base substance (such as starch, caseinate or gelatine) and a small proportion of polyol-like glycerol or propylene glycol to give a free-flowing powder that rapidly gives up the oil when dissolved in water, thus being valuable for the manufacture of the seasoning of soups.

2.10.1.3 Butter powders

Butter powder deserves special consideration when discussing fat powders. A considerable amount of development work has been carried out on this product in order to achieve an 80% butter fat product that performs well in bakery applications and can withstand higher ambient temperatures than butter itself.

The powder is made by way of standard spray-drier technology (Frede *et al.*, 1987) from an emulsion of milk powder solution and anhydrous milk fat that is homogenised as a 40% total solids solution and then spray dried. The mechanical stability of the powder is improved by the inclusion of trisodium citrate, possibly because of its influence on the fat–protein interface. Emulsifier, usually monodiglyceride, is included in the product, even though it has been found to affect adversely the mechanical stability of the powder, as it improves the aeration of the cake batter.

The risk of caking or clumping of the power can be reduced by using a higher melting fraction of butter fat and by cooling the powder immediately on leaving the spray tower as this creates many crystal nuclei, which help to retain the state over a wide range of temperatures. If the cooling is rapid enough the crystalline structure can also give benefits in bakery applications. Butter powder is useful as a bakery ingredient as it is dispersed easily and quickly in the mix and imparts the richness to the product associated with butter.

2.10.2　Applications of fat powders and powdered fats

At a time when the demand for convenience foods is increasing the use of pow-dered products has increased. Powdered products that can be easily reconstituted are now of considerable importance. The best-known examples come from the dairy industry where there is a considerable range of milk and milk proteins in powdered form being used as ingredients and nutritional supplements. There is also the opportunity for efficiency improvements in industrial situations where powders give easier handling and storage (Hogenbirk, 1984).

Powdered fat is now being used as a bread improver. This kind of product often includes emulsifiers. Flaked and powdered fats have a limited application because they have a high melting point; however, they play a major role as crystallisation accelerators in products such as fondant and as a stabilising agent for paste products such as peanut butter. In addition, lauric hard butters in powder form are used as easy adjusters for the fat content of compounded chocolate for coatings. Flaked and powdered lauric fats are used in the preparation of ice-cream mixes.

Fat powders can vary considerably in their fat content, from as low as 15–20% in microencapsulated powders, through coffee creamers and whippable toppings at 30–40%, to butter powder with greater than 80% fat content, used as a bakery ingredient.

The presence of nonfat dry solids in fat powders means that lower melting fats and fat blends can be used in fat powders than in powdered fats, which widens the range of applications compared with powdered fats, though, as mentioned above some constraint is placed on fat powders in that the carrier must be compatible with the other final recipe ingredients.

Instant desserts and whipped toppings that can contain up to 40% fat and emulsifier are popular convenience foods relying heavily on the added emulsifier system to ensure good aeration.

Prepared cake mixes for both catering and domestic outlets have been avail-able for many years. The traditional manufacturing method was to distribute a conventional boxed shortening or a pumpable shortening throughout the flour and other dry ingredients or to inject a softened shortening as small droplets into a haze of flour. In both cases the crystal structure of the fat is main-tained in order to ensure good bakery performance in the finished mix. The high-fat powders, which contain between 70% and 85% fat, now give the opportunity for bakery mixes to be prepared by mixing all the dry ingredients together. The difficulty that is found is that the fat will not necessarily be in the correct crystalline modification for optimum performance. Develop-ments in emulsifier systems in these products have substantially improved their performance.

It is not only necessary to compare the advantages and disadvantages of using fat powders and powdered fats; it is also important that these two forms of processed fats be compared with fats manufactured as plastic, fluid and liquid shortening.

In comparing powdered fat and fat powders it can be seen that because of the presence of a carrier in fat powders there is a greater microbiological hazard than in the all-fat powdered fat. It has been noted that fat powders deteriorate oxidatively more rapidly than does powdered fat because of the contact with the carrier. However, the comparison is confused by the fact that most fat powders are less saturated than are powdered fats.

Fat powders and powdered fats are easy to handle and store, as are boxed fats. However, liquid oils in bulk require expensive temperature-controlled storage installations. Where boxed fats are melted prior to use this gives a significant disadvantage because of equipment, energy consumption and handling of the melted fat.

Boxed fats have considerable advantages over fat powders and powdered fats in recipe versatility, although the method of incorporation can be more expensive than for powders in terms of energy consumption. The selection of the form a fat product is in depends on the final food product, the ease of handling, incorporation and, most importantly, the quality of the finished food.

2.11   Fat in biscuit baking

Biscuit manufacture is a major and highly specialised sector of the baking industry and is seen as being wholly separate from bread and confectionery baking. Biscuits appear in a wide variety of types, the majority of these requiring fat in the recipe, with some of the functionality described earlier for other baked products. However, biscuits differ from the majority of other baked products in that they are baked to have a much lower residual moisture content in order to ensure a long shelf-life; as a consequence the fats used must also exhibit good resistance to oxidation and to the development of off-flavours.

Biscuit manufacturing is probably the most mechanised and automated baking activity, yet many of the biscuit types that are still popular and so are regularly manufactured were made in the home before factory manufacture was introduced. In the early days, only fats such as butter and lard were used, which restricted the shelf-life of the higher-fat biscuits. The introduction of vegetable oils and hydrogenation improved this situation.

The function of fat in biscuits will be discussed in relation to the various biscuit types. However, biscuit coatings will not be considered here as chocolate and confectionery fats are dealt with elsewhere in this book.

A problem associated with fat in biscuits is the appearance of fat bloom, that can appear in biscuits containing particular fat blends and during storage in conditions where there is temperature cycling. The biscuits take on a dull unappetising appearance. This feature of how fat functions in biscuits will be discussed later.

2.11.1 The function of fats in biscuits

Fats are used in biscuit doughs, cream fillings and as surface sprays. Fat is also employed in biscuit coatings, the major example being chocolate. Before considering how fat functions in biscuit doughs it is necessary first to attempt to classify biscuit types. This is particularly difficult for a product that appears in so many forms with many overlapping qualities. Table 2.8 (Manley, 1983) shows a classification that attempts to take account of a range of parameters and properties in order to classify biscuit types. Those biscuits more influenced by the addition of fat will be considered here.

2.11.1.1 Short-dough biscuits

This is the largest group, spanning the fat-rich shortbread and the sugar-rich, fat-lean ginger snap, as well as digestive biscuits, and biscuits filled with cream. The biscuit dough is in principle made from fat, water, sugar and syrup, into which the flour is blended and, as in short pastry, the fat in the dough shortens and texturises the finished biscuit. Biscuits by design are crisp, and the presence of fat ensures they do not become hard.

In high-fat short-dough biscuits the fat prevents the sugar solution reacting with the flour to develop gluten. The fat also reduces the starch swelling and gelatinisation to give a soft-textured biscuit. The high level of fat also has a lubricating function that helps to control the amount of water used in the recipe.

In high-sugar biscuits the fat still functions to control gluten development but its principle function is to limit the extent to which the syrup solution develops to be a vitreous and brittle solid on cooling after baking.

The fats used in these doughs are plasticised, as in the case for short pastry, so that they can be smeared through the dough in small particles in order for the resulting dough to be cohesive and lacking in elasticity. The mixing times are closely controlled in order to control gluten development (these biscuit doughs

Table 2.8 Classification of biscuit types

	Crackers	Semi sweet	Short high-fat	Short high-sugar	Soft
Moisture in dough (%)	30	22	9	15	11
Moisture in biscuit (%)	1–2	1–2	2–3	2–3	⩾3
Temperature of dough (°C)	30–38	40–42	20	21	21
Critical ingredient(s)	Flour	Flour	Fat	Fat and sugar	Fat and sugar
Baking time (min)	3	5.5	15–25	7	⩾12
Oven band type	Wire	Wire	Steel	Steel	Steel

are usually allowed to stand for a period before they are rotary moulded or wire cut).

The oil blends used in the production of biscuit dough fats are structured to ensure that the solid fat content at the dough temperature is high enough that the fat can be smeared through the dough, but low enough at body temperature to avoid a waxy mouth feel when the biscuit is eaten. Further constraints are that biscuits are expected to have a long shelf-life, so that fats used to formulate the blend must exhibit good oxidative stability. Hence oils with high contents of polyunsaturated fatty acids are usually avoided, and it has been found that fat blends with steep melting curves can lead to the development of bloom. Typical blends are shown in table 2.9.

2.11.1.2   Semisweet and hard sweet biscuits

Although for this type of biscuit a dough fat similar to that used in soft dough biscuits will be used, semisweet and hard sweet biscuits have a well-developed gluten structure. They are sheeted and baked to give a smooth surface with a slight sheen and an open, light texture.

2.11.1.3   Laminated biscuits

In the manufacture of puff biscuits there is an analogy to the manufacture of puff pastry, though the layering is much less well defined owing to the fact that the layering fat distribution is nonhomogeneous during lamination. A tough extensible dough is made free of fat and then the fat is folded into it. The fat is distributed in fine lumps or flakes to cause flakiness in the areas where they are located.

Table 2.9 Short-dough biscuit recipes

	Biscuit		
Ingredient[a]	digestive	Lincoln	ginger snap
Shortening	121 (30.3)	100 (25.0)	73 (18.25)
Sugar	86 (21.5)	107 (26.8)	202 (50.5)
Molasses or syrup	10.5	20	88
Shimmed milk powder	20	13	0
Biscuit crumb	0	0	23
Salt	5	2	3
Colour or flavouring	q.s.	q.s.	q.s.
Water	57	51	57
Bicarbonates	8.5	1.75	4
Ginger	0	0	6
Flour	400 (100)	400 (100)	400 (100)
Oatmeal	14.25	0	0

q.s. *Quantum satis.*
[a]Amounts are in kilograms. Figures in parentheses are percentages relative to the flour content.

The fats used need to have a high solids content at the laminating temperature, but because the product is eaten cold the solids content at 35°C must be low to prevent waxiness. The fats used are palm kernel oil or hydrogenated rapeseed oil delivered into the dough mass as flakes, often after crystallisation, on a rotating chilling drum. This equipment consists of a horizontal revolving drum cooled internally by brine or liquid ammonia. The unit operates such that the melted fat is fed into a trough, which spreads the fat in a thin layer across the surface of the drum. As the drum rotates the fat is rapidly cooled and crystallised, to be scraped off as flakes by a scraper blade located near the feed trough.

Cream crackers are another form of laminated biscuit, but in this case a fermented dough is used. The dough fats discussed above can be used in these products as there is even less requirement for distinctive layering than in the case of the puff biscuits.

2.11.2   Biscuit filling creams

Biscuit filling creams are principally sugar, fat and flavourings. The fat in a filling cream is usually in the range of 25–30% of the cream. The fat used in the filling creams are selected to give the cream some specific characteristics:

- the cream must be firm enough at ambient temperatures to hold the two biscuit shells together and to avoid being squeezed out of the 'sandwich'
- the cream must give the consumer a firm bite yet melt quickly to give a cooling impression on the palate and release the sugar and added flavourings
- the cream must solidify sufficiently rapidly after spreading so that the two biscuit shells are held together during transport and packaging

The fats used in this application must have low solids content at body temperature and yet have a relatively high solids content between 15°C and 25°C to ensure the biscuit shells remain 'keyed' together. The fats preferred are blends containing natural and hydrogenated lauric fat and hydrogenated vegetable oils such as rapeseed oil and soybean oil. Table 2.10 shows a small selection of possible blends.

2.11.3   Spray fats

Fat is sprayed onto the surface of savoury crackers to give them an attractive sheen and also enables added salt to adhere to the surface. The requirements of the fat in this application is that it should be liquid at room temperature but must also show good resistance to oxidation because it is exposed to the atmosphere as a thin film. Blends containing lauric fats, palm oil, palm oleins and so on are favoured in this application, as shown in table 2.11.

Table 2.10 Cream fat blends for biscuit fillings

	Percentage
Blend 1:	
Hardened coconut oil	75
Coconut oil	25
Solid fat content	
20°C	48–58
35°C	1–5
Blend 2:	
Hardened palm kernel oil	50
Palm oil	20
Coconut oil	30
Solid fat content	
20°C	45–51
35°C	4[a]
Blend 3:	
Hardened rapeseed oil	50
Hardened palm kernel oil	45
Coconut oil	5
Solid fat content	
20°C	45–51
35°C	5[a]

[a]Maximum.

Table 2.11 Spray-fat blends

	Percentage
Blend A:	
Hardened rapeseed oil	35
Coconut oil	65
Solid fat content	
20°C	22–28
35°C	0.5[a]
Blend B:	
Palm oil	50
Palm kernel oil	50
Solid fat content	
20°C	18–22
35°C	0.5[a]

[a]Maximum.

2.11.4 Fat bloom

Fat bloom in biscuits becomes evident as a white powder on the surface of a biscuit, giving a dull appearance. The powdery substance on the biscuit surface consists of fat crystals. These crystals are caused by fractionation of the biscuit fat and migration of the liquid portion to the biscuit surface, where it recrystallises as fine powdery crystals.

Temperature cycling of biscuits during storage encourages the development of bloom. Fat bloom has also been associated with fats such as palm oil and lard, which show an ability to separate quickly into fractions. Fat bloom in biscuits can also be caused by the mixing of cream fats and dough blends, which in many cases leads to the production of a eutectic mixture in the fat system, and the new mixture of triglycerides more easily fractionates and migrates to cause bloom.

2.12   Conclusions

There will continue to be a demand for fats for use in baked products, though nutritional and life-style changes will continue pressure for change. Concerns about obesity and the impact of saturated and trans fatty acids on heart disease will influence the choices of the oils and fats manufacturer. Bakery fats now show a lower content of hydrogenated oils and increasing quantities of liquid vegetable oils, both to raise the levels of monounsaturated and polyunsaturated fatty acids and to reduce the levels of saturated and trans fatty acids consumed. These shifts in blend characteristics have led to the increased use of fractions and the growing use of interesterification, which is leading to products with lower solids contents and to an increasing interest in fluid shortenings.

There has already been a considerable amount of work done toward the use of reduced fat emulsions in bakery products, designed to maintain good functionality particularly with respect to eating quality. The greater use of emulsifiers and emulsifier–stabiliser systems is likely to be combined with changes in bakery recipes and processing methods in order to meet consumers' desire to have what appear to be traditional products but with lower fat content.

Some bakery products could be used as a source of functional food. Thus the fat manufacturer may be expected to supply fats and emulsions containing for example phytosterols or that are enriched with vitamin E, as well as to supply fat replacers such as inulin, which is a dietary fibre (de Dekere and Verschuren, 1999).

The change away from the traditional freshly baked product to the packaged product sold in a supermarket is likely to continue, so that bakery products could come into competition with confectionery count-line products (products retailed as individual items, such as chocolate bars), which will put further demands on the fat supplier for modified functional requirements and extended shelf-life.

The search for nutritional improvement could lead to the use of sucrose polyesters, which have been developed by Procter & Gamble, to the point where they are being used as frying media for certain snack foods in the

USA. It is not hard to visualise that by changing the fatty acids esterified to the sucrose molecule a sucrose polyester could be produced with the required functional properties. The same approach can be applied to other carbohydrate polyesters based on sorbitol, trehalose, raffinose and stachyose, which are also indigestible.

Medium-chain-length triglycerides are structured lipids, which have been used in spreads for many years. Medium-chain-length fatty acids do not increase plasma total cholesterol in the same way that lauric myristic and palmitic acids do, and they can contribute to weight reduction in obesity because of their lower energy content compared with longer-change-length triglycerides.

The impact of genetic modification will be felt in the production of bakery fats. The possibility of genetic modification of oilseed crops to produce oils with many of the desired characteristics already present in the oil will allow processors to simplify operations and hence supply a cheaper ingredient to the baker.

References

Anderson, A.J.C. and Williams, P.N. (1965) *Margarine* Press, 2nd edn. Pergamon Press, Oxford.

Bailey, A.E. (1950) *Melting and Solidification of Fats and Fatty Acids*, Interscience, New York.

Bent, A.J. (1997) Pastries, in *The Technology of Cake Making* (ed. A.J. Bent), Blackie Academic, London, pp. 275-289.

Birker, P.J.M.W.L. and Padley, F.B. (1987) in *Recent Advances in the Chemistry and Technology of Fats and Oils* (eds. R.J. Hamilton and A. Bhati), Elsevier, Amsterdam, Chapter 1.

Blenford, D. (1987) When the heat is on. *Food* (May), 19-23.

Carlin, G.T. (1944) A Microscopic Study of the Behaviour of Fats in Cake Batters. *Cereal Chem.*, **21**, 189-199.

Christie, W.W. (1973) *Lipid Analysis*, Pergamon Press, Oxford.

Cocks, L.V. and Van Rede, C. (1966) *Laboratory Handbook for Oil and Fat Analysis*, Academic Press, New York.

de Dekere, E.A.M. and Verschuren, P.M. (1999) in *Functional Foods* (ed. G. Mazza), Technomic, Basil, pp. 233-255.

Frede, E., Patel, A.A. and Bucheim, W. (1987) Zur Technologie von Butterpulver. *Molkerei Zeitung Welt der Milch*, **41**(51-52), 1567-73.

de Greyt, W. and Kellens, M. (2000) in *Edible Oil Processing* (eds. W. Hamm and R.J. Hamilton), Sheffield Academic Press, Sheffield, pp. 79-127 .

Haighton, A.J. and Mijnders, A. (1968) *Preparation of Liquid Shortening*. US Patent 3, 395, 023, assigned 30 July 1968, Lever Brothers Company, New York.

Hamaguchi, T. (1988) *A Fat or Oil Composition in Powdery or Granular form and a Process for Producing the Same*. European Patent 0289 069, assigned 2 November, 1988, Asahi Kasei Kogyo Kabushi Kaisha.

Hartnett, D. (1977) Cake shortenings. *J. Am Oil Chem. Soc.*, **54**(10), 557-560.

Hartnett, D. and Thalheimer, W.G. (1979) Use of oil in baked product, part 2–sweet goods and cakes. *J. Am. Oil Chem. Soc.*, **56**(12), 948-952.

Hegenbart, S. (1993) Examining the Roll of Fats and Oils in Bakery Products. *Food Prod. Design* (August), 109-126.

Hodge, D.G. (1986) Fat in Baked Products. *B.N.F. Nutrit. Bull.*, **11**(3), 153-161.

Hogenbirk, G. (1984) Powdered Fat Systems. *S.A. Food Rev.* (February/March), pp. 26-27.

de Jong, S. (1980) *Triacylglycerol Crystal Structures and Fatty Acid Conformations: A Theoretical Approach*, PhD thesis, University of Utrecht, Utrecht.

Joyner, N.T. (1953) The plasticizing of edible fats. *J. Am. Oil. Chem. Soc.*, **30**(11), 526-533.

Krog, N. (1975) Interactions between water and surface active lipids in food systems, in *Water Relations of Food* (ed. R.B. Duckworth), Academic Press, New York, pp. 587-611.

Krog, N. and Laurisden, B.J. (1976) Food emulsifiers and their association with water, in *Food Emulsions* (ed. S. Friberg), Marcel Dekker, New York, pp. 67-139.

Krog, N. and Nybo Jensen, B. (1970) Interaction of monoglycerides in different physical states with amylose and their antifirming effects on bread. *J. Food Tech.*, **5**, 77-87.

Lamb, R. (1987) Spray Chilling. *Food* (December), 39-43.

Larsson, K. (1980) in *Cereal for Food and Beverages: Progress in Cereal Chemistry and Technology* (eds. G.E. Inglett and L. Munck), Academic Press, New York, pp. 121-135.

Manley, D.J.R. (1983) Technology of biscuits, crackers and cookies, Ellis Harwood, Chichester, p. 157.

Mouries, M. (1869) French Patent 86480.

Mulder, H. and Walstran, P. (1974) *The Milk Fat Globule*, Purdoc, Wageningen.

Munch, E.W. (1986) *Deulsche Milch Wirtschaft* **37**(9), 567-569.

Opfer, W.B. (1975) Margarine products - A review of Modern Manufacturing Processes. *Chem. Ind.*, **18**, 681-687.

Pyler, E.J. (1973) *Baking Science and Technology*, Siebel Publishing, Chicago, IL.

Rossen, J.L. (1970) *Fluid Shortening*. US Patent 3, 528, 823, assigned 15 September, 1970, Lever Brothers Company, New York.

Schroeder, W.F. and Wynne, J.R. (1968) *Method for Making Fluid Oleaginous Suspension*. US Patent 3, 369, 909, assigned 20 February, 1968, National Dairy Products Corporation, New York.

Shepherd, I.S. and Yoell, R.W. (1976) Cake emulsions, in *Food Emulsions* (ed. S. Friberg), Marcel Dekker, New York, pp. 216-275.

Stauffer, C.E. (1996) in *Advances in Baking Technology*, pp. 336-367.

Timms, R.E. (1984) Phase behaviour of fats and their mixture. *Prog. Lip. Res.*, **23**, 1-38.

Waddington, D. (1986) Applications of wide-line nuclear magnetic resonance in the oils and fats industry, in *Analysis of Oils and Fats* (eds. R.J. Hamilton and J.B. Rossell), Elsevier, Amsterdam, pp. 341-401.

Willett, W.C., Stampfer, M.J. and Mansen, J.E. (1992) Intake of trans fatty acids and the risk of coronary heart disease among women. *Lancet*, **341**, 581-583.

Zurcher, K. and Hadorn, H. (1979) Erfahrung mit der Bestimmung der Induktionszeit von Speiseolen. *Gordian* **79**, 182.

3 Water continuous emulsions

H. M. Premlal Ranjith

3.1 Introduction

The mammary epithelium of the udder cells of mammals undergoes remarkable changes in order to secrete the white biological fluid termed milk. Humans utilised this wholesome food from early periods in history and developed techniques to produce specific food commodities using some of its major components, such as fat. Milk from sources other than cows have also been utilised in many parts of the world as the main source of food supply. For example, the milks of buffalo, goat, sheep, yak and horse have in some parts of the world been the main source of vital nutrients to supplement the native food supply. Milk exists in the form of an emulsion. The food emulsion exists in two basic forms: the oil-in-water type (O/W) or the water-in-oil type (W/O).

Creams, ice cream, mayonnaise and salad oils (dressings) are some of the two-phase O/W emulsions where water or the aqueous phase is continuous. These emulsions are constantly under threat from possible destabilisation caused by microbiological activity or by chemical or physical changes. In creams and ice creams, the pH is in the neutral range and that, together with the continuous aqueous material, provide an ideal environment for a large cross-section of microorganisms to thrive. In the case of salad dressings and mayonnaise, the pH is low (3–4), and such acid conditions are an unfavourable environment for many microorganisms, including pathogens. The aqueous phase in O/W systems contains a variety of nutrients, which can be easily digested by bacterial cells and therefore act as a good substrate for them to feed on.

An important variation on O/W emulsions is that found in ice cream. Its components—water, dissolved solids and salts—all exist in a frozen state. In addition to being able to withstand the freezing process, it should also hold fine air bubbles. Incorporation of air into ice cream and whipped cream means that these emulsions consist of a third phase.

3.1.1 The structure of water continuous emulsions

In defining an emulsion it is important to be aware of descriptions of other closely associated systems such as solutions, colloidal solutions, suspensions and foams. A genuine solution is the simplest of all physicochemical states, whether it be solid, liquid or gaseous. A simple example of a genuine solution is that of sugar dissolved in water. The solute particle size is less than $0.001\ \mu m$.

In a colloidal solution the particle size varies from about 0.001–0.1 μm. The complex calcium salts in milk exist in colloidal form together with albumin and casein. According to basic principles there is very little difference between an emulsion and a suspension. The most striking difference is that the particles of an emulsion are liquid, whereas those of the suspension are solid. In both systems the particle size (diameter) is greater than 0.1 μm and could be as large as 1 mm. The fat in milk exists in this form. Foam can be considered to be a suspension where the particle size (i.e. the diameter of the air cells in the foam) is about 30–100 μm.

An emulsion is defined as a mixture formed by combining two immiscible fluids in which one is uniformly distributed in the other without separation. As indicated, the particle size ranges from about 0.1 μm to visible size. In food emulsions the two main components are oil and water, with other semisolids or solids dispersed either in the continuous or in the dispersed phase. These systems are either O/W or W/O, and the semisolids and solids may also exist as crystals, as in ice cream or butter. Figure 3.1 shows the basic structure of O/W and W/O emulsions. In milk, the fat content, which is approximately 4%, is dispersed in an aqueous phase containing protein. Similarly, the higher fat milk emulsions such as creams also contain protein in the aqueous phase but to a lesser extent. However, milk, creams and dairy ice cream differ from other emulsions that have been formed by mixing vegetable oils and water. In milk the fat exists uniquely in globular form, with each globule surrounded by a membrane. The aqueous phase may be described as a colloidal phase because of the presence of proteins, salts carbohydrates, microcomponents and other aqueous-phase materials.

Milk and creams are considered to be natural O/W emulsions whereas ice creams, soups and salad creams are considered to be engineered O/W emulsions. Milk inside the cow udder is a stable emulsion with the dispersed phase (fat) uniformly distributed without separation. However, once the milk exits the udder

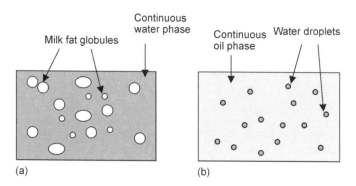

Figure 3.1 (a) Oil-in-water and (b) water-in-oil emulsions.

the tendency is for the fat globules to rise because of the difference in density between fat globules and the aqueous phase. This separation is also referred to as creaming and milk and cream storage silos and tanks are kept in gentle agitation at regular intervals to minimise this creaming effect. This natural separation of the fat phase in milk was used in early periods to prepare high-fat milk products.

3.1.2 Milk fat globule structure

A unique difference between milk fat and vegetable fat is that in its natural state milk fat is found in a globular form enrobed by a very complex membrane. This membrane protects the fat inside the globule from chemical and lypolytic action by the chemicals and enzymes present in the aqueous phase. The membrane also extends protection against mechanical damage during handling in the farm as well as in the dairy. This thin protective membrane also prevents the globules from flocculating and coalescing. The milk fat globule membrane is estimated to be about 5–10 nm thick (Mulder and Walstra, 1974). Scientists describe the fat globule as consisting of two basic layers: an inner core of fat surrounded by an emulsifier system with a hydrophobic tail linked to the fat, and a hydrophilic head attached to the surrounding aqueous layer consisting of proteins and other components. Therefore, the fat globule membrane surface is hydrophilic and maintains the fat in an emulsified state. The membrane is composed of a complex mixture of phospholipids, protein, vitamin A, carotenoids, cholesterol, high-melting-point glycerides and various enzymes. Figure 3.2 shows the layout of

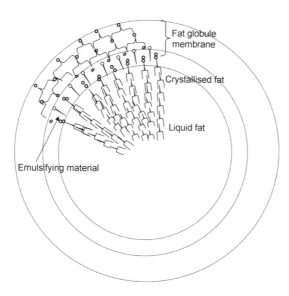

Figure 3.2 Milk fat globule system.

milk fat globule components to highlight the main layers, the inner layer being the triglycerides surrounded by the membrane. Chemical components such as lecithin hold the lipids and the aqueous material together. Other components such as casein units organise themselves around the aqueous layer to form the outermost layer. This biological membrane is formed during the milk secretion process in the mammary epithelium of the udder cells in the cow.

The fat content of milk varies more than the other major constituents and is influenced by environmental and other biological factors. As the fat fraction of milk tends to separate from the aqueous phase in most species the fat content varies throughout milking or suckling, generally increasing as the gland is emptied. Apart from this the fat content in all species is influenced by the stage of lactation, age, the breed type and diet. The fat in milk is present in the form of minute globules ranging in diameter from 0.1–20 μm (Walstra, 1969). It is estimated that the mean fat globule diameter is about 3 μm for cows' milk and about 2 μm for goats' milk. In different breeds the size of the globules tends to vary with the fat content of the milk. Therefore, the mean for Holstein milk is 2.5 μm and that for Channel Island cattle is 3.5 μm (Ling *et al.*, 1961). However, it was found that Ayrshire's are an exception as although the milk has a higher fat content than Holstein milk, the size of the globules is the same. As lactation advances the fat globules tend to become smaller.

The long-chain fatty acids of the phospholipid are buried in the fat and the hydrophilic part of the molecule is directed outwards. It is likely that the carotenoids, vitamin A and cholesterol associated with the membrane are in the phospholipid layer. The characteristic nature of the globule surface is such that at normal temperatures (8–20°C) the globules are grouped together in the form of clusters, which play an important part in the rising of cream.

Milk fat begins to melt at about 28°C but is not completely liquid until the temperature has reached about 33°C. The liquid fat sets over a similar but lower temperature range (24–19°C). At the time of secretion the fat globules are liquid. However, when milk is cooled crystallisation commences and this may take up to 24 h to reach completion. The reason for this lengthy period could be attributable to the complexities of the crystallisation process rather than to solidification. Also with the liquid fat being in the dispersed phase, there is no opportunity for the hastening effect of 'seeding', which usually operates when the liquid is in a continuous phase.

As with other fat material, milk fat is soluble in fat solvents such as petroleum and ether but it is not possible to extract milk fat by merely shaking it with these solvents. This is because of the protection it receives from the fat globule membrane. In quantitative solvent extraction the fat globule membrane is first removed by the action of acid or alkali. The membrane materials change constantly to keep in equilibrium with the components in the aqueous phase. This means that some membrane material may leave and join the aqueous phase, and vice versa. Therefore, the thickness of the membrane is variable.

Such clear demarcation of layers in the fat globule structure indicates that boundaries can be defined based on the components associated with each layer. The outer boundary is that associated with the components outside the fat globule and belongs to membrane materials. The inner boundary consists of the triglycerides in the core of the globule at the fat–water interface. However, the membrane undergoes rearrangement as a result of homogenisation and cooling.

3.2 Preparation of water continuous emulsions

3.2.1 Dairy creams

It is known from ancient writings that butter was used in food preparation in India between 2000 BC and 1400 BC (McDowall, 1953). It has also been reported that in 1480 cream was recovered by skimming the surface of milk and used for butter-like product manufacture in Italy. This therefore provides the evidence that, historically, the effect of differences in the density between oil and water has been utilised to recover cream and to produce dairy products from fats in milk. This also includes fat from mammals other than the cow.

Similarly, a denser material in milk will eventually settle to the bottom of the container holding the milk. The flotation or sedimentation velocity of particles in an emulsion (O/W) follows Stokes's law, as follows:

$$V_g = \frac{d^2(p_1 - p_2)}{18\eta} g \qquad (3.1)$$

where

V_g is the flotation or sedimentation velocity
d^2 is the particle diameter
p_1 is the particle density (kg m^{-3})
p_2 is the density of the continuous phase (kg m^{-3})
g is the gravitational pull of the earth
η is the viscosity of the continuous phase (kg m^{-1} s^{-1})

Equation (3.1) indicates that a larger particle will rise or sediment faster than will a smaller particle, that the velocity increases with increasing difference in density between the particles and continuous phase, and that the velocity increases with decreasing viscosity of the continuous phase. Equation (3.1) can be used to calculate the flotation velocity of a fat globule. For milk η is about 1.4 $\times$ 10^{-3} kg/m.s, p_1 is 980 kg m^{-3} and p_2 is 1.03 kg m^{-3}. The average globule diameter varies based on the mechanical and shear forces acting on the globule, leading to size reductions. However, the fat globules that form clusters or that aggregate rise faster because of the increase in overall diameter.

3.2.1.1 Clarification of milk

The cooled, stored raw milk is either filtered or clarified prior to processing and separation. A clarifier is used in almost all modern, medium-to-large dairy installations. A clarifier works on the same principal as a centrifugal separator. Centrifugal force is generated from the rotational movement of the material. The magnitude of the centrifugal force is dependent on the radius and speed of rotation and on the density of the material being rotated. Milk produced under good hygiene conditions will be substantially free from foreign matter as it will have passed through a filtering system in the farm. In all dairies milk passes through a filtering system (filter cloth, etc.) or through a centrifugal clarifier. Clarification is generally conducted in the cold and, unlike the case of a separator, the heavier and lighter fractions do not exit separately from the machine. The insoluble matter is thrown to the rotating bowl, where it is discharged at regular intervals.

3.2.1.2 Centrifugal separation

The dairy industry pioneered the development of centrifugal separation to produce cream fractions for the manufacture of high-fat dairy products. A dairy centrifugal separator is specially designed to separate fat phase in milk with minimum damage to the fat globules. In early designs, fat globules travelled a significant distance inside the unit prior to being separated by the centrifugal force. The centrifugal force throws the denser material away from the centre spindle, and the lighter phase moves towards the centre. In modern separators, however, the inner cavity of the separator bowl is fitted with a series of conical discs arranged one on top of the other to form a stack. This arrangement separates the milk into thin layers [figure 3.3(a)]. Such an arrangement increases the efficiency of the separation process. In each disc holes are arranged for the distribution of milk. The sediments and other heavy particles are also separated as sludge and are collected at the sidewall of the bowl. In small and medium throughput separators (up to about $5000 \, l \, h^{-1}$) the sludge is removed manually when the separator is dismantled for cleaning.

The operating principles of separators vary with design features present. There are three main categories of separator: the open design, the semi-enclosed type and the hermetic type. These names relate to the separator bowl design and to the arrangement of the entry and exit ports.

As already described, the denser material moves to the wall of the rotating bowl and the lighter fraction moves towards the centre spindle. The centre spindle is mechanically linked to the motor and gear system, which rotates the spindle at 7000–9000 rpm. In the dairy separator the open design introduces milk from the top of the bowl, usually via a float arrangement. The skimmed milk and cream exit the bowl from the top via an overflow into collecting spouts or trays. In this system, milk, skimmed milk and cream are exposed to the outside environment, hence the name 'open design'.

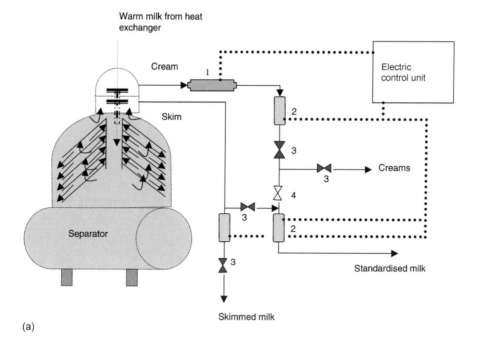

(a)

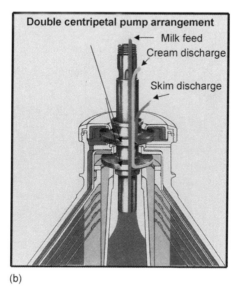

(b)

Figure 3.3 (a) A milk separation and standardisation circuit; 1, density transmitter; 2, flow transmitter; 3, flow-control value; 4, isolation valve; dotted line, control circuit. (b) Centrifugal method for extracting skimmed milk and cream: the centripetal pump system is fitted internally, with milk fed into the system from the top of the separator. Part (b) reproduced courtesy of Westfalia Separator, UK and Diotte Consulting & Technology, UK.

Semienclosed separators differ from the open type in that there are paring discs fitted at the top of the bowl to pump the cream and skimmed milk [see figure 3.3(b)]. These are also called centripetal pumps. The exit ports are fitted with throttle valves to maintain back pressure and to control the fluid flow.

All modern dairy separators designed to operate at a higher flow rate are able to isolate the separator rotating components and products inside the bowl (milk, cream and skimmed milk) from the outer environment. Such separators are of the hermetic type.

3.2.1.3 Control of fat content in cream

In early periods, upto the nineteenth century, cream recovered from milk was used mainly for the manufacture of butter-type products. The development of the dairy separator was an important process innovation which led manufacturers to launch added-value products with a high fat content, creams as well as low-fat and 'no-fat' milks such as semiskimmed and skimmed milk, respectively. The method of preparation included either using the skimmed fraction or whole milk to dilute the cream to obtain the desirable fat content in the final product. In these techniques the fat content of each fraction must be analysed in order to formulate the appropriate combination required. The dairy industry regularly standardises creams and milks.

To adjust the fat content accurately Pearson's rectangle is used. The basic method is demonstrated in figure 3.4. The example given illustrates that cream containing 50 g per 100 g fat may be diluted with skimmed milk in the ratio 4:1 to obtain cream containing 40 g per 100 g fat. In high-throughput separators, an in-line densitometer automatically makes the fat adjustment. A simple diagrammatic illustration of in-line automation for fat standardisation is given in figure 3.3(a).

The basic requirement is to use flow meters, densitometers and control valves to adjust the fat content by mixing the correct proportion of skimmed milk and cream. The actual fat content of the streams is continuously monitored by the

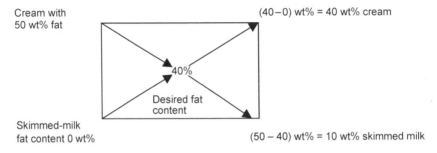

Figure 3.4 Pearson's rectangle for cream standardisation. Figures are in grams per hundred grams.

densitometer and the information is communicated back to the control unit for comparison with the preset parameters corresponding to the required fat content in the cream. The density transmitter is capable of detecting small changes in the density of the cream when the fat content changes. An increase in fat content in cream results in a decrease in density; the opposite is true when the fat content decreases. This means a relationship exists between fat content and density; that is, the fat content in cream varies inversely with density.

Factors such as entrained air in milk and cream influence the density measurement. If the processing circuit linking the separator has the tendency to suck air in because of poor flow control in the system then the cream produced will be lighter as air bubbles will be pushed towards the cream by the centrifugal force. This means that the process circuit must be airtight and, in addition, some installations include an in-line de-aerator to ensure that the oxygen content of raw milk is reduced to a uniform level prior to being separated.

It is possible to estimate the theoretical yield of cream of known fat content by using the following formula:

$$T_f = \frac{F_m S}{100} \tag{3.2}$$

where,

T_f is the total fat flowing through the system (kg h^{-1})
F_m is the fat content of the incoming milk (g 100 g^{-1})
S is the flow rate of the separator (kg h^{-1})

If the separator is adjusted or programmed to achieve 40 wt% fat cream, the cream yield Y (in kg h^{-1}), would be given by

$$Y = \frac{100 T_f}{40} \tag{3.3}$$

If T_f is the pure fat flowing per hour, then $Y - T_f$ is the skimmed-milk fraction flowing per hour included in the 40 wt% fat cream. These values calculated as shown can be used for approximate setting of the separator to obtain the desired cream fraction. For example, in separators where the flow is manually adjustable the skimmed-milk and cream exit ports are fitted with throttle adjustments to restrict the flow. They are adjusted so that the system pressure as well as the cream flow rate can be balanced as specified for a particular separator. The system pressure adjustment is in the skimmed-milk outlet and, when the cream exit port is set to the correct flow rate in the flow indicator, the fat content of the cream exiting the separator will be close to the desired value. Table 3.1 lists the categories of cream in use in the United Kingdom.

The efficiency of separation is reflected in the quality of skimmed milk obtained and the level of free fat in the cream. Efficient separation produces

Table 3.1 Cream categories in the United Kingdom, with minimum fat content (min. fat) measured in g per 100 g

Type	Min. fat
Half cream	12
Single cream	18
Whipping cream	35
Double cream	48
Sterilised cream	23
Clotted cream	55

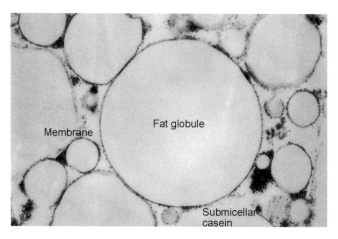

Figure 3.5 The microstructure of cream. Reproduce courtesy of Diotte Consulting & Technology, UK.

skimmed milk with less than 0.05 wt% fat and low amounts of free fat in the cream. Figure 3.5 shows the microstructure of single cream. The free fat level will increase if damage is caused to the fat globule membrane from mechanical and shear action in the circuit and pumps. Milk is usually warmed to at least 40°C prior to separation, and the feed to the separator should be adjusted to ensure that it is within the design specification of the manufacturer. A lower than optimum feed rate into the separator can cause the fat fraction to stay in the separator longer than necessary, causing further damage to the fat globule membrane as a result of shear. Once the globule membrane is damaged the free fat can escape into the aqueous phase. A high percentage of free fat leads to high quantities of free fatty acids (FFAs) in the cream, resulting from lipolysis by indigenous lipase enzyme. Not all FFAs are formed by this method; some are formed as a result of poor quality milk or poor handling practice. In poor-quality milk a high concentration of the microbial enzyme lipase is available to hydrolyse the triglycerides and thus increase the FFA content. Therefore, it is important to reduce the occurrence of FFAs in the raw milk that may originate from poor microbiological quality or handling methods.

3.2.2 Recombined creams

The formulations and procedure for recombined creams have been known to the food and dairy industries for quite some time. Buchanan and Smith (1966), Zadow and Kieseker (1975) and Towler and Stevenson (1988) studied the manufacture and properties of these products. The manufacture of such products was originally achieved by using butter oil, skimmed milk powder and water. However, the main drawback with this recombination process was that the resulting cream had a very poor overrun or no foaming at all: high-fat recombined cream should be pourable as well as thick enough to use as a dessert cream. Therefore, in formulating a recombined cream various ingredients are selected to enhance the desirable qualities and to make the process cost-effective. Recombined creams are popular in countries where real cream is in short supply or is not available and where cream is popular in bakery applications. For bakery applications high-fat products can be specially formulated by using stabiliser–emulsifier systems to minimise serum drain and to improve foam quality. Some details of emulsifiers and stabilisers are given in section 3.2.3.1. Emulsifiers, such as distilled monoglycerides, were found to produce acceptable overrun, and lecithin gives the desirable stiffness to the foam.

Zadow and Kieseker (1975) reported that use of anhydrous milk fat, nonfat milk solids and an emulsifier such as glycerol monostearate could produce good-quality whipping cream (35 wt% fat). A two-stage homogenisation step was used at pressures of 1.4–2.1 MPa at the first stage, and 0.7 MPa at the second stage. The optimum temperature for homogenisation was found to be about 48°C. In the formulation 0.1 wt% of the emulsifier was used and it was found the product could withstand process conditions used in ultra high temperature (UHT) treatment.

The steps involved in the preparation of recombined cream are as follows:

1. Dissolve the nonfat milk solids and emulsifiers in water at about 40–50°C with use of a high-speed mixer.
2. Melt the fat (e.g. anhydrous milk fat, butter, etc.) at about 40–45°C and add it slowly to the liquid mix.
3. Continue mixing to ensure that a homogeneous mix is formed that is free from lumps and oil droplets.
4. Heat treat (i.e. pasteurise or UHT) the mixture (see section 3.2.4).

3.2.2.1 Nondairy creamers or cream alternatives

The manufacture of these products is similar to that of recombined dairy cream. The most common ingredient that is substituted in making a nondairy cream is the fat source. Milk fat is replaced with vegetable oil in most nondairy preparations. Substitution of milk fat with vegetable oil is attractive for the manufacturer because of its substantially lower cost as well as because it is easy to obtain in various parts of the world. In some formulations fat replacers are used so that these products are suitable for those who prefer fat-free products. Other components such as nonfat milk solids have been replaced with alternative nondairy components. For example nonfat solids originating from soya powder or soya slurry extract has been used in varying proportions together with stabilisers and emulsifiers in nondairy cream formulations.

Development of such products tends in most cases to be protected by patents. A simple nondairy emulsion is a coffee whitener where the fat content is in the range 10–12 g per 100 g. Some of the components in coffee whitener are given in table 3.2.

Coffee whitener products became popular in the marketplace in the mid-1980s, their production being aided by the development of sophisticated emulsifier–stabiliser systems and much improved recombination technology. Similar to other recombined water continuous emulsions, nondairy creams were prepared with care to achieve the desirable physical and organoleptic characteristics. Products containing vegetable oil have been formulated to include a certain proportion of dairy fat and solids so that these products closely resemble real dairy creams. For example, butter milk or butter milk powder is frequently used in nondairy products as these ingredients provide the characteristic creamy taste as well as the emulsifying ability of O/W emulsions. The selection of emulsifiers is important to achieve desirable characteristics such as an increase in whitening ability of the coffee creamers and also to reduce the homogenisation pressure required to disperse the oil phase uniformly. The whitening power of a coffee whitener reflects the surface area created by the dispersed particles. The higher the surface area the greater the light reflectance from the dispersion and

Table 3.2 Constituents of coffee whitener (quantities are given in g per 100 g)

Constituent	Quantity
Vegetable fat	10–12
Carbohydrate[a]	8–12
Sodium caseinate	1–2
Emulsifier[b]	0.15–0.25
Stabiliser[c]	0.04–0.06
Polyphosphate	0.04–0.2
Flavouring	as required
Water	remainder

[a]Sources: dextrose monohydrate, polydextrose and maltodextrin.
[b]For example, monoglycerides.
[c]For example, carageenan.

thus the greater the whitening effect. However, overhomogenisation also has its drawbacks, as it disrupts the fat globules and leads to fat oxidation.

3.2.2.2 Cream liqueurs

Cream liqueurs were developed in the mid-1970s and are water continuous dairy emulsions of high added value. Cream liqueurs are simply a combination of milk-protein-stabilised cream emulsion with high alcohol content. These emulsions are now produced commercially in large volumes and they are a popular product, especially in Ireland, the United Kingdom and mainland Europe.

The manufacturing methods are described by Banks and Muir (1985). The procedure can progress in two ways: via single-stage or two-stage processing. In the single-stage process homogenisation is carried out after the addition of alcohol, whereas in the two-stage process the homogenisation step takes place prior to alcohol addition. A typical formulation for cream liqueur is given in table 3.3. In both methods the starting point is to prepare a cream base. It is prepared by mixing sodium caseinate powder and sugar into water. Cream is incorporated into this mix together with citrate by gentle agitation to form the cream base. Alcohol is added slowly with gentle agitation. This mixture is then homogenised twice at about 30 MPa at about 50–60°C and is then cooled. In the two-stage process the cream base is homogenised at 50–60°C, cooled to about 10–15°C and the alcohol is added slowly with gentle agitation. Banks and Muir (1988) observed that there are fewer large fat globules if the homogenisation is carried out in the presence of alcohol. As much as 97% of the fat globules in the emulsion has diameters less than 0.8 μm when the product is homogenised twice at 30 MPa. In view of the relatively small diameters of these fat globules the corresponding increase in total fat surface requires additional protein material to cover them adequately, and this protein is derived from the sodium caseinate incorporated in the cream base.

In some formulations either all or part of the sodium caseinate may be replaced with a suitable low molecular weight emulsifier. A heat-treatment stage is not required for cream liqueurs as they are protected from microbiological activity by the presence of the alcohol (14–17% v/v).

Table 3.3 Composition of a typical cream liqueur (quantities are in g per 100 g)

Constituent	Quantity
Cream (40 wt% fat)	30.0
Added sugar	18.0
Sodium caseinate	2.8
Alcohol (40% v/v)	38.0
Water	11.2

3.2.3   Ice-cream mix

An ice cream is both an emulsion and a foam containing ice crystals. Various components contribute to the stability of the system. Physico-chemically it is a very complex system, probably the most complex of all dairy products. The complexity is partly a result of the choice of ingredients and the use of sophisticated emulsifier–stabiliser systems.

A typical simple ice-cream mix contains the ingredients listed in table 3.4.

3.2.3.1   Sources and functions of ingredients
The desired chemical and physical properties and organoleptic quality can be achieved through the selection of appropriate ingredients. The selection of ingredients varies from manufacturer to manufacturer, from regions of the same country to another and from country to country. Such variation exists because the ingredients used in making up an ice-cream mix will vary in character and may fluctuate significantly from one formulation to another. Commercially, the composition is a vitally important factor, contributing to characteristics such as eating quality, demand and competitiveness of market price. Manufacturers will favour the combination that will best stimulate demand and at the same time be made at a favourable cost.

Fat.   Fat present in ice-cream products may be derived from animal or vegetable sources. In dairy ice cream, common sources of fat are whole milk, creams, butter, butter oil, sweetened condensed milk, evaporated milk, and milk concentrated by membranes (e.g. ultrafiltration and reverse osmosis). Other sources are milk powders, butter milk and whey. The minimum level of fat in ice cream varies according to national regulations. For example, in the United Kingdom ice cream is defined in the Food Labelling Regulations (1996), which state it should contain not less than 5% fat. This also applies to other European Union countries. Similarly, dairy ice cream should contain a minimum of 5% fat consisting exclusively of milk fat.

In some countries the description 'ice cream' may only be used if the fat is derived from cows' milk. Regardless of the origin of the fat it must be free from off flavours and undesirable taste characteristics otherwise these will carry

Table 3.4 Typical ingredients of a simple ice-cream mix (quantities are averages and are expressed in g per 100 g)

Ingredient	Quantity
Water	63
Sugar	15
Nonfat milk solid	11.5
Fat	10
Emulsifier–stabiliser system	0.5

through to the finished products. Vegetable oil has a neutral taste, whereas dairy fat provides a characteristic creamy flavour. In addition, the melting characteristics of fats are important to achieve stability during storage.

Nonfat milk solids. These may also be called simply nonfat solids. They play an important role in the final eating quality of ice cream and therefore must not be used as a filler to increase the total solids content as above an optimum amount nonfat solids can lead to a sandy texture resulting from the formation of lactose crystals. At the optimum level of addition they give body as well as enhance the aeration properties. Incorporation of nonfat solids supplements the use of proteins, milk salts (e.g. salts of calcium, magnesium, potassium and sodium, and chlorides, citrates and phosphates) and some vitamins (e.g. B, C and folic acid).

Sugars and sweeteners. The main function of added sugar in ice cream is to provide sweetness and therefore to increase the palatability. It also, for example, increases the food value and alters the physical properties, such as the freezing point, of the product. An increase in sugar content relative to the water content has a tendency to depress the freezing point of ice cream.

Sugar is derived mainly from cane and sugar beet. In formulations where sweetened condensed milk is used in significant quantities, it may not be necessary to add further sugar to the ice cream mix. Other sources such as malt sugar and corn sugar or corn syrup may be used to replace 'normal' sugars (sucrose from sugar cane or sugar beet). An important consideration is that the sugar, either in solid or in liquid form, must not present handling problems in commercial use.

The majority of consumers show a preference for relatively sweet ice creams; in some recipes the total sugar content may be as high as 20%. Most products generally contain about 15% sugar. The level of addition plays an important role in adjusting the required final total solids content.

Stabilisers. The function of stabilisers is to prevent large crystals from forming during freezing of the ice-cream mix. The most commonly used stabilisers are derived from two main sources: gelatine, of animal origin, and food hydrocolloids, such as alginates, carrageenan, carboxymethyl cellulose (CMC) and gums, of plant origin. The quantity of stabiliser used is very small and tends not to induce undesirable organoleptic characteristics. Desirable qualities expected from the use of stabilisers are a smooth body, good textural properties, the presence of small ice crystals throughout the storage period and an increase in product viscosity.

Emulsifiers. Food emulsifiers serve a number of purposes. They promote the emulsification of the oil and the aqueous phase without separation, they have a starch-complexing ability, they interact with proteins, they modify fat crystallisation and the viscosity characteristics of food ingredients, they control foaming and antifoaming, they disperse solids in water and they provide lubrication. Hence, in addition to providing basic emulsification, there are other extremely important functions performed by emulsifiers. These functional properties are key factors in determining their suitability for various applications. The use of an emulsifier increases the ease of formation and promotes the stability of emulsions by reducing the amount of work required to form a homogeneous mixture of two usually immiscible phases (e.g. oil and water). This function is possible if the chemical structure of the molecule consists of hydrophilic (i.e. water-soluble) and lipophilic (i.e. oil-soluble) groups. The emulsifier can then partially dissolve in each of the phases and thus unite those phases in the form of a homogeneous emulsion. Figure 3.6 illustrates O/W and W/O emulsions.

3.2.3.2 Hydrophilic–lipophilic balance

The hydrophilic–lipophilic balance (HLB) is used to classify emulsifiers. A number is assigned to each emulsifier that expresses the balance of the number and strength of its hydrophilic groups as compared with its number of lipophilic groups. In general terms, the HLB represents the oil and water solubility of an emulsifier. The HLB value can be calculated with knowledge of the chemical structure of the emulsifier, determined experimentally by comparison with emulsifiers of known HLB. For a homologous series of esters of polyhydric alcohols and fatty acids the simplest method is to calculate the value from analytical data by using the following formula:

$$\text{HLB value} = 20\left(1 - \frac{s}{a}\right) \tag{3.4}$$

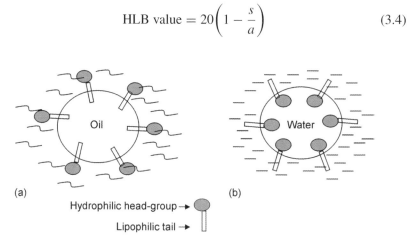

(a)   (b)

Hydrophilic head-group →

Lipophilic tail →

Figure 3.6 The role of emulsifiers in the formation of (a) oil-in-water emulsions and (b) water-in-oil emulsions.

where s is the saponification number of the ester, and a is the acid value of the fatty-acid radical. This method, which is based on weight percentages is satisfactory for emulsifiers that are nonionic. However, there are difficulties in determining the saponification value of some esters, and ionisation of the hydrophilic group in ionic emulsifiers will tend to exaggerate their hydrophilic character.

Table 3.5 shows that as the HLB value increases, the emulsifiers become more soluble in water and their function changes from being W/O emulsifiers to being O/W emulsifiers. Ice cream is essentially an O/W emulsion, and one would expect the most effective emulsifiers to have HLB values in the range 8–14. In fact, however, saturated monoglycerides with HLB values in the range 3–4 are by far the most widely used emulsifiers in ice cream. This apparent anomaly arises from the fact that the function of the added monoglycerides is not emulsification per se but interaction with the milk proteins present to form a protective hydrophilic layer of adsorbed protein around the fat globules. These layers prevent the globules from coalescing and thus stabilise the fat emulsion. In addition, some bonding takes place between neighbouring protein layers, causing the fat globules to clump. This clumping effect is responsible for the dryness, texture and stand-up properties of ice cream. The fat globules containing adsorbed protein also help to stabilise the air bubbles incorporated into the ice cream during the freezing and whipping process. Emulsifiers generally used in ice-cream manufacture are monoglycerides and diglycerides (saturated, and polysorbates). Monoglycerides are used in the range 0.25–0.5%. They function to control fat destabilisation to confer dryness at extrusion, to provide resistance to shrinkage, to ensure good melt-down properties and to control overrun.

Table 3.5 Common emulsifiers, categorised by hydrophilic–lipophilic balance (HLB) and application

HLB value 3–6: water-in-oil emulsions
    monoglycerides
    glycerol lactopalmitate
    propylene glycol monostearate
    sorbitan esters
HLB value 8–14: oil-in-water emulsions
    diacetyl tartaric acid esters
    polyoxyethylene sorbitan esters
    sucrose esters
    decaglycerol distearate
HLB value 14–18: detergents
    soaps
    lecithin
    decaglycerol monolaurate

3.2.3.3   Other ingredients

Egg products.   Egg yolk increases the nutritive value of ice cream but, equally, it is an expensive ingredient. Egg yolk also provides a characteristic flavour to ice cream and gives it body and texture. Egg yolk is rich in lecithin, which has emulsifying properties and therefore improves aeration and increases the viscosity. Therefore, in some ice-cream recipes it is used as a filler.

Starch.   As with solids from egg products, starch may be used as a filler in ice-cream recipes or in special ice creams. In some recipes starch may be used as a substitute for gelatine.

Flavours and colours.   Addition of such components to ice-cream mixes varies, depending on consumer preferences. There are many flavouring materials available. Commonly used flavouring materials in ice cream are vanilla, chocolate, fruit and fruit extracts, nuts and spices. Flavours may be harsh or delicate but, in general, the consumer will tolerate high concentrations of delicate flavours but may object to harsh flavours even in low concentrations. Therefore, finding the right flavour balance is important, especially when a product is made from mixed flavours. The intensity of the flavours should only be sufficient for the consumer to perceive it.

Colours are chosen in accordance with flavour. Fruit-flavoured ice creams may require only a small amount of added colour as the fruit itself may give sufficient colouring. Colours are not generally required for chocolate ice cream made with cocoa powder.

3.2.3.4   Creating a balanced ice-cream mix

As already mentioned, ice cream is a very complex product and requires careful selection of ingredients and close calculation of the proportions required to bring out the desirable organoleptic qualities and economic advantage for the manufacturer. Calculation methods are illustrated in detail by Arbuckle (1986). In the formulation of a balanced ice-cream mix it is important that initially the ratio between the fat and the nonfat milk solids or that between the fat and the total solids be calculated to establish that the minimum requirement is satisfied for the various ingredients used in the formulation. Further addition of specific ingredients is possible provided that the level of addition is not detrimental to the organoleptic quality of the ice cream.

In such calculations the adjustment of fat will be similar to that for creams, described in section 3.2.1.3, using Pearson's rectangle (figure 3.4). In modern calculations computers are used to simplify quantitative specifications for ice-cream mixes. In addition to formulation work, such methods are used for assessing and comparing costs across various recipes.

3.2.3.5 Preparation of ice-cream mixes, and freezing

Preparation. Ice-cream mixes are formulated from many ingredients derived from a variety of sources. The key components are fats and oils, liquid dairy ingredients, powdered dairy and food ingredients, stabilisers and emulsifiers. To achieve the best from each individual ingredient in the formulation, optimum conditions should be ensured. This can be achieved by taking care in the order of incorporation and by using appropriate timing for inclusion.

Figure 3.7 illustrates the procedure for mixing ice cream to produce an O/W emulsion. The ice-cream mix should be heat treated (i.e. pasteurised). In the batch method the warm ingredient mix is homogenised and collected into a batch pasteuriser (for minimum heat-treatment conditions, see section 3.2.4.1). In the continuous method, the preheating stage in the heat exchanger raises the temperature to 40–50°C. The mix is homogenised at 40–50°C and is then returned to the heat exchanger for pasteurisation. The normal pressure range for most formulations and homogenisers are given in Table 3.11 (page 104). Homogenisation can also be carried out after pasteurisation and cooled to about 45–50°C. In all the methods discussed here, the ice-cream mix must be cooled to below 10°C, preferably to less than 7°C. The cooled ice-cream mix is aged for at least 4 h or preferably overnight to allow fat crystallisation and the emulsifier–stabiliser system to take effect.

Freezing. Very small batches of ice cream can be frozen by using the batch method. However, all industrial operations use the continuous freezing method. A scraped-surface heat exchanger is used for freezing ice-cream mixes in the batch method and in the continuous method. The central shaft of the scraped-surface unit is fitted with scraper blades, and the shaft rotation speed can be altered in most industrial units to allow optimisation of product quality. Air is introduced into the mix either by suction of the feed pump or through injection into the freezer entry port. The scraped-surface freezer unit continuously whips the air into the ice-cream mix, with a concomitant reduction in temperature to freeze the water into small ice crystals. Figure 3.8 shows a cross-section of a scraped-surface freezer.

The freezing process thickens the consistency of the mix as the temperature is reduced to below about 0°C. The scraper blades, in addition to whipping air, continuously scrape the frozen layer of the product from the inner surface of the freezer barrel to facilitate good heat-exchange performance. The refrigerant material is circulated through the space between the product tube and the outer tube.

In the continuous phase there exist unadsorbed whey proteins, salts, high-molecular-weight polysaccharides and other macromolecules. After homogenisation, where the fat globules are reduced in diameter, new membranes are formed to cover the new fat globules. The membrane formation is

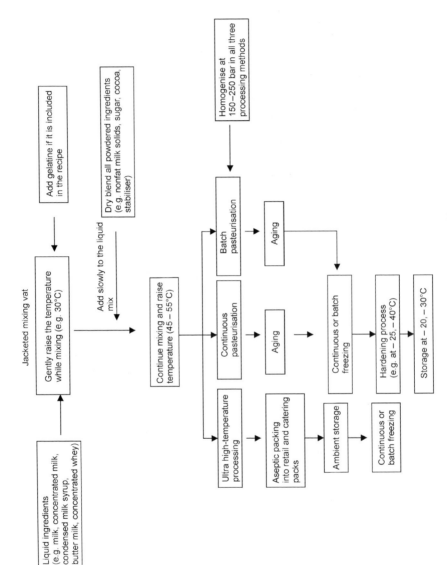

Figure 3.7 Flowchart for the manufacture of ice cream.

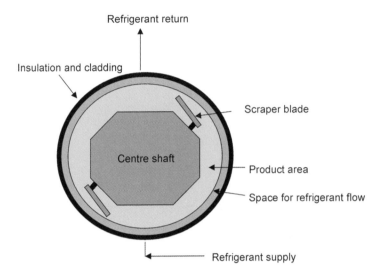

Figure 3.8 Cross-section of a scraped-surface freezer barrel.

completed during the ageing process, and some proteins are displaced by the emulsifiers.

In the freezing stage the air is uniformly distributed and forms a dispersed phase of air bubbles. Emulsifiers enhance the process of fat crystal formation due to nucleation during cooling and aging. They also enhance the whipping quality of the mix, the production of drier ice cream and facilitate moulding and various extrusion requirements. Dryness implies the absence of surface gloss and also a short nonsticky texture.

Ice cream exits the freezer unit at about −6°C to −8°C. At the exit from the freezer there is a back-pressure valve to generate the required pressure inside the scraped-surface unit. Ice cream exiting the unit is directed to the filling system for filling into individual servings and/or bulk packages. These containers go through a hardening process either in a batch process at −25°C to −30°C or in a continuous process by being passed through a tunnel. Hardened ice cream is stored at about −20°C to −25°C.

3.2.4 Heat treatment of emulsions

The primary objective of a heat-treatment process is to ensure food safety, to comply with hygiene requirements and to facilitate those ingredients that require heat to activate and initiate functional properties. This applies to water continuous emulsions. Milk separation may be carried out as part of the heat-treatment process (e.g. pasteurisation). In many continuous heat-treatment applications

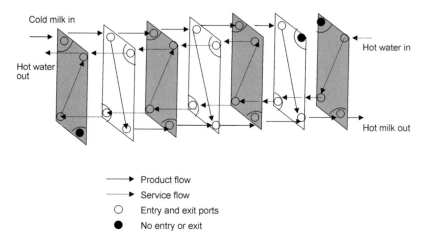

Figure 3.9 Flow arrangement in a plate heat exchanger.

on farms, a separator is linked to the heat-exchange unit either in the heating cycle or after the final heating stage and during the cooling cycle. In either arrangement the separation temperature used is in the region of 40–45°C. In the former method the separated fractions, skimmed milk and cream are further heat treated to pasteurise and are then cooled. The heat-treatment regimes designated, for example, in the Dairy Products (hygiene) Regulations (1995) are used for heat treatment of milks, creams and ice cream (see section 3.2.4.1).

3.2.4.1 Pasteurisation
Pasteurisation methods can be either of the batch type or of the continuous high-temperature short-time (HTST) type.

Batch method. In this method the product is heated to a temperature of not less than 65.6°C and held at that temperature for at least 30 min.

High-temperature short-time method. In this method the product is heated to not less than 72.0°C and held at that temperature for at least 15 s (continuous method), or it is heated to some other temperature for some other time regime, to have an equivalent effect to eliminate pathogens.

Ice-cream batch method. In this method the product is heated to not less than 71.1°C for not less than 10 min.

Ice-cream high-temperature short-time method. Here, the product is heated to 79.4°C for not less than 15 s and is cooled to less than 7.2°C within 1.5 h before freezing. If frozen ice cream reaches above $-2.2°C$ at any time during storage it must undergo pasteurisation again before sale.

In each method the ice cream must be cooled to less than $10°C$, preferably less than $7°C$. Figure 3.9 illustrates the product flow arrangement for a continuous plate pasteurisation system. Some important design features in a HTST heat-treatment plant are as follows.

- The pasteurisation temperature sensor is located in the early part of the holding tube.
- A divert valve is fitted at the end of the holding tube.
- A continuous recording method is used for probes monitoring the temperatures of pasteurisation, hot water and the final product cooler.
- A pressure differential measurement and indicating device should be fitted to monitor the raw and pasteurised products. If the pressure differential is not monitored a valid pressure-test certificate should be available for inspection by the licensing authority (every 12 months).

3.2.4.2 Ultra high-temperature treatment

Long-life creams and ice-cream mixes have been produced successfully for many years in Europe. Long-life ice-cream mixes have become popular in many other parts of the world. The ultra high temperature (UHT) process was originally applied successfully to milk. The same basic processing methods were later adapted for creams and ice-cream mixes. The UHT process regimes used in the United Kingdom are defined in the Dairy Products (Hygiene) Regulations of 1995.

The minimum heat treatment required is defined as follows:

- heat the product to not less than $140°C$ and hold that temperature for not less than 2 s

or

- heat the product under other temperature and time regimes having an equivalent lethal effect on vegetative pathogens and spores

The normal heat-treatment regimes used in commercial operations are 136–145°C for 2–6 s. For ice-cream mixes it is necessary to raise the temperature to not less than $148.9°C$ for at least 2 s. The regulations may vary according to the food process control measures introduced in individual countries.

Process plants have been developed and mechanised to be able to withstand heat-treatment duty and to satisfy hygiene and food safety requirements and the organoleptic quality preferences of the consumer. Heat-exchange systems are divided into three main types based on the method of construction: plate, tubular and scraped-surface. The most commonly used in the dairy industry is the plate-type heat exchanger. Plate heat exchangers are efficient in terms of energy usage and some plants have been designed to operate with 95%

energy efficiency. The method of heat transfer is also divided into two main groups based on how the transfer is carried out: direct or indirect. The direct method may involve steam injection, in which high-pressure steam is injected into a stream of product, steam infusion, in which the product is injected into a chamber containing steam under pressure, or it may involve electrical heating, in which a high voltage passes between two electrodes placed inside a stainless steel tube carrying the product, the resistance to electrical conductance producing the necessary heat. Commercial plant using electrical heating are called simply 'ohmic heaters' and are supplied by the APV Company, England (for a description of this process see Murray, 1985). In the indirect method heat-conducting material such as thin plates or tubes separates the product and the heating medium. The heating medium may be steam, superheated water or electrical energy.

The time–temperature profiles of typical commercial UHT plants are given in figure 3.10. Ranjith and Thoo (1984) described a procedure for producing fresh-tasting milk and milk products after UHT. This process is in commercial production in the United Kingdom. Burton (1988) has documented a comprehensive account of UHT processing. The heat treatment given to a product was originally quantified in terms of lethality values based on work carried out in the food canning industry. For example, the lethality value given to food products is described in terms of F_o values with reference to the death rate of the organism *Clostridium botulinum*. An F_o value of 1 is given when a product receives a heat treatment of 121.1°C for 1 minute. A simplified formula used in the calculation to derive F_o values in high-temperature processes is as follows:

$$F_o = 10\left(\frac{T - 121.1}{z}\right)t \qquad (3.5)$$

where

  T is the temperature (°C) of the process
  t is the time (min)
  z is the change in temperature (°C) required for the thermal death time
    to transverse one $\log_{10}$ cycle

Kessler and Horrak (1981a) described alternative dimensionless values for quantifying the lethal effects of a heat treatment: the B^* and C^* values. A B^* value of 1 refers to a heat treatment when the spores of *Bacillus stearothermophillus* were reduced by $10^9 \log_{10}$ cycles. A C^* value of 1 refers to a heat treatment where vitamin B_1 (thiamine) is reduced by 3%. For milk and cream processing it is necessary to achieve a higher B^* value and a C^* value as low as possible. These values can be calculated by using graphical methods (Kessler, 1981; Kessler and Horrak, 1981a).

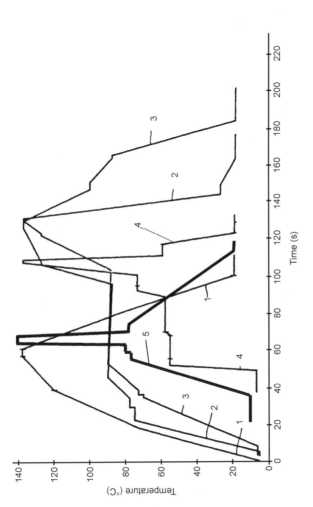

Figure 3.10 Temperature–time profiles for some commercial high-temperature plants. 1, Indirect plate, 85% regeneration; 2, indirect plate, 95% regeneration; 3, indirect tubular plate, 60% regeneration; 4, indirect plate, 70% regeneration; 5, indirect steam injection. Reproduced by courtesy of the Milk Marque Product Development Centre, Nantwich, UK.

Ultra high temperature (UHT) products are always stored and filled under aseptic conditions, without which a long shelf-life at ambient temperatures would not be possible. An aseptic system in commercial production is shown in figure 3.11. A key feature in such a system is that every point where a valve isolates the sterile product from a nonsterile environment (air, water, cleaning material) a steam tracing device exists to protect the product from contamination. In addition, where the product is diverted to aseptic storage tanks and aseptic filling machines, an aseptic valve cluster is installed to facilitate product diversion, steam barrier functions and cleaning in place (CIP). The sequence of valve operations and monitoring of events are very complex in an industrial installation and require a programmable controller. The program itself is carefully prepared to control a specific installation so that coordination of events and all necessary precautions are taken to ensure that recontamination does not take place.

3.2.4.3 Extended shelf-life processes

A systematic investigation into the poor keeping quality of milk and creams commenced in the United Kingdom in the early 1970s by the Milk Marketing Board (MMB, England and Wales). At that time it was well known that the storage life of cream, milk and other fresh liquid milk products was very short. For example, when these products are stored at ambient temperature (e.g. 10–20°C) for a few hours the microbiological quality can reach unacceptably high levels even when they are subsequently stored under refrigerated conditions. The shelf-life of such products can be prolonged by storing the products under refrigerated conditions (5°C–8°C) immediately after heat treatment. Such low-temperature storage conditions prolong the shelf-life from perhaps 1 or 2 days to perhaps 4 or 6 days, but prolongation of shelf-life by this extent is of limited value industrially. The deterioration in the quality of milk, cream and other fresh liquid milk products is due to microbiological activity that generally develops within a few days of storage to such a level that the product takes on unacceptable flavour characteristics and frequently undergoes unacceptable physical changes. The microbiological activity that gives rise to these unacceptable changes is not prevented by conventional pasteurisation treatment. The bacteria in refrigerated bulk raw milk consist mainly of Gram-negative psychotrophs. Some of these organisms synthesise extracellular proteases and lipases that are resistant to heat (Griffiths et al., 1981; Law, 1979) and therefore control of their numbers in milk is important. Pasteurisation of dairy products is a thermal treatment to destroy pathogens that may be present. However, it also destroys a large proportion of the nonpathogenic bacteria. Thus the thermal treatment makes the product safe and improves the keeping quality but it does improve the organoleptic quality of the product. Recently, the E-coli 0157 strain has been found in heat-treated milk in the United Kingdom and elsewhere. This was not due to survival from heat treatment but was mostly due to post-process contamination. This

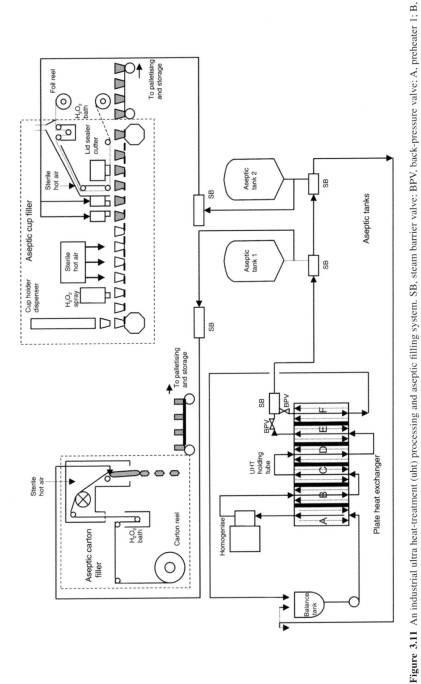

Figure 3.11 An industrial ultra heat-treatment (uht) processing and aseptic filling system. SB, steam barrier valve; BPV, back-pressure valve; A, preheater 1; B, preheater 2; C, uht heater; D, cooler 1; E, cooler 2; F, sterilising cooler (Diotte Consulting & Technology, UK).

highlights the problems faced in delivering milk with good keeping qualities and 'freshness'.

Microbiological quality of raw milk. Cows' milk is an almost perfect food for human beings and is the perfect food for calves. Unfortunately, it is also a good source of food for microorganisms. Milk from the udder of a healthy cow contains very few organisms (not more than about $300 \, ml^{-1}$) and these are of no danger to the consumer. Therefore, milk is contaminated generally by post-production handling, including milking equipment and the general hygiene of operatives. The keeping quality of raw milk is determined mostly by the initial number of microorganisms present in the milk and by the temperature at which it is retained after production. Hygienic milk production has advanced in most countries, and in the European Communities the standards are defined for the maximum allowable total viable counts per millilitre of raw cows' milk. Raw milk viable counts are becoming an important factor in modern dairy processing as the time interval between milk production and processing has increased. Other factors such as alternate day collection and changes to factory operation methods further extend this time interval. Delay in processing means that psychotrophic bacteria can proliferate. It is the psychotrophic spores in large numbers that can undermine a heat-treatment regime, as the microbial inactivation follows first-order reaction kinetics. For example, if the heat-treatment regime were capable of reducing the number of colony-forming units (cfu) by a factor of 10^4, an initial count of $10^5 \, cfu \, ml^{-1}$ would leave $10 \, cfu \, ml^{-1}$ after processing. This is the scenario with heat-resistant organisms, but not with vegetative pathogens. Thermodurics are defined as those organisms that survive a temperature of $63°C$ for 30 min; endospores can survive a temperature of $80°C$ for 10 min (Lewis, 1999). Therefore, the quality of raw milk will vary depending on general production hygiene, the equipment used, the environment, and organism population and type.

High-temperature pasteurisation. Pasteurisation of milk and milk products by the continuous flow method applies a heat-treatment regime of $72°C$ for 15 s. These conditions destroy pathogens in milk, in particular *Mycobacterium tuberculosis*. However, a pathogen *Mycobacterium paratuberculosis* was found to survive this pasteurisation regime, and many milk processing dairies have already extended the holding time from 15 s to 25 s in order to destroy this organism.

One drawback of the present pasteurisation process is that the whole system is vulnerable to post-process contamination. Thus the shelf-life of the product may vary from one day to another, depending on the total counts in the final product after heat treatment and depending on fluctuation in storage temperature. It is difficult to guarantee the hygiene standard of milk holding tanks, filling machines, packaging materials and the packaging environment. If contamination is high then the shelf-life may be reduced significantly. Therefore, a method to extend the shelf-life of creams and milks is desirable in order to minimise losses through microbiological spoilage.

Consequently, there is still a need for a heat-treatment process that will inhibit the microbiological activity in fresh milk and milk products to an extent that will permit the product to be stored under refrigeration for prolonged periods of time (e.g. in excess of 4 weeks), and at the same time avoiding the difficulties of an unacceptable 'sterilised' flavour. The method described here is based upon careful selection of temperature and time combinations suitable for milk and milk products to inhibit microbiological activity when they are stored at 5–10°C for periods of up to 7 weeks or longer. Any heat-treatment process showing poor organoleptic characteristics were considered to be unacceptable.

The microbiological development of milk after a heat-treatment regime in the range 74–81.5°C (using an indirect plate heat exchanger) is shown in figure 3.12(a). Contrary to the general belief that by increasing heat-treatment temperature, shelf-life will be prolonged, results indicate that bacteriological counts increase during storage. For example, in this experiment, cream was treated at the lowest temperature had a shelf-life of 20 days whereas that treated at the highest temperature (81.5°C) was unacceptable after 13 days at 7°C.

In flash pasteurisation heat-treatment temperatures in the range 80–90°C are used commercially with a holding time of 1 s or less. Figure 3.12(b) shows microbiological development in heat-treated cream stored at 7°C. The end result shows that the microbiological quality obtained with this treatment was similar to that obtained with the higher temperature pasteurisation as illustrated in figure 3.12(a). Comparing the results in figures 3.12(a) and 3.12(b), one can see that the holding time plays a significant part in the shelf-life of milk and milk products. For example, the cream heat treated at 80°C for 1 s had a longer shelf-life than that treated at 79°C and 81.5°C for 15 s. It can be seen from figure 3.12(b) that 20 days shelf-life was achieved for heat treatment at 80°C for 1 s, whereas creams treated at 82.5°C, 85°C and 90°C for 1 s were unacceptable after 9 days.

The pasteurisation effect P^* of each heat-treatment regime was calculated according to Kessler and Horak (1981a) and the results are listed in table 3.6. The recommendations of the International Dairy Federation (IDF) for minimum conditions for pasteurisation (72°C for 15 s) was given as $P^* = 1$. The pasteurisation effects at 72°C for 15 s and at 80°C for 1 s were close, and the total microbiological development after storage at 7°C for 20 days was comparable. However, the pasteurisation effect was 15-fold more at 81.5°C for 15 s, yet the shelf-life was limited to 13 days. Similarly, at 82.5°C up to 90°C for 1 s, the pasteurisation effect was more severe than at 72°C for 15 s but the shelf-lives were limited to 9–13 days at 7°C.

The reason for this contradiction is associated with the microbiological development of those organisms that are tolerant of high temperatures. The aseptic packing techniques in the sample production ensured that other organisms did

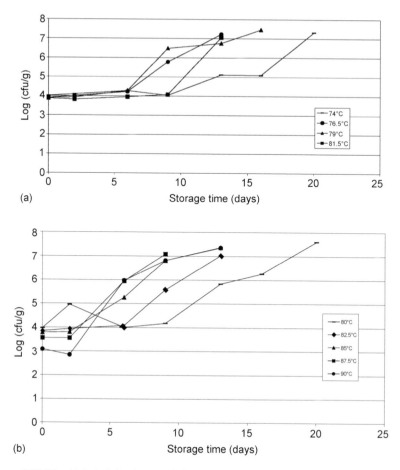

Figure 3.12 Microbiological development in heat-treated cream stored at 7°C: (a) pasteurisation at temperatures in the range 74–81.5°C (holding time 15 s); (b) flash pasteurisation at temperatures in the range 80–90°C (holding time 1 s). Note: cfu, colony-forming units.

Table 3.6 Pasteurisation effect (P^*) of various heat-treatment regimes

Time (s)	Temperature (°C)							
	72	75	79	80	81.5	82.5	85	90
15	1	2.4	7.5	10.0	15.4	20.5	–	–
1	< 0.1	0.16	0.5	0.7	1.03	1.4	2.8	11.9

not enter the system to cause post-process contamination downstream of the highest temperature in the system. Therefore, bacteria present in the cream had survived the heat treatment and were able to grow at 7°C. The results of thermoduric and spore counts at various heat-treatment regimes are listed in

Table 3.7 Thermoduric and spore counts of heat-treated cream stored at 7°C (Brown *et al.*, 1980)

Heat treatment		Thermoduric count/g	Spores count/g	Shelf-life (days)
temperature (°C)	time (s)			
74.0	15	1.02×10^4	1.85×10^3	20
76.5	15	2.75×10^4	1.71×10^4	13
79.0	15	1.08×10^4	1.00×10^3	16
81.5	15	2.36×10^4	1.55×10^4	13
80.0	1	1.07×10^4	8.55×10^2	20
82.5	1	2.44×10^4	2.60×10^4	13
85.0	1	1.48×10^4	1.75×10^2	13
87.5	1	4.70×10^3	2.70×10^2	9
90.0		7.8×10^3	1.69×10^4	13

table 3.7. Out of the thermoduric group of organisms, *Bacillus cereus* is able to grow rapidly at 7°C and therefore would limit the shelf-life of the product.

Scientists have known about the activation of thermoduric organisms such as *Bacillus cereus* by high-temperature pasteurisation for a long time (Brown *et al.*, 1980; Kessler and Horak, 1984; Schroder and Bland, 1984). It was believed activation is caused by the heat shock on the spores resulting from high-temperature treatment regimes. However, investigations by Barrett *et al.* (1999) into the lactoperoxidase system (LPS) in milk found that the high-temperature inactivation of LPS also plays an important role in the poor keeping quality. The antimicrobial activity of LPS was attributable to its oxidation product, with a hypothiocyanite group, being able to oxidise the sulfhydryl (—SH) groups of bacterial cell walls (Reiter and Harnulv, 1981). Therefore, if the LPS were inactivated by a heat-treatment regime, the level of antibacterial activity could also be reduced. Barrett *et al.* found that inactivation of LPS was temperature sensitive, with z values of about 4°C. The inactivation appears to take place close to 80°C, and Lewis (1999) reported that when hydrogen peroxide and thiocyanate were added to heat-treated milk (to enhance the LPS) it increased the keeping quality of milk heat treated in the region 72–76°C for 15 s when kept at 30°C, but no such increase was detected in the region 78–90°C. This showed that at the higher temperatures the LPS had been inactivated.

In the absence of antimicrobial activity from LPS, the preservation of milk and milk products requires more severe heat treatment to reduce the thermodurics and spores. Table 3.8 shows the microbiological development in cream heat treated to 115°C, 117.5°C, 120°C, 122.5°C and 125°C.

A heat-treatment temperature range of 120–125°C for 1 s tends to increase the keeping quality of cream to an acceptable standard when tested after storage for 49 days at 7°C. As before, post-process contamination was avoided by using aseptic filling techniques.

Table 3.8 Microbiological development in heat-treated cream stored at 7°C (total counts per gram)

Age of cream (days)	Heat-treatment temperature (°C)[a]				
	115	117.5	120	122.5	125
0	9	7	2	< 1	1
2	10	8	1	1	< 1
6	7	3	2	3	< 1
9	3150	700	105	3	1
13	$> 3 \times 10^6$	$> 3 \times 10^6$	20	4	3
20			15	10	10
41			40	15	15
49			< 10	< 10	< 10

[a]For a duration of 1 s.

Table 3.9 Organoleptic quality of milk and creams after heat treatment and storage at 7°C

Product	Shelf-life (days)	Mean acceptability score[a]		
		120°C, 1 s	125°C, 1 s	74°C, 15 s
Milk	> 37	7.7	7.6	7.8
Single cream	> 49	7.7	7.6	6.75
Whipping cream	> 37		6.9	7.6
Double cream	> 37		6.8	6.9

[a]A value of 1 indicates the product is unacceptable; a value of 10 indicates the product is good and fresh.

An acceptable product with very low surviving thermodurics and spores was obtained by the high-temperature heat treatment, but is was also desirable that the final product be similar to pasteurised milk and cream in terms of organoleptic characteristics. The organoleptic quality of milk and cream after heat treatment are shown in table 3.9. Freshly prepared pasteurised milk and cream were used for organoleptic assessment. The organoleptic characteristics for high-temperature heat-treated milk and creams stored for more than 37 days (more than 49 days for single cream) were similar to those of pasteurised products. A heat-treatment temperature range of 120–125°C for 1 s tends to increase the keeping quality of cream to an acceptable standard when tested after storage for 49 days at 7°C. As described before, the use of aseptic filling techniques prevented post-process contamination.

The results of total counts and organoleptic quality assessment provide sufficient information to design a high-temperature pasteurisation method to extend the shelf-life of milk and milk products with use of a standard dairy heat exchange system. Table 3.10 compares the lethal effects of high-temperature

Table 3.10 Lethal and chemical effects of some UK heat-treatment methods

Heat treatment	B^*	C^*	Lactulose (mg l^{-1})
Extended shelf-life	0.006	0.013	< 40
Ultra heat-treatment			
direct	3.55	0.27	80–100
indirect	1.20	0.19	200–500
Sterilisation	1.00	5.50	550–750

Note: $B^* = 1$ for $\log_{10}$ 9 reduction of thermophilic spores; $C^* = 1$ for 3% reduction of thiamine (vitamin B_1).

pasteurisation with some commercial heat-treatment methods. The values increase as the severity of the heat treatment increases. The aim is to achieve a high B^* value and low C^* and lactulose values to produce a safe product with good keeping quality and acceptable organoleptic characteristics.

To guarantee an extended shelf-life of more than, for example, 30 days in chilled distribution, it was necessary to include aseptic filling to prevent post-process contamination.

A method based on high-temperature pasteurisation was first commercially operated in the United Kingdom in 1980. The extended shelf life (ESL) process together with aseptic filling was also robust, withstanding fluctuations in chilled temperature distribution in commercial operations.

Microfiltration. In membrane filtration technology, microfiltration is the specified area where the particle size ranges from about 0.09 μm to about 9 μm. This size range is able to reject molecular weights greater than 100 000. In the ESL process the membranes have a nominal pore size of about 1–2 μm and are capable of reducing the bacterial cells and spores in the product by more than 99%.

In the microfiltration process the milk is first separated at about 60°C, and the skim fraction is then cooled to about 50°C. The membrane filters in the circuit then separate the permeate and retentate in a form of concentration process. The microbial cells and spores are also rejected and get collected along with the retentate. This is a very small fraction and is either discarded or heat-treated to destroy the cells and spores before the fraction is added back to the microfiltered skimmed milk. The microfiltered skimmed milk, permeate and cream can be mixed in appropriate proportions to formulate various milks and creams and can then be pasteurised.

Bactofugation. This process was developed based on the principles of centrifugal separation, where the dense bacterial cells and spores are collected as a single fraction after separation. A hermetic centrifuge called a bactofuge is employed as the main unit to carryout the separation of the bacterial cells and their spores. There are three main methods of bactofugation:

- two-phase bactofuge with continuous discharge of bactofugate
- single-phase bactofugate with intermittent discharge of bactofugate
- double bactofugation with two single-phase bactofugates in series

Byland (1995) gives details of the third system. In essence, it can be described as being a two-phase bactofugation with continuous discharge and sterilisation of bactofugate. The milk is separated into cream and skimmed milk, the skimmed fraction being rich in bacterial cells and spores. The skimmed fraction is subjected to separation by a two-phase bactofuge. The bactofuge separates the bactofugate, which is rich in spores and other microbial cells, as it is denser than the rest of the skimmed milk. This bactofugate is subjected to sterilisation by steam infusion. The sterilised fraction is mixed with part of the skimmed fraction to cool prior to adding it to the main stream skim flow in the circuit. The skimmed milk resulting from bactofugation and heat treatment is mixed with cream to formulate standardised milk and creams. These products can be pasteurised according to the standard methods described in section 3.2.4.1.

3.2.4.4 Homogenisation of emulsions

A homogeniser is simply a high-pressure pump usually designed to operate with a three-piston arrangement. It was invented in 1899 by a Frenchman, August Gaulin. The general method of homogenisation of creams and other dairy products involves pumping the liquid by a positive displacement pump arrangement into a homogeniser valve chamber or head. The head encloses the valve arrangement, with a very narrow gap allowing the product to exit. When the product is allowed to exit through the narrow slit the product particle velocity undergoes a sudden increase, which breaks down the coarse material and incoming fat globules (up to about $20\,\mu m$ in diameter) to a much finer particle size ($< 1\mu m$). The term 'homogenisation' is also used by equipment suppliers for equipment that carries out high-shear mixing of products in batch operations. For example, high-shear mixers and colloid mills are used to incorporate powders and other ingredients into liquids to ensure uniform distribution of particles. This type of operation does reduce the particle size and disperse the ingredients uniformly, but the size-reduction capability is not comparable with a high-pressure homogeniser. Particle-size reduction by ultrasonic waves has also been used but practical, economic and commercially viable units are not as yet available in the food and dairy industries.

With the inclusion of a homogenisation step in the process one would expect to achieve a stable emulsion as a result of particle size reduction, a smoother mouth feel as a result of the smaller fat globules, the need to use less stabiliser, a shorter ageing time, better overrun and a decreased tendency to churn fat in the freezer in ice-cream mixes. Therefore, even small variations from the optimum process conditions can lead to significant deterioration of consistency and texture. The design of the homogenising valve has been improved over the years, and Phipps (1985), Stistrup and Andreasen (1966) and White (1981) have reported on the performance of various commercially available systems.

There are four main valve designs, the bell-flow, flat, conical and liquid whirl. The particle sizes vary significantly for the different valves. In the flat-valve design the pressure must be almost double that of the liquid-whirl design to achieve the same particle sizes. It is important to note that the homogenisation procedure may be either single-stage or two-stage, depending on the design of the head. The second stage involves simply routing the product through a similar path as the first stage.

The main components in a homogeniser are the piston driving unit, the high-pressure head and the homogenising valve housing. The movement of the pistons out of the high-pressure head shuts the exit valves and opens the inlet valve and allows the product to enter the chamber. When the pistons move into the head the inlet valve shuts and opens the exit valve to allow the product to reach the homogenising valve (or valves in a two-stage process) under pressure. The product exits through a restriction and then through a second restriction (called the first stage and second stage, respectively) or, in the case of the liquid-whirl design, two steps in the same valve allows two homogenisation stages. In both methods the increase in pressure is achieved by manually operating a plunger (forcer) or by hydraulic pressure, to close the product exit gap between the two faces of the homogenising valve (or to restrict the exit orifice of the product). When the product is forced to flow through a very narrow orifice, the particles (mainly fat globules) must go through a size-reduction stage, their new size reflecting the size of the gap from which they escape. The second-stage homogenisation pressure is lower than the first-stage pressure, and its main purpose is to break up the clusters of fat globules formed after the first-stage homogenisation. For example if the total pressure specified for homogenisation of cream were 200 bar, then the first-stage pressure could be set at 17 MPa, and the second-stage pressure at 3 MPa. Table 3.11 indicates some homogenisation conditions used for fresh creams and ice creams with use of a flat-valve homogeniser. In table 3.11, a range of values is given, reflecting the conditions used in the dairy industry and taking into consideration variables such as the throughput of the homogeniser, the temperature used at different installations and the efficiency of the valves in the head.

The product itself can determine the optimum conditions, as the ratio of nonfat milk solids to fat is important to ensure sufficient solid material is available to reduce interfacial tension. Sommer (1944) reported that for cream a ratio greater than 0.85 would prevent fat clumping. A ratio in the range 0.6–0.85 could lead to some clumping. A ratio less than 0.6 significantly increases the clumping of fat globules. The ideal homogenisation conditions for creams are best established

Table 3.11 Homogenisation pressures and temperatures used in the commercial production of creams and ice creams

Product	Temperature (°C)	Homogenisation pressure (MPa)	
		stage 1	stage 2
Cream			
12% fat	45–70	15–20	3–6
18% fat	45–70	10–32	3–8.5
Ice cream	50–75	16–20	3–5

by conducting trials to select the most appropriate parameters to produce the desired product characteristics.

In long-life milks, creams and ice-cream mixes the homogenisation can be done after the sterilisation stage and the cooling section of the process (downstream). To be able to homogenise in this downstream position, the pistons and pressure-adjustment devices must be fitted with steam tracing to protect the product from post-process contamination.

Double homogenisation. Information on double homogenisation or multiple homogenisation is limited. Geyer and Kessler (1989) reported that double homogenisation of 12% fat cream showed improvement in physicochemical properties. Stistrup and Andreason (1966) were among the early investigators to establish the effect of single-stage, two-stage and double homogenisation of ice-cream mixes. Their results showed that fat-globule dispersion in ice cream was better with the liquid-whirl valve design compared with the flat-valve and conical valve designs in single-stage homogenisation. Two-stage homogenisation did not improve the degree of dispersion in comparison with single-stage homogenisation, whereas double homogenisation gave a higher degree of dispersibility than single-stage or two-stage homogenisation. Four linear relationships were found to exist between the logarithm of optical dispersion (a measure of light scattering) and the logarithm of pressure applied, for all methods.

Homogeniser care. Basic care of the components of the homogeniser not in contact with food is as equally important as for the homogenising head and valve arrangement. Components such as drive belts, crankshaft, oil pump and seals should be regularly checked for wear and tear to ensure that running efficiency is optimal. In the homogenising head the chevron seals can easily be damaged by abrasive food components such as cocoa powder. This is also true for the pistons. Ordinary pistons are hardened and coated with a thin layer of chromium oxide to withstand excessive wear. Other materials such as ceramic-coated pistons have been used successfully for processing emulsions containing abrasive material.

The inlet and exit valve seat surfaces in the homogeniser head should be free from pitting and crevices. They required regular inspection, and, if necessary, surface cutting and grinding should be carried out according to the supplier's instructions. Similar checks are necessary for homogenising valves. One problem is the appearance of extensive wear on the face of valves, forming a specific pattern of erosion rings caused by separation and cavitation (Phipps, 1985). A sharp-edged inlet arrangement increases flow separation, and at low homogenising pressure the valve seating develops flow patterns. A suitable valve with the correct inlet angle should be used for efficient homogenisation.

The seals in the homogenising valve chamber must be checked regularly, as faults in these seals can allow unhomogenised product to leak out. The seals when fully compressed tend to leak under high pressure. The result of such a leak is the formation of a fat ring on top of UHT milk when it has been left standing for a few days.

3.2.5 Preparation of dressings

Salad dressings are designed to form an O/W colloidal macroemulsion but, unlike creams and ice cream, they are acidic food products with a pH in the range of 3–4. There are two basic categories of dressings: the mayonnaise-type salad dressings made with cooked starch paste, and the French-type dressings.

The basic ingredients in mayonnaise are oil, water, stabilisers or thickeners (eggs, edible gums), flavouring and colouring materials. Some regulations specify 65% as the minimum oil requirement for mayonnaise, but many commercial products contain 75–80% oil. Table 3.12 gives an example of ingredients used for high-fat and medium-fat mayonnaise formulations.

Salad dressings contain a slightly lower oil content compared with mayonnaise (30–40%). The ingredients are cooked starch, emulsifiers and stabilisers (i.e. gums to provide stability and thickness). There are two types of French

Table 3.12 Formulation of mayonnaise (quantities are in g per 100 g)

Ingredient	Oil content	
	80%	50%
Vegetable oil	80.0	50.0
Stabiliser	0.0	1.0–4.0
Emulsifier	9.0	6.0
Salt	0.5	1.0
Sugar	0.5	4.0
Vinegar[a]	6.0	15.0
Mustard	0.5	1.0
Seasoning	optional	optional
Water	remainder	remainder

[a] 10% acetic acid.

Table 3.13 Formulation for a French dressing (quantities are in g per 100 g)

Ingredient	Quantity
Sugar	2.5
Dried mustard	2.5
Salt	2.5
Worcestershire sauce	1.5
Vegetable oil	60.0
Vinegar	31.0

dressing, the separating type and the nonseparating type. The separating type is a temporary O/W emulsion containing oil, vinegar, lemon juice and seasonings. Table 3.13 lists an alternative formulation for a typical French dressing. The main ingredients in nonseparating French dressings are egg yolk and/or other emulsifying ingredients (to keep the oil in suspension) and stabilisers such as gums to provide extra stability to the emulsion.

Mayonnaise is used to enhance the flavour and texture of some foods and also functions as a spread. Applications for salad dressings are much more diverse and the products are supplied in a variety of flavours and thicknesses. Therefore, a wide range of products also exists based on the modifications to physical characteristics. Such perceived physical characteristics (viscosity, body and thickness) have a direct relation to the ingredients used as stabilisers and to oil-phase volume in the recipe.

Mayonnaise is a stiff product whereas salad dressings are somewhat thinner and can be made to a pourable or spoonable consistency. Consumer demand is responsible for such a wide variety. Reducing the oil content and substituting oil with fat replacers also produces the low-fat and no-fat varieties. Thickeners are added to compensate for the loss of oil and to restore the required physical characteristics.

3.2.5.1 Manufacturing procedure

In the manufacture of mayonnaise and salad dressings, salad oils are used for the oil phase and, typically, vegetable oils such as soybean, sunflower, corn, rapeseed, olive and cottonseed oils are used. Soybean has been the most cost effective and widely used oil but other oils, typically rapeseed oil (also known as canola), can be a more economic alternative. The term 'salad oil' is generally reserved for those products that remain substantially liquid at refrigerator temperatures (4–7°C). This property is referred to as resistance to graining at chilled storage. They have varying degrees of flavour and oxidative stability, depending on the main oil and manufacturing standards. Most oils are refined, bleached and deodorised to remove flavour and are sometimes lightly hydrogenated. An exception to this concerns olive oil, which is generally not deodorised because it lends a special flavour quality desirable for dressings. Those salad oils that have high melting oil fractions that solidify in refrigerated storage are put through a process called 'winterisation' to remove these fractions. Sometimes, crystal inhibitors such as polyglycerol esters and methylsilicone are added to further protect the oils from grain formation or crystallisation.

Salad oils contain a high proportion of unsaturated fatty acids and are easily oxidised. The oils contain natural vitamin E in small concentrations and provide protection against oxidative rancidity.

Flavour in salad dressings and mayonnaise is derived from the oils used, the spices added as solids finely dispersed in the emulsion, water-soluble salts, alcoholic products or from flavoured vinegars and vegetable preparations such as tomato products. The aqueous phase in these products is usually the vinegar, which is a weak solution of acetic acid (at least 2.5% by weight). Highly flavoured products such as cider, wine or malt vinegar are also used. Equally, other well-established flavouring components are citric acid and lemon juice.

For mayonnaise and salad dressing production in industry (the batch method), equipment such as jacketed stainless-steel tanks fitted with temperature-control devices and agitators are used, together with colloid mills or high-shear mixers. In the production of mayonnaise, initially the egg and flavour ingredients are mixed thoroughly to form a base to which oil and vinegar are added by continuous agitation. Once the ingredients are mixed, the mixture is passed through a high-shear mixer such as a colloid mill. The final consistency and texture of the emulsion is affected by the rate of addition of the oil during the premixing stage. For example to obtain fine oil droplets the oil is dispersed by using a high-shear mixer, increasing the viscosity of the product. The mixing and colloid mill treatment is carried out at about $10°C$.

In the manufacture of salad dressings, gums are mixed with an acidic aqueous phase; this process may include a heating profile to hydrate the solid particles. Gums are added to increase the viscosity, as the oil content is less than that for mayonnaise.

In the manufacture of mayonnaise and salad dressings a heat-treatment stage is not included. The acidity of the final product is sufficiently low to inhibit the proliferation of microorganisms.

The manufacture of mayonnaise and salad dressings can also be automated as well as produced by a continuous production method for large-scale industrial operations. Figure 3.13 illustrates an industrial layout for mayonnaise production.

Originally, salad dressings and mayonnaise products were packed in glass containers. Glass containers provide protection against attack by oxygen. However, polyethylene (PE) based containers have been introduced more recently. PE is not as good as glass as a barrier material against oxygen, but it provides greater protection for it allows antioxidants to be incorporated into the product.

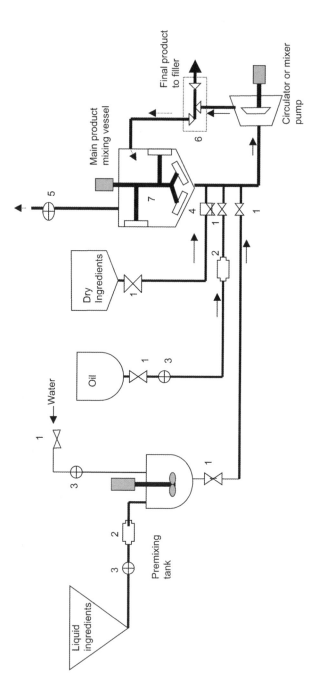

Figure 3.13 A batch industrial mayonnaise production circuit. 1, isolating valve; 2, flow meter; 3, product pump; 4, Ventury power injector; 5, vacuum pump; 6, product divert valve; 7, scraper agitator.

The manufacture of a nonemulsified (separating) type of salad dressing is a simple procedure. It involves only mixing, and the mixture is then filled into containers.

3.3   Factors affecting water continuous emulsions

Water continuous emulsions, where the pH is in the neutral range, provide a good medium in which microorganisms can grow. Therefore, products such as milks, creams and ice-cream mixes provide a good environment for the multiplication of bacteria, yeasts and moulds unless some control measures are taken to limit their number. Salad dressings and mayonnaise-type products are acidic food emulsions, the low pH inhibiting the growth of most microorganisms.

In ice cream the microbiological problems are reduced to some extent because of the low temperature at which it is stored until it is consumed or until the end of its shelf-life ($-25°C$ to $-30°C$). It is important to ensure that the storage temperature is maintained throughout its shelf-life period. In these emulsions, microbiological activity leads to changes in pH, mainly lowering it depending on the extent of the activity. In milks and creams, such a change in pH (e.g. to a pH less than about 6) would bring about emulsion instability as well as undesirable organoleptic characteristics.

3.3.1   Emulsion stability of high-fat creams

At normal storage temperature, the fat globules in milk show a tendency to form clusters compared with those in a nonrefrigerated environment. Milk silos are kept under slow agitation at regular intervals to minimise cluster formation and fat separation. The size of the fat globules affects the efficiency of the separation of cream from milk, and the optimum temperature for separation will vary depending on globule size. A large proportion of smaller fat globules ($< 2\,\mu m$) can reduce skimming efficiency by allowing the fat percentage in the skimmed milk to rise. Larger fat globules with a distinct yellow colour (compared with that of milk from, for example, a Friesian herd) is generally present in milk from Jersey herds. Rothwell (1966) and Foley *et al.* (1971) investigated various processing parameters that influence the emulsion stability of creams. These investigations highlighted the influence of milk separation temperature, cream pumping, pasteurisation, cooling and fat percentage on the physical properties of cream and on damage to the integrity of the fat globules.

3.3.1.1   Physicochemical defects in fresh creams
In many industrial installations equipment designed for products more robust than cream (i.e. milk, skimmed milk, juice drinks) is employed for handling creams. As stated before, the handling and processing circuits need to be very gentle when handling cream in order to minimise shear and mechanical damage.

In the preparation of high-fat creams [e.g. 48% fat cream (double cream in the United Kingdom)] there is a tendency for a solid plug of cream to develop and, in some instances the viscosity may rise to unacceptably high levels under optimum process conditions. Defects in cream are associated with the fat percentage of cream and, depending on the use to which the product is put, these defects may come under strong criticism from the end user.

Defects such as oiling off, formation of a cream plug and age thickening have been associated directly with the level of free fat (or solvent-extractable fat) present in the cream, known to increase with fat content, especially above about the 40% fat level. Section 3.2.1.2 highlights the incidence of fat in relation to the use of a centrifugal separator in the preparation of creams.

An O/W emulsion prepared with increased levels of free fat will result in a product likely to show instability immediately after preparation or in the early stages of the shelf-life. The instability has been described in many ways, the most familiar descriptions relating to creaming, flocculation, coalescence and disruption.

The process of fat globules rising to the surface as a result of differences in density between the aqueous phase and the fat globules and the consequent formation of a thick, fat layer is termed 'creaming'. The presence of free fat tends to aggravate this as the crystallised and liquid fat exits the globules as a result of damaged or missing globule membranes and forms a solid fat layer on the surface of the cream.

The fat globules also come together and form floccules, making them much larger particles. These then begin to rise faster than individual globules as the floccules are less dense than the aqueous phase. However, the floccules are fairly redispersible. In floccules, individual fat globules exist but these are bound together with neighbouring globules by weak forces.

Fat globules coming together and bound by strong forces form clusters. These fat globules unite at their contact points and share interfacial layers. The clusters can be redispersed with application of mechanical energy, as in the case in the second stage of the homogeniser. The formation of clumps indicates serious destabilisation of the emulsion, as the clumps are not usually considered to be redispersible. The fat globule membrane material of individual globules come together and form a continuous membrane round them, and the fat forms a continuous mass, completing the clump formation. It is reported that clumps are formed from partly solid globules (Mulder and Walstra, 1974). These clumps will coalesce into one large globule when all the fat becomes liquid (owing to a rise in temperature). Once the cream is cooled some of the high melting fat fraction will have become crystallised, the crystal formation generating physical forces within the globules. If the cold cream is subjected to mechanical forces as a result of harsh handling methods, the fat globules will tend to become damaged. This leads to further cluster formation. Te Whaite and Fryer (1975) found that gel formation in cream is linked directly with the formation of free fat in creams.

3.3.1.2   Physicochemical defects in ultra heat-treated creams

The most widely manufactured UHT creams in the United Kingdom are the half-creams (not less than 12% fat) and single creams (not less than 18% fat). The general term 'coffee cream' is used for creams with a fat content in the range of 10–20% and is used mainly in catering applications. Other high-fat creams (whipping cream, 35–38% fat; double cream, 48–50% fat) are produced but the demand tends to be seasonal. Since the mid 1980s demand for low-fat whipping cream has increased in Europe and the USA. Mann (1987) reviewed whipping cream, including low-fat whipping cream. Anderson and Cawston (1975) reviewed the progress of research work under the subheading 'The milk fat globule membrane', covering details of the milk fat globule membrane and its composition.

Stability of UHT coffee cream.   A common problem associated with some coffee creams is an instability known as 'feathering', which shows coagulated curd-like flocs floating on the surface when the product is added to hot coffee. This phenomenon was reported as far back as the 1920s and 1930s by Doan (1929, 1931). Anderson *et al.* (1977a, 1977b) and Geyer and Kessler (1989) have also investigated the stability of UHT coffee creams with reference to shelf-life, extending the shelf-life and the influence of manufacturing methods on feathering.

The stability of cream to hot coffee is affected by homogenisation pressure and temperature, homogenisation position (i.e. upstream or downstream position in the circuit), the hardness of the water used to make coffee, the acidity of the coffee and the temperature of the coffee. This shows that feathering is partly a result of changes brought about by alteration of the structure of milk proteins caused by processing and also, partly, by the harsh conditions of coffee preparation. Usually, these creams are stable to hot coffee immediately after processing (i.e. they resist feathering); the tendency is for the product to become susceptible to feathering on storage at ambient temperature. The exact transformation of various components in the O/W emulsion leading to feathering was not initially fully understood. Early researchers suggested the possibility of changes to the fat–water interface, but Anderson *et al.* (1977a, 1977b) confirmed that an increase in the ratio of fat-phase casein to calcium was associated with the feathering in UHT coffee cream, suggesting that casein is the more important factor. They also observed that susceptibility to feathering is associated with the tendency for adjacent fat globules to become linked by bridges of casein (see figure 3.14). Most of the fat globules have casein micelles associated with them, and numerous submicellar casein particles appear to be attached to the surface. Therefore, feathering is accompanied by an increase in calcium and casein levels and in the ratio of casein to calcium in the fat phase of the cream. I have investigated single cream prepared by various methods (table 3.14), including a new method of producing cream involving making adjustments to

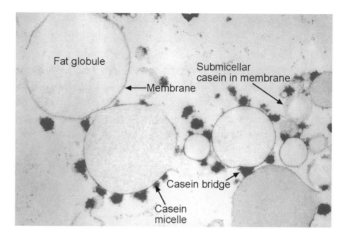

Figure 3.14 Ultra heat-treated cream exhibiting severe feathering (Diotte Consulting & Technology, UK).

the diffusible-calcium (calcium in the aqueous phase) content (Ranjith, 1995). In this method, the milk is first subjected to membrane filtration (ultrafiltration) so that the retentate has a total solids content of 35–38 wt%. The permeate is subjected to an ion-exchange process to remove calcium and magnesium (diffusible divalent ions). Deionised permeate is added back to the retentate to reconstitute the original whole milk. The whole milk is then separated to obtain diffusible-calcium-reduced (DCR) cream (single cream or coffee cream).

Each cream preparation was divided into two equal portions and subjected to UHT treatment by the indirect plate method, with one portion being homogenised in the upstream position at 170/34 bar (170 bar at first stage, 34 bar at second) by using an aluminium pressure valve (APV) Manton Gaulin homogeniser. The cream was preheated to 55°C prior to homogenisation. In the downstream method after UHT treatment at 140°C for 3.5 s it was cooled to about 60°C prior to homogenisation (170/34 bar). The creams were cooled, in both methods, to 20°C before aseptically filling into 150 ml plastic containers and sealing with an aluminium-foil lid.

When tested, the cream prepared by this method was found to resist feathering for up to six months at ambient storage temperature. More importantly, this cream was very stable to alcohol in the alcohol stability test. It was stable to 95 vol% ethanol when tested after six months at ambient storage. A cream liqueur prepared with use of this cream was stable to retort sterilisation, and the emulsion was stable for more than two years.

The viscosity of coffee cream made by the methods summarised in table 3.14 were tested, and the results are given in figure 3.15. All cream samples made by adjustment of calcium content (except batch 3, containing added caseinate) showed a slight reduction in viscosity after six months at ambient storage. The

Table 3.14 Methods used for the preparation of coffee cream (fat content 20 g per 100 g product)

Cream	Composition
1	Control cream
2	Cream with calcium content reduced to c. 20% of original level
3	As for cream 2, with 1.5 g per 100 g sodium caseinate added
4	Diffusible-calcium-reduced cream

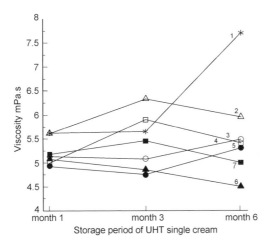

Figure 3.15 The change in viscosity of ultra heat-treated single cream as a function of storage time. Cream 1, control cream, made by downstream homogenisation; creams 2 and 3, calcium-reduced cream, made by upstream and downstream homogenisation, respectively; creams 4 and 5, calcium-reduced cream with added sodium caseinate, made by upstream and downstream homogenisation, respectively; creams 6 and 7, diffusible-calcium-reduced cream, made by upstream and downstream homogenisation, respectively.

control cream made by upstream homogenisation developed a thick plug after two months. The control sample made by downstream homogenisation showed thickening after six months (the viscosity increased by 36%) but did not form a plug.

The sample containing added sodium caseinate showed a gradual increase in viscosity, although the total calcium level was reduced by about 20% of the original value. It is possible that calcium adjustment reduces the likelihood of casein bridges forming between fat globules, thereby keeping the viscosity low or even reduced during long-term storage and helping to maintain a uniform, homogeneous O/W emulsion. The increase in casein content could be responsible for the increase in viscosity during storage, with casein bridges being built between the fat globules.

Other experimental work has indicated that the problem of feathering can be minimised by immobilising the calcium in cream by using chemical additives such as phosphates, citrates and carbonates.

Stability of UHT whipping cream. The primary objective of UHT is to achieve a long shelf-life for products at ambient storage with minimum heat damage and minimum changes to organoleptic characteristics compared with the fresh product. However, a homogenisation step is essential in the manufacture of high-fat liquid dairy products to minimise fat separation. An optimum homogenisation pressure is desirable as high pressure tends to be somewhat disadvantageous when one wishes to create a stable foam having about 100% overrun (see equation 3.6). Kieseker and Zadow (1973) found that milk separation at 43°C was ideal for manufacturing UHT whipping cream (36% fat) with desirable whipping properties. Whipping properties are overrun, serum leakage or seepage, free-fat content and whipping rate (assessed by using a scale of 1 to 5, where 1 is very poor, and 5 is excellent). The overrun can be calculated as follows:

$$O = \frac{W - W^{\text{whip}}}{W^{\text{whip}}} \times 100 \qquad (3.6)$$

where,

O is the percentage overrun
W is the weight of a specific volume of cream
W^{whip} is the weight of same specific volume of whipped cream

The control of overrun in industrial whipping equipment is possible as such equipment is provided with devices to adjust the cream feed rate, the air injection rate and the rate of rotation of the worker unit.

The overrun may not be the same for different cream preparations and each batch of cream must be treated differently from previous batches, and fine-tuning and adjustments are necessary to optimise the control parameters to achieve a stable foam.

Excessive shear in the worker unit can be detrimental to the cream and leads to poor overrun and other whipping properties. Table 3.15 lists the parameters affecting the production of good-quality UHT whipping cream. The methods described for whipping cream can also be applied to the manufacture of double creams and other high-fat creams.

3.3.1.3 Foam formation and stability

The formation of air bubbles when handling milk, skimmed milk and creams is a common problem in dairies and food factories. Nevertheless, based on this foaming ability, aerated products such as whipping cream and ice cream have been developed and became established in the markets. Recently, milk for cappuccino coffee has become popular in many countries. As its popularity

Table 3.15 Effect of preparation and processing on the whipping characteristics of UHT whipping cream

Parameter and its effect on product properties

Milk quality
Poor microbiological quality causes lipolysis of fat as a result of lipase, originating from bacteria surviving the UHT treatment

Milk separation and cream preparation
A temperature above about 40°C is suitable for milk separation to ensure fat is in the liquid form
Milk supply to the separator must be appropriate for required throughput of the separator to reduce the production of free fat through shear action. The fat content of the cream separated should be as close as possible to the final fat content of the product. This tends to reduce excess free fat in cream.
Permitted stabilisers can be added at this stage by incorporating them into the skimmed milk by using a high-shear mixer; this fraction is then added to the cream with gentle agitation. Stabilisers tend to reduce seepage but also reduce the overrun; therefore optimum level of stabilisers is desirable

Cream pumping
A positive displacement pump is desirable to ensure minimum disruption to fat globules

UHT processing and homogenisation
UHT treatment by indirect or direct methods are suitable, but shear damage to fat globules as a result of steam injection may affect whipping properties
Homogenisation is usually carried out above about 40°C in the preheating stage, but the downstream process (after UHT-treatment) may have a slight advantage over this in terms of whipping properties. Two- stage homogenisation with 20–25 bar in the first stage and 7–12 bar in the second stage is a suitable specification for most homogenisers to produce good whipping and storage properties
One-stage homogenisation tends to encourage cluster formation and, depending on the homogenisation pressure, gel formation in the cream is a possibility during storage

Cream cooling methods
Whipping cream is generally manufactured to have low viscosity.
To obtain a low viscosity the cream is rapidly cooled in the heat exchanger to below 10°C.
To obtain a slightly thicker consistency, the cream is cooled to about 25–30°C
and filled aseptically into containers and held for 2–4 h at ambient temperature
before overnight storage at a chilled temperature below 10°C. This
treatment allows the crystallisation of liquid fat and develops the viscosity

increases, the consumer will expect a good quality foam when consuming the beverage. For example, the foam in a cappuccino coffee is expected to stay stable until at least half the coffee is consumed. Similarly, whipped cream must produce stable foam after whipping to produce an overrun of about 100%. Additives to improve aeration are not permitted in milk in many countries.

Sometimes milk fails to produce an acceptable level of foam in cappuccino coffee. Whipping cream also sometimes produces a poor overrun or the foam collapses soon after it has formed. This means that the fundamental mechanism of foam formation must be understood so that appropriate steps can be taken to ensure the desired properties of foam are achieved. In whipped cream the air bubbles in the foam are held in a three-dimensional matrix of partially coalesced fat globules.

The microstructure study of whipped cream also clearly shows partly destabilised fat globules adsorbed at the air–water interface. The gas phase in foam provides specific textural character; for example, the lightness in whipped cream or the scoopability in ice cream. In dairy creams and alternatives the emulsions are stabilised by the proteins. These proteins should be soluble in the emulsion and rapidly diffuse into the oil–water or air–water interface. This is an important functional property of protein, and poor solubility characteristics means poor emulsion stability as well as poor foaming ability of the emulsion. Other equally important characteristics of the protein is that it should reorient its structure until some degree of unfolding of the molecule is achieved, sufficient to bring about intermolecular interactions leading to the formation of a coherent film. This continuous, cohesive film brings about considerable mechanical strength and viscoelastic properties, which plays an important part in the stability of emulsions and foams.

During aeration the protein diffuses into the air–water interface and gets adsorbed. When more and more proteins are adsorbed, the surface tension at the interface reduces. The protein structure orientation at the interface is important at this stage, as its hydrophobic part must unfold towards the air whereas the hydrophilic part must associate with the aqueous part, or water phase. Interaction and association between proteins is important to the integrity of the film. Emulsion temperature, acidity and ionic strength all affect the protein–protein interactions. The whey proteins, especially β-lactoglobulin, is well known for its foaming properties in milk and milk products. It increases its adsorption rate with increase in ionic strength. This protein exists in globular form and during pasteurisation unfolding of the structure causes more charged sites to become exposed to the emulsion, which enhances the adsorption process. Anderson *et al.* (1987) examined the structure of whipped cream by surface electron microscopy and the photographs taken show evidence that the inner surface of air bubbles consists of a continuous air–serum interface through which individual fat globules protrude. In the whipping process fat globules penetrate into the air–water interface and are then attached to the air bubbles. At the same time fat globules clump and form a network spreading some of the fat on the air bubble surface. A network of air bubbles and fat clumps finally entrap the liquid from the emulsion to provide rigidity and stability to the foam. Therefore, the foaming ability of milks and creams depend on:

- the solubility of proteins, and protein–protein interactions
- the ability of proteins to reorient and become adsorbed at the air–water interface

- the formation of a strong and viscoelastic film to entrap air bubbles
- the formation of fat globule clumps and network over air bubbles and film
- the ability of the film to hold serum and provide rigidity to the foam

A high free-fatty acid content in the emulsion tends to give antifoaming properties and causes poor overrun as well as serum drain. Stabilisers help to minimise serum leakage from the foam but they also reduce the overrun of the foam.

3.3.2   Defects in ice cream

In commercial terms the composition of an ice-cream mix is an important factor for two main reasons. The composition influences and stimulates consumers to eat the product and creates demand. The demand is also a reflection, to some extent, of the pricing strategy of the product. A further concern for the manufacturer is the cost of producing a specific volume of ice-cream mix. Therefore, the choice of ingredients together with the cost of production, if optimal, stimulate demand for generating a commercially viable product or process system. Defects in the final product originate from flavour, body and texture, melting characteristics, colour, packaging, microbiological quality and composition.

3.3.2.1   Compositional issues
Too low a fat content affects the palatability and food value of ice cream. A very high fat content is also found to be difficult to assimilate and, owing to its large content of heat units, can become especially objectionable in warm weather. Fat content in the range of 8–10% appears to be popular for most markets. It is equally important to optimise the level of nonfat milk solids as in many formulations it is responsible for good overrun because of the foaming characteristics of casein and whey protein. Therefore, about a 5% level of nonfat milk solids in the ice-cream mix has been used for most formulations as the lower limit, but about 8–12% is frequently used by many manufacturers in commercial products. Figure 3.16 illustrates the structure of ice cream.

The ingredients used to provide the nonfat milk solids could be responsible for one particular defect, which is described as 'sandy ice cream'. Sandy ice cream arises from the crystallisation of the milk sugar (lactose). Crystallisation of milk sugar is possible under certain conditions, such as those experienced after the mix has been frozen. The other possibility is that sugar crystals already exist in some of the dairy ingredients in the formulation. The level of nonfat milk solids is responsible for a sandy texture, and a level of less than 9% tends to prevent this problem from occurring. Other factors, such as aqueous phase material, could influence the solubility of lactose, but

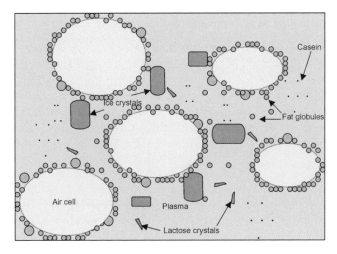

Figure 3.16 The structure of ice cream (not to scale).

this does not appear to be a significant factor in causing a sandy texture. However, long-term storage of ice cream also allows sufficient time for crystal formation. Therefore, storage period may be an important factor for some formulations. Closely associated with texture defects is the coarse taste resulting from large ice crystals. Gelatine and alternative stabilisers have been used in formulations, as they are capable of influencing the water crystal size during freezing. A smooth texture can be achieved by incorporating, for example, about 0.5% gelatine, but higher levels, apart from being costly, could become organoleptically objectionable, as the ice cream does not melt readily on the tongue.

3.3.2.2 Shrinkage in retail containers

Shrinkage of ice cream is a problem that is visible in retail containers from time to time without any other obvious defects. The mechanism of shrinkage is attributed to the collapse of lamellae, or the walls between the air cells, caused by changes in the internal pressure together with temperature fluctuation. The exact reason for ice-cream shrinkage is poorly understood and its manifestation is sudden and persists for a period and very often this problem disappears without any known cause.

Goff (2001) reported on the control of ice-cream structure by examining fat–protein interactions. Figure 3.17 shows the effect of adsorbed proteins on the structure of ice-cream mix [figure 3.17(a)], ice cream [figures 3.17(b) and 3.17(c)] and melted ice cream [figure 3.17(d)]. Goff's work highlights the importance of a network of partially coalesced fat globules in the formation of ice-cream structure.

3.3.3 Defects in mayonnaise and salad dressing

The formulations of mayonnaise and salad dressings are carefully carried out to ensure the dispersed oil phase is kept uniformly distributed throughout the period of its shelf-life. The only exception to this is for the separating-type French dressings. Oils that contain a high degree of saturated triglycerides are subjected to a process called 'winterisation' to remove the high melting fractions. If these are not removed they solidify at refrigeration temperatures.

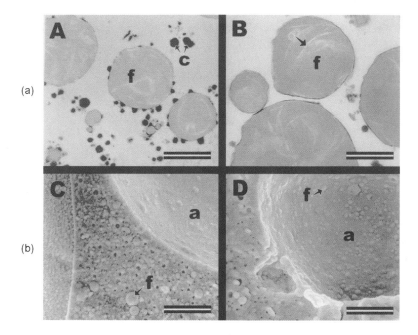

Figure 3.17 The effect of adsorbed protein on the structure of ice-cream mix, ice cream and melted ice cream. (a) Ice-cream mix with (A) no surfactant and (B) added surfactant, as viewed by thin-film transmission electron microscopy (TEM; for methodology, see Goff *et al.*, 1987); double bar = 1 μm; there are high levels of adsorbed protein, especially casein micelles, in the matrix illustrated in part (A). (b) Ice cream with (C) no surfactant and (D) added surfactant, as viewed by scanning electron microscopy (SEM; for methodology, see Caldwell *et al.*, 1992); double bar = 4 μm; the lack of a surfactant impedes the adsorption of fat. (c) Ice cream with (E) no surfactant and (F) added surfactant, as viewed by thin-section TEM with freeze substitution and low-temperature embedding (for methodology, see Goff *et al.*, 1999); double bar = 1 μm; the added surfactant can be seen to inhibit the partial coalescence of the fat. (d) Melted ice cream with (G) no surfactant (double bar = 1 μm) and (H) added surfactant (double bar = 5 μm), as viewed by thin-section TEM (for methodology, see Goff *et al.*, 1987); the absence of surfactant shows rapid meltdown with recovery of mostly intact fat globules. a, air bubble; c, casein micelle; f, fat globule; fc, fat cluster; fn, fat network; arrow, crystalline fat. Reproduced by courtesy of D. Goff, University of Guelph, Canada.

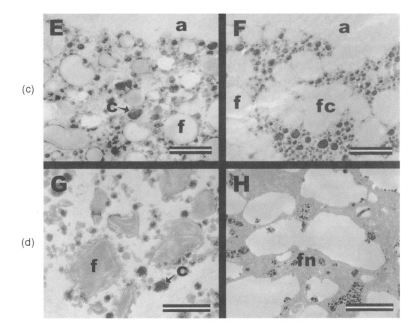

Figure 3.17 (continued)

The main problem in salad dressings and mayonnaise is the development of rancidity. The oils in these products contain unsaturated bonds and are easily oxidised by contamination with traces of transition metals, in particular iron and copper, as they act as catalysts in the reaction. Other factors such as the viscosity of the aqueous phase, particle size and influence of water tend to support the oxidation process.

References

Anderson, M. and Cawston, T.E. (1975) Review of the progress of dairy science: the milk fat globule membrane. *J. Dairy Res.*, **42**, 459-483.

Anderson, M., Brooker, B.E., Cawston, T.E. and Cheeseman, G.C. (1977a) Changes during storage instability and composition of ultra-heat-treated aseptically-packed cream of 18% fat content. *J. Dairy Res.*, **44**, 111-123.

Anderson, M., Cheeseman, G.C. and Wiles, R. (1977b) Extending shelf life of UHT creams. *J. Soc. Dairy Technology*, **30**, 229-232.

Anderson, M., Brooker, B.E. and Needs, E.C. (1987) The role of proteins in the stabilisation/ destabilisation of dairy foams, in *Food Emulsions and Foams* (ed. E. Dickinson), The Royal Society of Chemistry, pp. 100-109.

Arbuckle, W.S. (1986) *Ice Cream*, 4th edn, AVI Publishing, Westport, CT.

Banks, W. and Muir, D.D. (1985) Effect of alcohol content on emulsion stability of cream liqueurs. *Food Chemistry*, **18**, 139-152.

Banks, W. and Muir, D.D. (1988) Stability of alcohol containing emulsions, in *Advances in Food Emulsions and Foams* (eds. E. Dickinsons and G. Stainsby), Elsevier, London, pp. 257-283.

Barrett, N., Grandison, A.S. and Lewis, M.J. (1999) Contribution of lactoperoxidase to the keeping quality of pasteurized milk. *J. Dairy Res.*, **66**, 73-80.

Brown, J.V., Wiles, R. and Prentice, G.A. (1980) The effect of different time–temperature pasteurisation conditions upon the shelf life of single cream. *J. Soc. Dairy Technol.*, **33**, 78-79.

Buchanan, R.A. and Smith, D.R. (1966) Recombined whipping cream. *Proc. XVII Int. Dairy congress*, **E/F**, 363-367.

Burton, H. (1988) *Ultra high temperature processing of milk and milk products*, Elsevier Applied Science, London.

Bylund, G. (1995) *Dairy Processing Handbook*, Tetra Pak Processing Systems AB, Bylund, Lund, Sweden.

Caldwell, K.B., Goff, H.G. and Stanley, D.W. (1992) A low-temperature scanning electron microscopy study of ice cream. 1. Techniques and general microstructure. *Food Structure*, **11**, 1-9.

Doan, J.F. (1929) Some factors affecting the fat clumping produced in milk and cream mixtures when homogenised. *J. Dairy Sci.*, **12**, 211.

Doan, J.F. (1931) The homogenising process. *J. Dairy Sci.*, **14**, 527.

Food Labelling Regulation (1996) Statutory Instruments No. 1499, HMSO, London.

Foley, J.F., Brady, J. and Renolds, J. (1971) The influence of processing on the emulsion stability of cream. *J. Soc. Dairy Technol.*, **24**, 54-58.

Geyer, S. and Kessler, H.G. (1989) Effect of manufacturing methods on the stability to feathering of homogenised UHT coffee cream. *Milchwissenschaft*, **44**, 423-427.

Goff, H.D. (2001) Ice cream under control. *Dairy Inds. Int.*, **66**, 451-454.

Goff, H.D., Liboff, M., Jordan, W.K. and Kinsella, J.E. (1987) The effects of polysorbate 80 on the fat emulsion in ice cream mix: evidence from transmission electron microscopy studies. *Food Microstructure*, **6**, 193-198.

Goff, H.D., Verespej, E. and Smith, A.K. (1999) A study of fat and air structures in ice cream. *Int. Dairy J.*, **9**, 817-829.

Griffiths, M.W., Phillips, J.D. and Muir, D.D. (1981) Thermostability of proteases and lipases from a number of species of psychrotrophic bacteria of dairy origin. *J. Appl. Bacteriol.*, **50**, 289-303.

Kessler, H.G. (1981) *Food Engineering and Dairy Technology* (translated by M. Wotzilka), A. Kessler, Freisling, pp. 173-201.

Kessler, H.G. and Horak, F.P. (1981a) Objective evaluation of UHT milk heating by standardisation of bacteriological and chemical effects. *Milchwissenschaft*, **36**, 129-133.

Kessler, H.G. and Horak, F.P. (1981b) Effect of heat treatment and storage conditions on keeping quality of pasteurised milk. *Milchwissenschaft*, **39**, 451-454.

Kessler, H.G. and Horak, F.P. (1984) Effect of heat treatment and storage conditions on keeping quality of pasteurised milk. *Milchwissenschaft*, **39**, 451-454.

Kieseker, F.G. and Zadow, J.G. (1973) Factors influencing the preparation of UHT whipping cream. *The Australian J. Dairy Technol.*, **15** (December), 165-169.

Law, B.A. (1979) Enzymes of psychrotrophic bacteria and their effect on milk and milk products. *J. Dairy Res.*, **46**, 573-588.

Lewis, M.J. (1999) Microbiological issues associated with heat treated milks. *Int. J. Soc. Dairy Technol.*, **52**, 121-125.

Ling, E.R., Kon, S.K. and Porter, J.N.G. (1961) The composition of milk and the nutrition value of its components, in *Milk: The Mammary Gland and its Secretion*, (eds. S.K. Kon and A.J. Cowie), Academic Press, New York, Vol. 11, pp. 195-263.

McDowall, F.H. (1953) *The Butter Making Manual, Volume 1*, New Zealand University Press, Wellington.

Mann, E.J. (1987) Whipping cream and whipped cream. *Dairy Ind. Int.*, **52** (9), 15-16.

Mulder, H. and Walstra, P. (1974) *Milkfat Globule: Emulsion Science as Applied to Milk Products and Comparable Foods*, Commonwealth Agricultural Bureaux, Farnham, pp. 101-226.

Murray, S.A. (1985) Thermal processing of particulate containing liquid foods, *A septic processing and packaging of foods* (proceedings), The Swedish Food Institute, Lund University, Sweden, pp. 81-99.

Phipps, L.W. (1985) Technical bulletin 6: the high pressure dairy homogeniser, National Institute for Research in Dairying (NIRD), Reading, UK.

Ranjith, H.M.P. (1995) *Assessment of Some Properties of Calcium-reduced Milk and Milk Products from Heat Treatment and Other Processes*, Ph D thesis, Department of Food Science and Technology, University of Reading, Reading, UK.

Ranjith, H.M.P. and Thoo, Y.C. (1984) *Int. Patent* 85306220.6.

Reiter, B. and Harnulv, B.G. (1981) The preservation of refrigerated and uncooled milk by its natural lactoperoxidase system. *Dairy Inds Int.*, **47**, 13-19.

Rothwell, J. (1966) Studies on the effect of heat treatment during processing on the viscosity and stability of high-fat market cream. *J. Dairy Res.*, **33**, 245-254.

Schroder, M.A. and Bland, M.A. (1984) Effect of pasteurisation temperature on keeping quality of whole milk. *J. Dairy Res.*, **51**, 569-578.

Sommer, H.H. (1944) *The Theory and Practice of Ice Cream Making*, Sommer, Madison, WI.

Stistrup, K. and Andreasen, J. (1966) Homogenisation of ice cream mix. *Proc. XVIIth International Dairy Congr., Munchen*, **E/F**, 375-386.

Te Whaite, I.E. and Fryer, T.F. (1975) Factors that determine the gelling of cream. *New Zealand J. Dairy Sci. & Technol.*, **10**, 2-7.

The Dairy Products (hygiene) Regulations (1995) Statutory Instruments No. 1086, HMSO, London, UK.

Towler, C. and Stevenson, M.A. (1988) The use of emulsifiers in recombined whipping cream. *New Zealand J. Dairy Sci. and Technol.*, **23**, 345-362.

Walstra, P. (1969) Preliminary note on the mechanism of homogenisation. *Netherlands Milk and Dairy J.*, **23**, 290-292.

White, G.C. (1981) Homogenisation of ice cream mixes. *Dairy Inds Int.* (February), 29-36.

Zadow, J.G. and Kieseker, F.G. (1975) Manufacture of recombined whipping cream. *The Australian J. Dairy Technol.* (September), 114-169.

4 Hydrogenation and fractionation

Albert J. Dijkstra

4.1 Introduction

Edible oils and fats are natural products and their physical properties are determined by their agricultural origin. Since many oils and fats are by-products such as lard or cottonseed oil, or co-products such as soybean oil (where the protein from the meal is an equally important product), their availability and price are not only governed by demand. Since food manufacturers require specific physical and/or chemical properties for the edible oils and fats to be used in their products, it is up to the refiner to provide fully refined oils, fats or fat blends meeting these specifications at the lowest cost. To meet these goals with use of the raw materials available on the market, the refiner may be required to modify the properties of the oils by using one or more of the processes listed below in order of increasing cost per tonne of oil processed (Kellens, 2000).

- Blending: this is by far the cheapest process and it is therefore widely used to produce, for instance, margarine fat blends by mixing a hardstock with a liquid oil.
- Fractionation: this process requires dedicated investment and uses some energy but it does not entail a yield loss nor does it require auxiliary products. It separates the fat into a higher melting and a lower melting fraction, redistributing the triglycerides that provide the fat with its consistency as expressed by its solid fat content profile.
- Interesterification: although the investment required for the interesterification process is low, the process itself is more costly than the processes listed above because of yield loss and catalyst usage. The process is used to redistribute the fatty acids between triglycerides in such a way as to alter the melting profile [i.e. solid fat content (SFC)] of the fat, changing for instance a blend with too high a melting point into a fat blend with acceptable mouth feel (i.e. into a fat with a melting point below body temperature).
- Hydrogenation: this is the only process that creates consistency (i.e. increasing the percentage SFC) in oils and fats by converting low melting into higher melting triglycerides. It was developed at the beginning of the twentieth century by Normann in 1902 (see Kaufmann, 1939) to counteract a shortage of solid fats such as edible tallow used in margarine and shortenings. Because it is the only process that converts a liquid oil

into a solid fat, and despite its high cost (because of the need for hydrogen, catalyst and investment), the hydrogenation process powered the growth of the margarine and shortening industry and is still considered essential in meeting the growing demand for semisolid fat products.

4.2 Hydrogenation

During the hydrogenation of edible oils many different reactions occur simultaneously:

- Hydrogen is added to the double bonds in the unsaturated fatty acids so that a triene is saturated to a diene, a diene is saturated into a monoene and, with continuing hydrogenation, a monoene is converted into a fully saturated fatty acid. A consecutive addition is generally assumed. Since the melting point of a triglyceride depends on the melting points of its fatty acids and since these rise with decreasing unsaturation, hydrogen addition leads to higher triglyceride melting points (a process also referred to as hardening).
- Double bonds isomerise geometrically so that *trans*-isomers are formed. Since *trans*-fatty acids can be straightened, they are more like saturated fatty acids than are the *cis*-isomers and thus have a higher melting point than their corresponding and nonlinear *cis*-isomers. They therefore also cause the triglyceride melting point to rise.
- Double bonds can also isomerise positionally and thus shift along the fatty-acid chain; this geometrical isomerisation in practice hardly affects the triglyceride melting point.

The purpose of the hydrogenation process is twofold: to convert a liquid oil into a product with greater consistency (i.e. with a higher SFC), and to produce a more stable product by eliminating or reducing the polyunsaturated fatty-acid content of the starting material. In Europe, this second aspect has led to the hydrogenation of fish oil, and, formerly, whale oil, and thereby to the utilisation of these oils in food products. However, in the USA the Food and Drug Administration (FDA) affirmed only in 1997 (Anon., 1997) that partially hydrogenated menhaden oil with an iodine value (IV) between 86 and 119 is generally recognised as safe (GRAS).

Also, in the USA partial hydrogenation of soybean oil is carried out to reduce its linolenic acid content because, in the USA, this acid is regarded as a major cause of flavour reversion and lack of stability. For soybean oil intended for use as salad oil, this so-called 'brush hydrogenation' is followed by a fractionation process to eliminate the higher melting triglycerides formed during hydrogenation (Coppa-Zuccari, 1971; Evans *et al.*, 1964). A process incorporating a directed interesterification reaction to concentrate the saturated fatty acids and *trans*-isomers into higher melting triglycerides after hydrogenation and before fractionation gives a better yield of salad oil (Baltes, 1975). Also, soybean-oil-based margarine fat blends in the USA have tended to be entirely hydrogenated and thus virtually free from linolenic acid. In Europe the situation is different in that nonhydrogenated soybean oil and rapeseed oil with their original linolenic acid contents are sold as salad oils in competition with for instance sunflower seed oil, olive oil or corn oil, which contain no linolenic acid. Similarly, margarine fat blends in Europe generally contain a liquid oil component that may be nonhydrogenated soybean oil or, if cheaper, rapeseed oil. Such blends are considered to be sufficiently stable.

The reason for this difference between Europe and the USA in oil stability is not quite clear but may well stem from differences in industrial deodorisation practice. Deodorisation temperatures in the USA tend to be higher and thus lead to a higher loss of tocopherols, the residual concentration of which has been demonstrated to affect oil stability (Lampi *et al.*, 1997). The difference may also be a result of the way in which oil stability is assessed. In the USA, accelerated tests which expose oil to oxygen under conditions that are very different from the way food products are normally stored are used to measure oil stability, whereas in Europe the value of such methods is actively doubted (Dijkstra *et al.*, 1996). Lacoste *et al.* positively demonstrated (1999) the failure of the Rancimat test to predict the onset of rancidity by observing that the Rancimat value of a sunflower seed oil did not change over a two-year period, during which the oil became clearly rancid.

4.2.1 Kinetics and mechanism

The hydrogenation process that is used industrially all over the world involves straight oil or fat as its substrate, molecular hydrogen as the reagent and supported metallic nickel as the catalyst. Other catalysts such as a copper/chromium and noble metal catalysts (palladium) have been studied on a laboratory scale, but these catalysts are not used industrially (Koritala *et al.*, 1973). Accordingly, the reaction involves three different phases, which means that mass transfer also affects the rate of the hydrogenation reaction and complicates the picture. In addition, the analytical problems involved in following what happens during a hydrogenation reaction are very complex because of the many different fatty-acid isomers formed during this reaction. Moreover, research in the USA on the kinetics and mechanism of the hydrogenation process has focused on the study of soybean oil; because this is a more complex substrate than for example model compounds, this work has not facilitated a clear understanding of the reaction either. Consequently, a century after the hydrogenation process was invented (Normann, 1902), its kinetics and mechanism are still not fully understood, let alone quantified. Accordingly, brochures of commercial hydrogenation catalysts illustrate catalyst activity only through examples of hydrogenation experiments but do not quantify the intrinsic kinetic parameters.

Nevertheless, the reaction is sufficiently well understood for the industrial hydrogenation process to be controlled to such an extent that different products can be made in a reasonably reproducible manner; there is, however, still considerable need for improvement. In fact, nearly all industrial hydrogenation is carried out as a batch process because the continuous process is too unpredictable. Several causes for this unpredictability have been identified, including variations in catalyst quality, variable amounts of catalyst poison in the substrate and an inability to measure the concentration of the hydrogen that is dissolved in the oil.

To describe the course of a hydrogenation reaction in qualitative terms, several selectivity concepts have been introduced. Despite their lack of theoretical foundation, they can serve a useful purpose. Linoleic acid selectivity for instance indicates to what extent the hydrogenation of linoleic acid to a mono-unsaturated fatty acid is favoured over the saturation of monoenes leading to stearic acid. The various selectivity concepts will now be discussed.

4.2.1.1 Positional selectivity
The literature contains quite a number of papers that try to determine whether or not an unsaturated fatty acid at the outer (1,3) position of the glycerol moiety in a triglyceride reacts at the same rate as the same fatty acid esterified to the secondary (2-position) hydroxyl group of this moiety. In my Chang Award Address, I reviewed this literature (Dijkstra, 1997) and concluded that it did not demonstrate the existence of this positional selectivity: the reactivity of an unsaturated fatty acid in a triglyceride molecule apparently does not depend on its position in this molecule.

4.2.1.2 Triglyceride selectivity
The concept of triglyceride selectivity has been poorly defined in the literature, and early authors did not even use the term. When Bushell and Hilditch hydrogenated mixtures of mixed triglycerides in 1937 they noted that molecules with three unsaturated fatty acids (tri-unsaturated glycerides) 'are attacked more rapidly than di-oleo-glycerides and the latter somewhat more so than the mono-oleo-compounds'. In other words, they looked at the rate at which triglyceride molecules were hydrogenated and not so much at the rate at which the fatty acids reacted. They concluded that a higher degree of unsaturation of the triglyceride caused it to react faster, but they did not conclude that the rate of reaction was proportional to the degree of unsaturation.

This proportionality was subsequently put forward by Bailey and Fisher (1946) as a very carefully worded suggestion (the 'common pool' concept), but, sadly enough, their suggestion soon became accepted as representing what happens during the hydrogenation of triglycerides and the concept started to lead a life of its own. Accordingly, nearly all kinetic studies of the hydrogenation reaction published subsequently have been based on this 'common pool' concept even if the authors do not mention this specifically, which they seldom do.

Nevertheless, a few authors have studied triglyceride selectivity, which now can best be regarded as a departure from the 'common pool' concept. These authors have shown that triglyceride selectivity does indeed exist. In other words, they have proved that the 'common pool' concept is invalid. Beyens and I (Beyens and Dijkstra, 1983) stumbled across triglyceride selectivity when studying steric effects during hydrogenation. We found that a triglyceride molecule containing two linoleic acid moieties did not react twice as fast as a molecule containing only one linoleic acid moiety and two short-chain (C_8 and C_{10}) fatty acids. Originally, we interpreted this observation on the grounds of the easier accessibility of the linoleic acid surrounded by short-chain fatty acids, but subsequently (Dijkstra, 1997), the observation was found to tie in with other studies of triglyceride selectivity.

Coenen (1978; Linsen, 1971), for instance, noted that the use of certain, narrow-pore catalysts led to the formation of more trisaturated triglycerides than expected from the 'common pool' concept, and he attempted to express this tendency mathematically (Coenen, 1976). Again, triglyceride selectivity was not regarded as a fundamental aspect of the hydrogenation of triglycerides but it was tied to certain catalysts. However, its existence was clearly demonstrated.

The only author to study triglyceride selectivity systematically is Schilling (Schilling, 1968, 1978, 1981). Because this study requires the determination of the molecular composition of the reaction products and not just of the fatty-acid composition, Schilling used model mixtures to ease his analytical problems. When hydrogenating a mixture of trilinolenin and monolinolenin, Schilling (1978) concluded that these molecules reacted equally fast instead of at rates differing by the factor of 3 predicted by the 'common pool' concept. Subsequently, Schilling (1981) modified his former conclusions by introducing weighting factors in an attempt to simplify the mathematics involved. He did not have enough experimental data to elaborate his proposals fully and, sadly, his valuable approach has not been followed up. Accordingly, the existence of triglyceride selectivity has only been demonstrated but has not been quantified. Consequently, it has not been possible to incorporate the extent of the effect of triglyceride selectivity into other studies of the kinetics of the hydrogenation.

4.2.1.3 Linolenic and linoleic acid selectivities

Linolenic acid selectivity and linoleic acid selectivity are the most commonly used examples of fatty-acid selectivity, constituting an attempt to express the difference in reactivity (or extent of reaction) between various unsaturated fatty acids. There is an obvious need for such an expression, for instance, to distinguish between a hydrogenation experiment exhibiting a certain drop in IV and giving a product with a certain melting point or SFC profile and another experiment whereby the same drop leads to a different melting point and SFC profile. Since analysis of these different products shows that their fatty-acid compositions differ and thereby explain the differences in physical properties, it is only logical to try and use the difference in reactivity of these acids—or, rather, of their precursors—to characterise the hydrogenation process.

Accordingly, Albright and Wisniak (1962) introduced and defined the concept of linoleic acid selectivity as the ratio of the rate constants of the rate of disappearance of linoleic acid and the rate of appearance of stearic acid. In doing so, they assumed these rates to be proportional to the concentration of the substrates (linoleic acid and monoenes) and to depend on the hydrogen concentration in an identical manner. If an oil containing no linolenic acid is hydrogenated, the rate at which linoleic acid reacts equals its rate of disappearance, but in the case of soybean oil, the rate of disappearance of linoleic acid has to be corrected by the amount formed by the hydrogenation of linolenic acid (i.e. by its disappearance).

However, when introducing linoleic acid selectivity, authors have based its definition on a number of assumptions (Dijkstra, 1997). One assumption (absence of triglyceride selectivity) subsequently turned out to be incorrect (*vide supra*), others (absence of positional selectivity, absence of shunt reactions whereby several double bonds within the same fatty acid react at the same time) are likely to be correct; a further assumption of equal reactivity of fatty-acid isomers such as conjugated and methylene-interrupted polyunsaturated fatty acids was already known to be incorrect (Thompson, 1951) at the time the definition was published. The final assumption (of identical dependence of rates of reaction of monounsaturated fatty acids and polyunsaturated fatty acids on the concentration of dissolved hydrogen) also turned out to be incorrect.

Several authors have reported on the effects of changing process parameters on product properties and composition and subsequently these were summarised by Coenen (1978) in a now well-known table (see table 4.1).

Table 4.1 shows the effect of process parameters such as rate of agitation on selectivity (linoleic acid selectivity and *trans*-selectivity). It also shows

Table 4.1 Effect of the increase in the value of various process parameters on linoleic acid selectivity and isomerisation index

Process parameter	Linoleic acid selectivity	Isomerisation index
Rate of agitation	−	−
Temperature	+	+
Pressure	−	−
Amount of catalyst	+	+
Activity of catalyst	+	+

that a parameter change that causes the hydrogen concentration in the oil to increase (faster agitation, less catalyst, etc.) causes the linoleic acid selectivity to decrease. According to its definition (Albright and Wisniak, 1962) the linoleic acid selectivity is the ratio of two rate constants and therefore it must be constant itself. Increasing the hydrogen concentration by increasing the rate of agitation should not affect rate constants nor their ratio.

To illustrate that the linoleic acid selectivity is not constant and varies during a hydrogenation run, a hydrogenation experiment was carried out during which samples were taken at regular intervals (Dijkstra, 1997). The samples were analysed for fatty-acid composition, and values for the linoleic acid selectivity were calculated over these intervals. The results have been summarised in table 4.2.

In the early stages of the experiment, hardly any stearic acid is formed and consequently a high linoleic acid selectivity is calculated. Later on, when the residual linoleic acid content has dropped to below 20%, the rate of formation of stearic acid increases markedly, and this is reflected in a much lower linoleic acid selectivity. The experiment was isothermal, used a fixed amount of catalyst and neither the rate of agitation nor the rate of hydrogen supply were changed. Accordingly, several parameters that are known to affect the linoleic acid selectivity were kept constant.

However, the experiment was carried out in an autoclave with the vent closed. Therefore the pressure could and did vary. In the early stages, the pressure was around atmospheric since the rate of hydrogen supply and the rate of agitation had been balanced in such a way that the rate of hydrogen supply and the demand for hydrogen were in dynamic equilibrium at around atmospheric pressure.

Table 4.2 Fatty-acid compositions during the hydrogenation of sunflower seed oil and the linoleic acid selectivities calculated for subsequent time intervals

Time (min)	Stearic acid (wt%)	Monoenes (wt%)	Dienes (wt%)	Linoleic acid selectivity
0	4.15	18.5	77.35	
80	4.40	45.0	50.60	59
100	4.60	52.2	43.20	40
120	4.80	60.5	34.70	55
140	5.10	69.5	25.4	68
160	5.50	77.5	17.00	72
180	7.00	83.8	9.2	33
200	16.6	80.5	2.90	10

Subsequently, at about the point in time when the linoleic acid selectivity increases, an increase in hydrogen pressure was noted.

This observation can be explained by taking the concentration of the hydrogen dissolved in the oil into account. At the start of the experiment, the reactivity of the reaction mixture is high because of its high linoleic acid content. In the course of the experiment, this linoleic acid content decreases, thereby also decreasing the reactivity of the mixture. Since a fourfold decrease in reactivity (from 77% to less than 20% linoleic acid) does not lead to a significant pressure increase in the autoclave, the concentration of the hydrogen in solution must have been very low in comparison with its solubility.

This conclusion can be deduced by considering which variables govern the rate of dissolution of hydrogen into oil. At a given temperature and within a given autoclave these variables are: the rate of agitation and the difference between the hydrogen solubility and its actual concentration. During the experiment, the rate of agitation was not changed. Hydrogen was supplied at a constant rate and the pressure in the autoclave did not change noticeably either. Since the hydrogen solubility is proportional to this pressure, the solubility remained almost the same and, since the rate of dissolution was constant, this would lead to the conclusion that the concentration must also have remained constant. However, since during this experiment a slight increase in agitation immediately led to a pressure decrease, it follows that the rate of hydrogenation increases with an increase in hydrogen concentration. During the experiment, the linoleic acid content and thus the reactivity of the reaction mixture decreased, and, since the rate of reaction remained constant, this would lead to the conclusion that the hydrogen concentration must have increased to compensate for the loss of reactivity.

This paradox can be resolved by assuming a hydrogen concentration that is very low in comparison with its solubility. Then, the large relative increase in hydrogen concentration needed to maintain a constant rate of reaction will only lead to a small absolute increase and thus to a negligible decrease in difference between hydrogen solubility and hydrogen concentration.

The low hydrogen concentration during the early stage of this experiment (Experiment A) is further demonstrated by the following observation: near the end of Experiment A, the pressure in the autoclave reached practically one atmosphere gauge. During a similar experiment (Experiment B), whereby the pressure was allowed to remain constant (atmospheric) and the rate of agitation was increased in order to compensate for the diminishing reactivity of the reaction mixture, the fatty-acid composition was identical to the composition observed in the Experiment (A) at a point in time when the agitator in Experiment (B) ran at its maximum speed. Then, the hydrogen concentration should have been much closer to its solubility. Since the pressure in Experiment A, with constant agitation, maintained a fixed difference between hydrogen solubility and concentration, and the pressure almost doubled in absolute terms, it must have been very low to start with.

Accordingly, it can be concluded that the hydrogen concentration can vary widely during a hydrogenation run. It can also be concluded that the hydrogen concentration seriously affects the course of the hydrogenation and the linoleic acid selectivity, as observed. These are very important findings for process and product control during industrial hydrogenation processes; they also have implications for the kinetics and mechanism of the hydrogenation reaction itself.

4.2.1.4 Trans-*selectivity*

In order to distinguish between hydrogenated products with varying amounts of *trans*-isomers, the concept of *trans*-selectivity has been introduced; this is also referred to as the isomerisation index. It has been defined as the ratio between the percentage increase in *trans*-isomer content and the decrease in IV during a hydrogenation run. Instead of the decrease in IV, the decrease in the percentage of double bonds is also used in the definition. In practice, this definition turns out to be very handy since almost all hydrogenation of *trans*-isomer-free starting materials display an increase in *trans*-isomer content that is perfectly proportional to the decrease in IV, up to quite high values. This is quite remarkable since:

- the rate of *trans*-isomer formation is apparently zero order in terms of IV or double bonds
- the rate of *trans*-isomer formation is nevertheless linked to the rate of hydrogenation of the double bonds, whereby a decreasing total number of double bonds goes on producing *trans*-isomers at a constant rate
- the number of *cis*-double bonds decreases even faster, but *trans*-isomers can re-isomerise to yield *cis*-isomers

The way to explain this remarkable phenomenon is by making a distinction between polyunsaturated fatty acids and monounsaturated fatty acids. During the hydrogenation of, for instance, sunflower seed oil or even soybean oil the polyunsaturated fatty-acid content at the start of the reaction is high and the monounsaturated content is fairly low. At that time, the linoleic acid selectivity is also quite high (see table 4.2). Consequently, the monounsaturated fatty-acid content will increase at the expense of the polyunsaturated fatty-acid content as and when the reaction proceeds. Therefore, the assumption that monounsaturated fatty acids are more prone to isomerise than are polyunsaturated fatty acids can go a long way towards explaining the proportionality between increase in *trans*-isomer content and decrease in IV.

This assumption was proven experimentally (Dijkstra, 1997, figure 4.1) by the hydrogenation of a number of oils with widely different fatty-acid compositions. A high-oleic sunflower oil was chosen as the starting material, with a high monoene-to-diene ratio, whereas normal, high-linoleic sunflower seed oil was used to include a low monoene-to-diene ratio. This high-linoleic oil was also diluted with an inert solvent (a C_8–C_{10} triglyceride oil), and palm

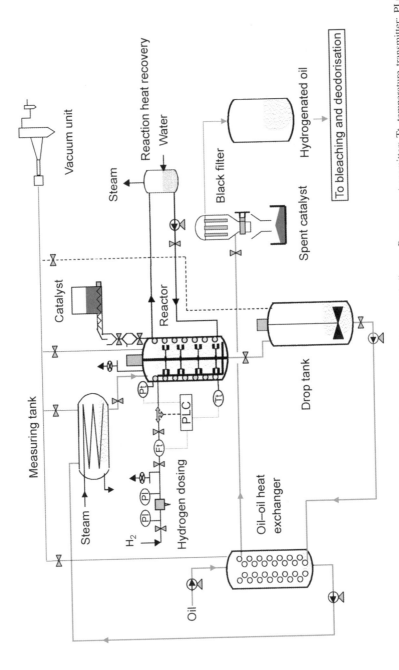

Figure 4.1 Flow diagram of a hydrogenation plant. Ft, flow transmitter; PI, pressure indicator; Pt, pressure transmitter; Tt, temperature transmitter; PLC, programmable logic controller. Reproduced by courtesy of N.V. Extraction De Smet S.A.

olein provided an intermediate ratio. It was then observed that the isomerisation index of the monoenes increased linearly with the monoene-to-diene ratio and was independent of the actual values of their concentrations, thus showing that polyenes suppress the isomerisation of monoenes.

This conclusion was also reached by Coenen and Boerma (1968) on the basis of an experiment in which high-erucic-acid rapeseed oil was selectively hydrogenated. It was an elegant experiment because it allowed a distinction to be made between *trans*-isomers formed by isomerisation (brassidic acid, the *trans*-isomer of erucic acid) and *trans*-isomers formed by the hydrogenation of C_{18}-dienes. Unfortunately, the authors also concluded that monoenes are isomerised only when they are also being hydrogenated, and they went too far when stating that the catalyst surface is monopolised by polyenes. This statement is clearly contradicted by the observations described in the previous paragraph.

With respect to *trans*-isomer formation it can therefore be concluded that such isomers originate from the hydrogenation of polyenes as well as from the isomerisation of monoenes and that the monoene source becomes more and more important as and when the monoene content of the reaction mixture increases. The study of any hydrogenation mechanism and its ensuing kinetics will have to take this conclusion into account.

4.2.1.5   Hydrogenation mechanism

As mentioned above, the kinetics and mechanism of the hydrogenation process are still not fully understood let alone quantified. Nevertheless, it has been possible to arrive at a qualitative mechanism that takes the observations and conclusions described above into account. These can be summarised and extended as follows:

- industrial hydrogenation runs are mass transfer limited; consequently, the concentration of the hydrogen dissolved in the oil is very small in comparison with its solubility, especially during the early stages of such runs
- because this concentration is so small in comparison with the hydrogen solubility, concentration changes do not manifest themselves macroscopically to any significant extent
- this concentration results from a dynamic equilibrium between the hydrogen supply and its demand
- the hydrogen supply (the rate of hydrogen dissolution) is governed by the agitator design, the rate of agitation and the difference between the hydrogen concentration in the oil and its solubility therein
- hydrogen solubility is proportional to the hydrogen pressure in the autoclave

- hydrogen demand is governed by the activity of the catalyst, the fatty-acid composition of the reaction mixture, the hydrogen concentration in the oil and the oil temperature
- actions that increase the hydrogen concentration such as use of increased agitation, less catalyst and a higher pressure or a lower temperature lower the linoleic acid selectivity and the isomerisation index
- in the very early stages of an industrial hydrogenation run, the hydrogen concentration may start at a fairly high level and then decrease as and when the catalyst becomes fully activated and displays its full design activity
- subsequently, the hydrogen concentration will increase as and when the linoleic acid content and thus the reactivity of the reaction mixture decreases
- this gradual decrease in linoleic acid content also gradually promotes the isomerisation of monoenes
- this isomerisation is reversible and can lead to an equilibrium, the position of which is governed by its thermodynamics: $\Delta H = -4.1\,\text{kJ mol}^{-1}$ (Veldsink et al., 1997)
- at commonly used hydrogenation temperatures, an equilibrium mixture will therefore contain around 75% trans-isomers, and publications claiming values greater than 90% (e.g. Baltes, 1972) should therefore be disregarded
- the increase in hydrogen concentration resulting from the decrease in reactivity of the reaction mixture eventually also causes monoenes to be saturated
- accordingly, monoene isomerisation precedes monoene hardening
- saturation of monoenes occurs only when the hydrogen concentration is sufficiently high
- the decrease in linoleic acid selectivity during a hydrogenation run indicates that the order in hydrogen for linoleic acid saturation is lower than the order in hydrogen for the saturation of a monoene

The mechanism that has been suggested (Dijkstra, 1997) is in line with the above observations and is based on the Horiuti–Polanyi mechanism (Horiuti and Polanyi, 1934) which assumes consecutive additions of the two hydrogen atoms to the double bond and therefore assumes a semihydrogenated intermediate. It also assumes the addition of the first hydrogen atom to be reversible. In scheme 4.1, this consecutive addition of hydrogen atoms has been indicated for dienes (D) and monoenes (M).

According to the Horiuti–Polanyi mechanism, double bonds in dienes and monoenes react according to the same mechanism so that it does not explain why these substrates apparently exhibit a different order in hydrogen. Consequently, the Horiuti–Polanyi mechanism must be amended to explain this difference.

$$D + H \longrightarrow DH \qquad\qquad (4.1)$$

$$DH \longrightarrow D + H \qquad\qquad (4.2)$$

$$DH + H \longrightarrow M \qquad\qquad (4.3)$$

$$M + H \longrightarrow MH \qquad\qquad (4.4)$$

$$MH \longrightarrow M + H \qquad\qquad (4.5)$$

$$MH + H \longrightarrow S \qquad\qquad (4.6)$$

Scheme 4.1 Hydrogenation reaction scheme. D, diene; M, monoene; S, stearic acid.

This amendment (Dijkstra, 1997) involves the assumption that in the case of dienes, D, reaction (4.1) in scheme 4.1 (the formation of the semihydrogenated intermediate, DH) is the rate-determining step, whereas for the monoenes, M, the rate-determining step is the addition of the second hydrogen atom to the semihydrogenated monoene, MH, via reaction (4.6) in scheme 4.1. This means that the observed rate of hydrogenation of monoenes depends on the hydrogen concentration in reactions (4.4) and (4.6), whereas the rate of hydrogenation of dienes depends only on the hydrogen concentration in reaction (4.1).

According to this assumption, the reactions taking place in an oil containing both linoleic acid and oleic acid (e.g. sunflower seed oil) will change in emphasis in the course of the hydrogenation run. At the start, when the high reactivity of the reaction mixture causes the hydrogen concentration to be low, linoleic acid (D) will preferentially react [reaction (4.1)] to form the semihydrogenated intermediate, DH, which will readily react [reaction (4.3)] further to form a monoene, M. Of course, this intermediate (DH) can also dissociate again and form *trans*-isomers and/or positional isomers of dienes, but this dissociation requires a very low hydrogen concentration. Indeed, the extent of 'diene conjugation (positional isomerisation) was found to increase as the temperature and the catalyst concentration increased and as the degree of hydrogen dispersion decreased' (Feuge *et al.*, 1953) since these factors lead to a decrease in hydrogen concentration. In industrial practice, artificially low hydrogen concentrations are usually avoided because they slow down the hydrogenation process.

During these early stages, monoenes (M) also participate in the hydrogenation process by forming semihydrogenated intermediates (MH) via reaction (4.4), but the hydrogen concentration is still too low for these intermediates (MH) to react further via reaction (4.6) to form stearic acid (S). Accordingly, the intermediates tend to dissociate via reaction (4.5), as can be seen from the formation of *trans*-monoenes (also denoted as M in scheme 4.1). Therefore, the linoleic acid selectivity remains at a high and almost constant level.

When the reaction proceeds and the reactivity of the mixture decreases, the hydrogen concentration increases and eventually becomes sufficiently high for the intermediates (MH) to react with a further hydrogen [reaction (4.6)] to form stearic acid (S); this is observed as a decrease in linoleic acid selectivity. The *trans*-isomer content may still increase but, if the hydrogenation is pursued, this content will eventually decrease to zero when full hydrogenation (IV=0) has been attained. By then the isomerisation index has lost its original meaning by having attained a negative value.

The above mechanism qualitatively explains the observations that have been itemised above. It is based on the generally accepted Horiuti–Polanyi mechanism but contravenes a number of conclusions that have been published in the literature. The mechanism:

- like Hashimoto *et al.* (1971) and Ahmad *et al.* (1979) assumes a higher order in hydrogen for the hydrogenation of monoenes than for the hydrogenation of dienes, but in doing so it contravenes the definition of linoleic acid selectivity (Albright and Wisniak, 1962)
- stresses the importance of the increase in hydrogen concentration during an industrial hydrogenation process, whereas many authors of research papers (Jonker, 1999) aim to study the process at near saturation; this is understandable since in doing so they eliminate an unknown variable; however, this choice of experimental conditions should prohibit authors from drawing conclusions (Veldsink *et al.*, 1999) about what happens industrially
- contravenes the conclusion (Coenen and Boerma, 1968) that monoene isomerisation necessarily coincides with monoene saturation

4.2.2 Industrial hydrogenation processes

In principle, industrial hydrogenation processes are merely scaled-up versions of laboratory experiments in that they are carried out in an agitated autoclave provided with a hydrogen supply. However, the scaling-up process, the need for product quality assurance, safety requirements and process economics have affected the design of industrial hydrogenation plants and their process control:

- the hydrogenation reaction is highly exothermic in that a drop in IV by 1 unit raises the oil temperature by 1.6–1.7°C; this necessitates temperature control of the autoclave content and also permits heat recovery
- hydrogen forms explosive mixtures with air; accordingly, the autoclave is preferably situated outside so that the wind will disperse any hydrogen escaping from the autoclave; if it is situated inside a building, extensive safety precautions, including hydrogen detectors, are necessary

- the autoclave is a relatively expensive vessel so that savings in investment can result from using cheaper auxiliary vessels for operations that do not require an autoclave
- for process control reasons, both the batch weight and the amount of hydrogen supplied to the autoclave must be determined sufficiently accurately
- to minimise exposure of the operators to nickel, fully enclosed and automated catalyst filters can be used, but their large heel makes this type of filter unsuitable if frequent changes of oil type are foreseen; in this case, a classical frame and plate type filter is preferred

A flow diagram of a modern hydrogenation plant is given in figure 4.1. It incorporates an oil-to-oil heat-exchange vessel in which the oil to be hydrogenated flows over spiral tubes before being pumped to a measuring vessel in which the oil can be further heated by steam coils if necessary. From this measuring vessel the hot oil can be dropped into the autoclave, where the hydrogenation proper takes place. Accordingly, the autoclave is provided with a means to feed the required amount of catalyst, a hydrogen supply, an agitator that maintains the catalyst in suspension and that dissolves the hydrogen into the oil, and cooling coils. After the reaction, the autoclave is emptied into a drop tank, which is agitated to prevent the catalyst from settling. The hot hydrogenated oil is then pumped through the coils of the heat-exchange vessel to the catalyst filter.

4.2.2.1 *Hydrogenation process conditions*

Industrial hydrogenation reactions employing a nickel catalyst require a starting oil temperature preferably greater than 150°C. If the previous batch had a final temperature of about 220°C, this starting temperature can be reached in the heat-exchange vessel. However, if a series of lightly hydrogenated products has to be produced (e.g. brush hydrogenated soybean oil), the final temperature is too low to bring the next batch up to starting temperature. Then additional heat must be supplied in the holding vessel. Products with an IV drop in excess of 40 generally require cooling to avoid the batch temperature exceeding 220°C. In standard practice, the temperature is allowed to increase to this final level before active cooling is instigated. For products with a lower than normal *trans*-isomer content, a low batch temperature throughout the process is required (see table 4.1).

The hydrogenation pressure is determined by what the autoclave can withstand and by the product properties required. Normally, the pressure varies (increases) during a run. In the beginning, the highly unsaturated oil tends to be highly reactive and tends to prevent the autoclave pressure from rising. As and when the reactivity decreases, the pressure may increase and be controlled at a certain, safe level (e.g. 5 bar gauge). As will be discussed below, pressure control during the latter stages of a hydrogenation run can effectively ensure batch-to-batch reproducibility. In addition, high pressure can be applied on purpose to generate hydrogenation products with a lower than normal *trans*-isomer content.

4.2.2.2 Catalyst usage

The amount of nickel catalyst to be employed merits some discussion. Some operators prefer to use fresh catalyst for each hydrogenation batch. In the USA, this has been a common procedure for a long time. In Europe, new plants also tend to use the catalyst only once. This necessitates cleaning the oil as thoroughly as possible by a pre-bleaching step. Then, 0.03–0.05 wt% catalyst (at approximately 25 wt% nickel) will generally suffice for partial hydrogenation. Full hydrogenation to give an IV less than 3 benefits from a somewhat higher catalyst loading. Fish oil and rapeseed oil, which contain sulfur catalyst poisons that cannot be removed easily by bleaching, also need more catalyst to ensure reproducible processing.

Another possibility is to reuse the spent catalyst until its catalytic activity is almost exhausted or until catalyst filtration becomes too time-consuming. According to this approach, the catalyst filter cake is slurried in oil when the filter has to be cleaned. This requires additional investment in a reslurry system and a catalyst slurry dosing system (for a flow diagram, see Coppa-Zuccari, 1971). Because the catalyst is reused until it is almost exhausted, higher catalyst loadings do not immediately raise the process costs. Therefore, higher loadings tend to be used since they make the hydrogenation process more robust (i.e. less susceptible to variable concentrations of catalyst poisons). Because of the steady decrease in catalytic activity of the catalyst lot (as determined by the filter capacity), catalyst loadings are preferably increased each time the lot is reused, until the loading has become so high that discarding the lot represents a saving. If for one reason or another a lot turns out to have been contaminated, it should also be discarded.

Reuse of catalyst is certainly preferable on cost grounds when a high catalyst loading is required as for instance for partially poisoned catalysts used to produce partially hydrogenated products with a *trans*-isomer content close to equilibrium, as used for confectionery fats. Then, a high catalyst loading is essential to further promote geometrical isomerisation by keeping the hydrogen concentration at a low value.

4.2.2.3 Hydrogen dissolution systems

It has already been mentioned that industrial hydrogenation processes are mass-transfer limited. Accordingly, the hydrogen dissolution system represents an essential process characteristic. Because laboratory autoclaves often use a sparging ring situated beneath the agitator impellers to disperse and dissolve the hydrogen in the oil, industrial autoclaves have also been fitted with this type of dissolving system. However, this system has the disadvantage that a sluggish reaction will cause the concentration of the dissolved hydrogen to become rather high. This decreases the driving force for dissolution, so that hydrogen will collect in the roof of the autoclave until a safety pressure switch cuts off the hydrogen supply and no more hydrogen will dissolve.

This disadvantage has been overcome by extracting hydrogen via a cooler from the roof and then recycling it to the autoclave via the sparging ring by means of an additional hydrogen pump. Another way to overcome this disadvantage is by using an agitator that sucks hydrogen from the head-space into the oil (Weise and Delaney, 1992). Such an agitator obviates the need to use a sparging ring and allows the hydrogen to be fed directly into the autoclave head-space.

The BUSS loop reactor is yet another system for dissolving the hydrogen into the oil. In this system (Duveen and Leuteritz, 1982) an external pump circulates the oil over the autoclave via a Venturi tube that sucks in hydrogen gas and dissolves it into the oil. The loop also contains a heat exchanger for oil temperature control. Improvements to the Buss loop reactor have been described by Urosevic (1986).

4.2.2.4 *Process and product quality control*

Process control in hydrogenation should aim for reproducibility and thus ensure that subsequent batches of the same grade have almost identical compositions and thus properties. This means that not only the extent of hydrogenation (drop in IV) has to be controlled but also the various selectivities that characterise a hydrogenation run. In addition, the control has to be quite accurate since small changes in IV and *trans*-isomer content can have a significant effect on the SFC, as illustrated by equation (4.1) (P.J.A. Maes, personal communication):

$$\Delta N_{20} = -1.2 \, \Delta \, V(\text{IV}) + 0.5 \, \Delta \, (\% \, \textit{trans}) \tag{4.1}$$

Equation (4.1) shows that for a partially hydrogenated soybean oil (melting point 35°C) a drop in IV of 1 unit will increase the SFC at 20°C (N_{20}) by more than 1.2%, since this IV drop will be accompanied by an increase in *trans*-isomer content.

The process control should also deal with variations in catalyst activity and catalyst activation and with a variable and often unknown content of catalyst poison in the oil. Of course, the effect of catalyst poison can be reduced by bleaching the oil to be hydrogenated, but this treatment constitutes an additional cost element since the hydrogenated oil also has to be bleached to remove residual nickel.

Given the hydrogenation mechanism described above, it is clear that product reproducibility can result only if the batch temperature and the hydrogen concentration in the oil follow standard profiles with regard to the extent of hydrogenation, which can be determined by calculating the accumulated hydrogen consumption from measurements of its flow.

According to a method described by Colen *et al.* (1990), good product reproducibility can be achieved if pertinent batch parameters (temperature and hydrogen concentration) are controlled with respect to the IV achieved; they replace the time variable by degree of hydrogenation. To this end they constructed a database comprising:

- the extent of hydrogenation as calculated from the measured hydrogen flow
- temperature
- rate of hydrogenation, by calculating the rate of drop in IV
- hydrogen concentration in the oil

The dissolved hydrogen concentration is calculated from the mass-transfer equation (Koetsier, 1997):

$$J = k_{L}a(C_{max} - C_{bulk}) \qquad (4.2)$$

where

J is the rate of hydrogen solution, that equals the rate of hydrogenation
$k_{L}a$ is the volumetric liquid-side mass-transfer coefficient which is an equipment parameter determined separately in a test run (Stenberg and Schöön, 1985)
C_{max} is the hydrogen solubility at the prevailing pressure and temperature, whereby

$$C_{max} = m(T)p \qquad (4.3)$$

where

$m(T)$ is Henry's law constant
p is pressure
C_{bulk} is the concentration of dissolved hydrogen

Having obtained a dataset that is characteristic of a certain hydrogenation grade, Colen *et al.* (1990) then ensured that subsequent runs aiming to produce that grade followed the same temperature and hydrogen concentration profiles. They controlled the temperature by adjusting the extent of cooling and they achieved the same hydrogen concentration by varying the pressure in the autoclave. They also mentioned that control is more critical in the final stages of the hydrogenation than at the beginning.

Thus, if for one reason or another a hydrogenation run turns out to be slow (J is lower than normal), this will be signalled by a higher than normal hydrogen concentration, as shown by rewriting the mass-transfer equation as follows:

$$C_{bulk} = C_{max} - \frac{J}{k_{L}a} \qquad (4.4)$$

Accordingly, Colen *et al.* (1990) then decreased the autoclave pressure to decrease the driving force for hydrogen dissolution, as a result of which the dissolved hydrogen concentration fell. Of course, this slows down the rate of reaction even further, which goes against the grain of the operator, who would tend to increase the pressure to speed up the reaction and thus gain time lost. However, decreasing the pressure will ensure that the various selectivities are controlled at around the values shown by the dataset and thereby lead to a similar final product.

To ensure that the final product properties are even closer to specification, it is advisable to interrupt the hydrogenation before the expected end-point and take a sample for quick analysis. One method of quick analysis is to determine the IV and possibly the *trans*-isomer content by Fourier transform Near Infra Red analysis (Cox *et al.*, 2000). Another method is to take a fast SFC measurement by pulsed nuclear magnetic resonance (NMR) (Rutledge *et al.*, 1988), involving the crystallisation of the sample in liquid nitrogen. In both cases, tables based on past performance should be used to calculate how much more hydrogen needs to be added to the autoclave to obtain the target product.

4.3 Fractionation

The fractionation process and the winterisation process both aim at the separation of higher melting constituents from triglyceride oils. In Europe, the term 'winterisation' refers only to processes aiming at wax removal from, for instance, sunflower seed oil and involving a filtration step. If some of the waxes are removed during a low-temperature degumming step, the term 'de-waxing' is preferred. In the USA, the term 'winterisation' is also used for salad oil preparation, involving the removal of high melting triglycerides from, for example, cottonseed oil (Neumunz, 1978; Porter Lee, 1939) or brush hydrogenated soybean oil. In Europe, such processes are called dry fractionation processes.

Among the various fractionation processes that have been or are still being used industrially, the dry fractionation process is by far the most important. For the detergent fractionation process (Haraldsson, 1979, 1981; Reck, 1973) no new plants are being built because developments in the dry fractionation process resulting in a higher olein yield have made dry fractionation more economical, and the expensive solvent fractionation process is in practice limited to a few high-added-value products such as cocoa butter equivalents. For palm oil, a comparison of the olein and stearin fractions is given for the three processes by Kreulen (1976).

Detergent fractionation has also been used for wax removal (Gibble and Rhee, 1976; Pallmar and Sarebjörk, 1981) but was found to be insufficiently effective in that a subsequent filtration step was found to be necessary to obtain satisfactory cold stability (Denise, 1987). A recent review by Timms (1997) not only discusses the theoretical aspects but also describes the various fractionation processes and applications of the products.

4.3.1 Theoretical aspects

All fractionation processes are based on the difference in solubility in the oily mother liquor (dry fractionation and detergent fractionation) or in the oil miscella (solvent fractionation) between the various triglycerides constituting the oil or fat to be fractionated. Winterisation is based upon the same principle in that waxes are poorly soluble in cold oil and thus make salad oil cloudy if they are not removed by winterisation.

In all processes, the oil or fat being fractionated or winterised is first of all fully melted and kept at some 10–20°C above the melting point to obliterate 'crystal memory' and then cooled to induce crystallisation of the least-soluble components. This process may be continuous, as in solvent fractionation and winterisation plants, or may be carried out in batch crystallisers, as invariably used for dry fractionation. Since the cooling and the removal of the latent heat of crystallisation imply heat transfer, crystallisers are almost always agitated. In batch crystallisers the agitator also prevents the crystals from settling and ensures temperature uniformity throughout the batch. Crystallisation without agitation is used only for the fractionation of lauric fats (coconut oil and palm kernel oil), which are allowed to solidify in trays, after which the solidified blocks are loaded into a press where the olein is expelled (Kokken, 1991; Rossell, 1985).

The crystallisation process requires the formation of crystal nuclei which should then continue to grow as long as the solution or melt is supersaturated (van Putte and Bakker, 1987; Timms, 1991). Since the rate of nuclei formation is more temperature-dependent than is the solubility of the least-soluble components, rapid cooling will promote nuclei formation and thus lead to the formation of many small crystals. Larger crystals result from a slow rate of cooling from above the cloud point of the material being processed. In this context, seeding this material can also promote the formation of larger crystals, whereby the crystals may be added as seeds (Iida et al., 1981; von Rappard and Plonis, 1979) after they have attained the desired crystal morphology (Dieffenbacher, 1986) or they may be added continuously to the crystalliser while still molten (Maes et al., 1994).

The rate of crystallisation depends on many different factors. Since poorly soluble constituents have to diffuse towards a nucleus or growing crystal, the rate of agitation certainly affects the rate of crystallisation. Another factor is the extent of supersaturation, which is, after all, the driving force of the crystallisation process. Yet another factor is the number of nuclei or growing crystals, or, in other words, the number of active sites at which crystallisation takes place.

Partial glycerides also affect the rate of crystallisation. Saturated partial glycerides tend to be poorly soluble and may thus attach themselves to a growing triglyceride crystal. However, in doing so they disturb the regularity of the lattice and this inhibits the crystallisation of further triglycerides at that formerly active site. Since crystallisation is a dynamic process, whereby molecules on the outside of the crystal also redissolve, crystal growth will restart when such a partial glyceride has gone into solution again. For palm oil it has been noted (Goh and Timms, 1985) that free fatty acids and diglycerides are concentrated preferentially in the olein whereas monoglycerides are concentrated in the stearin. Among the diglycerides, dipalmitate has a preference for the stearin (Siew and Ng, 1995). Like partial glycerides, phosphatides also affect the rate of crystallisation and the resulting crystal morphology (Smith, 2000). If crude palm oil is fractionated, the free fatty acids concentrate in the liquid fraction (Taylor, 1976).

All fractionation and winterisation processes include a separating step which is nearly always a filtration, and it is this step that determines the selectivity of the process rather than the composition of the crystals formed during the cooling stage. This composition depends only on the temperature and is not affected by the presence of a solvent (Coenen, 1974; Hamm, 1986). Accordingly, the temperature during the separation step determines the composition of the liquid fraction (the olein), and the separation efficiency determines the composition of the solids fraction (the stearin) as well as the yield of each fraction.

Therefore, the reason that the solvent fractionation process is more selective than most dry fractionation processes is only because the liquid entrained by the crystals during the separation stage is diluted by solvent and thus contains less olein. This olein content is then further reduced by washing the crystals with fresh solvent (Gee, 1948).

4.3.2 Wax removal by winterisation

Wax removal from edible oils is necessary only for oils to be sold as salad oil and that are likely to be kept in a refrigerator. Therefore, winterisation in Europe is limited to sunflower seed oil and corn germ oil. Olive oil, that also contains some wax, is not winterised since it will partially solidify in a refrigerator anyway. In Asia, rice bran oil, which has a much higher wax content than for example sunflower seed oil (1–3% compared with less than 0.15%), may also be winterised.

Industrial processes that aim at wax removal by winterisation comprise a crystallisation step and a filtration step. Since wax crystals are small and tend to generate a poorly permeable filter cake, filter aids such as diatomaceous earth are generally used. However, this use constitutes a cost element in that the filter aid has to be purchased and that the filter cake contains oil and thus lowers the yield of the process. Disposing of the filter cake rarely constitutes a cost element since it can be sold to animal feed compounding plants.

Accordingly, the use of a filter aid should be minimised by careful control of the wax crystallisation process. In this context it has been found (J. De Kock, personal communication) that the rate at which the oil is cooled is an important parameter, especially when the cloud point of the oil is about to be reached. Cooling the oil to about 5°C above the cloud point can be fast without affecting the filterability of the wax crystals, but from then onwards a slow rate of cooling (e.g. 5°C h^{-1}) is essential to obtain wax crystals that require a minimum of filter aid during filtration. Asbeck and Segers (1990) even omitted all filter aids when using tubular microfilters for wax removal. Therefore a process whereby warm oil is fed into a crystalliser where it is shock-cooled to below its cloud point is to be avoided. It leads to the sudden formation of many nuclei that can only develop into small wax crystals requiring inordinately large amounts of filter aid.

Several industrial plants using a continuous crystalliser also contain maturation vessels between this crystalliser and the filtration unit, the idea being that larger crystals will be formed in these maturation vessels by Ostwald ripening. However, preventing small crystals from being formed by slow cooling from above the oil cloud point is a more effective way than trying to remedy harm already done. Besides, Rivarola *et al.* (1985) do not observe any change in crystal habit even with residence times of 24 h in the crystallisation tank. Accordingly, the entire winterisation process, comprising rapid cooling to above the cloud point, controlled cooling during crystallisation followed immediately by filtration, need take only 6–8 h.

It is also unnecessary to heat the oil containing the wax crystals just prior to the filtration step from the final crystallisation temperature of 5–8°C to 20–15°C to reduce its viscosity and thereby facilitate its rate of filtration, provided proper crystals have been formed during the crystallisation process.

4.3.3 Industrial fractionation processes

Only dry fractionation processes will be discussed in any detail here since the detergent fractionation process has become obsolete and the solvent fractionation process, which is quite expensive to operate (Hamm, 1986), is in practice used only for speciality fats. The reasons the solvent process is expensive are manifold:

- since the solvents used are inflammable, the entire plant must be X-proof, which adds to investment costs
- dilution of the oil with solvent increases the bulk of the material to be handled, which also increases investment

- consequently, the fractionation cost depends to a large degree on plant occupation
- lower crystallisation temperatures and a larger volume to be cooled than in dry fractionation require more energy in cooling, except when the temperature is lowered by evaporation of the solvent (Smorenburg, 1972)
- solvent evaporation and, in the case of acetone solvent, rectification are energy-intensive processes
- accordingly, variable operating costs are high

However, the solvent process has some cost advantages over the dry fractionation process in that:

- cooling in a scraped-surface heat exchanger is faster than in the dry fractionation process
- the presence of a solvent allows a higher proportion of feed to be crystallised while maintaining crystal slurry pumpability and may thus avoid multistage fractionation.

Whereas the equipment and installations for most edible-oil processes (degumming, neutralisation, bleaching, etc.) are supplied by independent contractors such as Extraction De Smet, Alfa Laval/Crown, and Lurgi, solvent fractionation plants tend to be in-house developments. Only Construzioni Mecchanice Bernardini offered at one stage a hexane fractionation process (Bernardini, 1978; Bernardini and Bernardini, 1975). Accordingly, different companies use different solvents (Thomas and Paulicka, 1976) such as acetone (Unilever and Fuji Oil Co.), 2-nitropropane (Durkee, Karlshamns; see Andrikides, 1960; Kawada and Matsui, 1970), azeotropic mixtures (Luddy and Longhi, 1983) and hexane (CMB plants). For the same reason, little has been published about the process characteristics involved.

4.3.3.1 Cooling profiles in dry fractionation

When palm oil production started to increase in Malaysia, fractionation developed with it. First, a fair number of detergent fractionation plants were in operation (Kheiri, 1985) but, subsequently, dry fractionation gained in both relative and absolute importance, with Fractionnement Tirtiaux providing both plant and processing know-how. A flow diagram of a modern Tirtiaux plant is given in figure 4.2.

As shown by the diagram, this plant incorporates a membrane filter, but former plants used a vacuum belt filter, the so-called Florentine ('Florent' Tirtiaux, 1976). Since the performance of this filter is highly dependent on crystal morphology, great attention was given to the crystallisation process (Deroanne et al., 1976) and how to control it and to the design of the crystallisers. As a result, a slow batch crystallisation process was adopted whereby the temperature difference between the oil and the coolant (water) is used as the

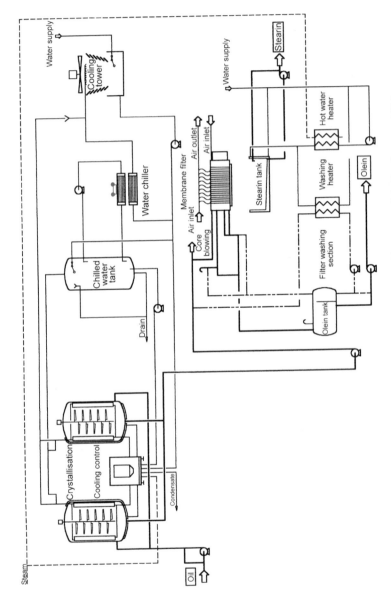

Figure 4.2 Flow diagram of fractionation plant. Reproduced by courtesy S.A. Fractionnement Tirtiaux.

steering parameter, and supercooling is strictly limited. For the production of a palm mid-fraction from the first olein, a process has been described that involves heating this olein to 65°C, cooling it to 38–48°C while keeping it at that temperature for at least 4 h, heating it again to 60–65°C for up to 2 h and then cooling it to the fractionation temperature of 14–17°C (Pike *et al.*, 1978). This particular process implies that a relatively large number of crystallisers is needed for a given throughput and this also holds for the general slow cooling. It also means that the crystallisation process is quite reproducible and yields fractions of predictable properties. Its applications in palm oil processing and butter fat fractionation were reviewed by Deffense in 1985 and 1993, respectively.

Perhaps the large number of crystallisers required by the slow crystallisation process developed for the Florentine filter prompted other companies to develop a faster crystallisation process. This became feasible after the development of another type of filter: the membrane press (Willner *et al.*, 1989; Willner and Weber, 1994) that is far less sensitive to crystal morphology. Accordingly, Extraction De Smet now offers a crystallisation process that comprises a crystalliser consisting of concentric annular vessels that are agitated by a common rotating beam provided with downward pointing agitating blades. The use of these nesting annular vessels separated by double walls through which a cooling medium is circulated ensures a large cooling surface per volume of oil and thus allows fast cooling. The resulting crystal morphology is such that the slurry often cannot be filtered on a laboratory Buchner filter and requires a membrane press to squeeze the olein from between the often irregular crystals.

Accordingly, there are two opposite approaches to crystal growth during dry fractionation, with the slow process aiming at the development of easily filterable crystals, and the fast process relying on the subsequent separation stage. As shown in figure 4.2, it is also possible to combine the characteristics of the two processes by aiming at easily filterable crystals and then squeezing them in a membrane press. This combination entails a relatively high investment but also ensures a high olein yield and a dry filter cake. Which process to choose depends on the demands placed on the fractionation products and on their prices. If stearin properties are important, attention to the crystallisation process may well be advantageous. If only a small amount of stearin has to be removed, and if stearin properties are not that important, a fast crystallisation process may well be preferable on cost grounds.

4.3.3.2 Separation in dry fractionation

It has already been mentioned that the separation efficiency during dry fractionation determines the stearin properties and the yield of both the stearin fraction and the olein fraction. As listed by Hamm (1986), the separation efficiency during vacuum filtration leads to a crystal content of the filter cake of only some 30–45%, the remainder (55–70%) being entrained olein. A much higher separation efficiency can be attained by the use of a membrane press (55–70%).

Such presses are plate and frame presses (figure 4.3) with alternating recessed plates and membrane plates, both of which are covered with filter cloth. When a crystal slurry is fed to the membrane press, crystals accumulate between the filter cloths whereas olein flows through these cloths. After the press has been filled a pressure medium (air for pressures less than 5 bar, and water for higher pressures) is introduced into the membrane plates to compress the cakes between the filter cloths and force more olein through the cloths. The pressure is then released and the press is opened to allow the cakes to be discharged. Because of the pressure applied, the filter cloths tend to clog and need to be cleaned periodically. This can be done *in situ* with hot oil, commonly heated olein (see figure 4.2).

Modern membrane presses are fully automated and whereas early membrane presses suffered high maintenance costs because of frequent membrane

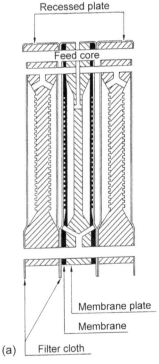

Figure 4.3 Operation stages of a membrane filter press: (a) details of the filter press; (b) the crystal slurry is fed through the press (crystals accumulate between the filter cloths whereas the olein passes through); (c) air pressure compresses the cloths (forcing more olein through the cloths); (d) the pressure is released and the cloths are discharged. Reproduced by courtesy S.A. Fractionnement Tirtiaux.

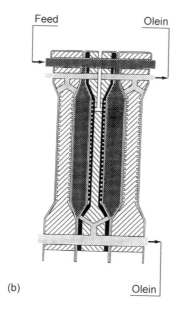

(b)

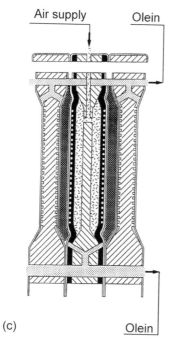

(c)

Figure 4.3 (continued)

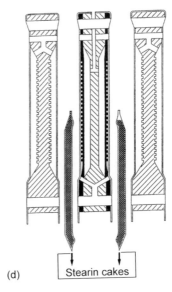

(d)

Stearin cakes

Figure 4.3 (continued)

failure necessitating costly plate replacement this aspect has also been improved. Nevertheless, the investment in membrane presses and maintenance costs are high, and other separation systems merit consideration.

In this context the use of a nozzle separator (clarifier) can be mentioned (Wilp, 2000). If such a separator is fed with a crystal slurry, the fat crystals having a higher density than the olein collect at the periphery of the rotating bowl and, since this has been fitted with nozzles, the crystal mass is extruded through these nozzles. This process is in fact a variant of a process that employs a regular discharge from a self-cleaning separator (Little, 1963). Operating such a clarifier requires careful matching of nozzle diameter, crystal content of the slurry and feed rate. Too small nozzles, too much crystalline matter and/or too high a feed rate cause the emerging olein to contain residual crystals. Similarly, too large nozzles, few crystals and/or too low a feed rate cause olein to be extruded together with the crystal mass, which is merely compacted as in the membrane filter press.

A more interesting process is a patented process (Maes and Dijkstra, 1983) describing the use of a conical sieve centrifuge provided with a scroll. This scroll both scrapes the screen surface, thereby preventing build-up of solids, and retains crystals within the centrifuge, thus ensuring adequate residence time and exposure to the centrifugal force that drives the olein away from the crystals. Whereas the membrane press and the clarifier compact the crystal mass, the sieve centrifuge replaces the interstitial cavities between the crystals by atmosphere and thus leads to very dry, fluffy cakes (high separation efficiencies) indeed. This is why the process is used for the production of high-grade cocoa butter equivalents and replacers (Deffense, 2000), which hitherto could only be produced by solvent fractionation. Subsequent authors (Breeding and Marshall, 1995), although not being aware of the above patent (Maes and Dijkstra, 1983), have confirmed the superiority of centrifugal filtration on a laboratory scale.

Like the vacuum band filter, the conical sieve centrifuge requires easily filterable crystals and thus carefully controlled crystallisation; this has repercussions on the investment required. However, it has the advantage of providing a fully continuous process, without hold-up, so that switches between products are much easier than with a membrane press, and the centrifuge is cheaper than a membrane press for a given throughput. The centrifuge also protects the products being processed from exposure to the atmosphere and allows cleaning-in-place (CIP). Accordingly, it is expected that in due course the conical sieve centrifuge will replace the vacuum band filter and become the preferred separation system in dry fractionation, at least for easily filterable crystals.

Therefore it is unlikely that the conical sieve centrifuge will be used in the dry fractionation of palm kernel oil. Accordingly, this oil will continue to be processed by pressing bags filled with partially crystallised oil, where labour costs are low, and by detergent fractionation where these costs prohibit use of this process (Timms, 1986). On technical grounds, the use of a membrane press should also be possible, but then this press should also be used for the further crystallisation of the partially crystallised feed, and this may well turn out to be a too expensive use of this equipment.

4.3.3.3   Dry fractionation products

The predominant application of the dry fractionation process concerns the production of palm olein with palm stearin as a by-product. As pointed out by Hamm (1995), quality demands for palm olein have become more stringent so that nowadays the first olein (cloud point 7–10°C) is refractionated to provide a palm superolein with a lower (cloud point 3–4°C). For a still lower cloud point (less than 0°C), this superolein can be refractionated again.

Because the processing costs during fractionation are fairly low, the economics of refractionation are determined mainly by the yields and the market values of the final products, such values depending very much on the properties of the product. Since the separation efficiency achieved during the filtration stage determines both the yields of the fractions and the properties of the stearin, the choice of the type of filtration equipment can strongly affect the process economics. During the fractionation of a palm olein of IV 57, a high separation efficiency not only produces more superolein (IV about 62) than a separation process with a low efficiency, but also can lead to a palm mid-fraction that can be used, for instance, as a confectionery filling fat and thus can command a higher price than can a soft palm mid-fraction.

Although several palm oil fractionation schemes have been published (Deffense, 1995; Hamm, 1995; Kellens, 2000), some of which include the recycling of intermediate fractions, most schemes in use industrially are proprietary. They have resulted from an internal optimisation by taking plant characteristics (capacity, separation efficiency and yield) and market values of the resulting products into account. These schemes also differ between companies because of differences in product emphasis. What may be the target product for one company may be more of a by-product to be disposed of on the open market for another company.

Anhydrous milk fat (AMF) or butter oil is another raw material for which quite intricate fractionation schemes have been developed (Deffense, 1993). In Europe, AMF fractionation developed when what was then the European Economic Community (EEC) was confronted with a 'butter mountain' and subsidised the use of AMF or its fractions in certain applications such as ice cream, bakery products and chocolate (Hartel, 1996). Consequently, the use of milk fat stearin in puff pastry became well established, since its longer plastic range in comparison with butter or AMF allowed industrial processing by obviating the need for intermittent cooling of the dough. This established the use of butter-fat stearin in its own right so that even after the subsidies were abolished AMF fractionation continued in Europe and also gained acceptance in other countries such as the USA (Kaylegian, 1999), New Zealand and Japan.

The olein fractions arising from the stearin production found their way into spreadable butter (Deffense, 1987). Butter as produced from cream by churning has the disadvantage that it is too hard to spread easily on bread after it has been stored in a household refrigerator. Seasonal variations in the triglyceride composition of the butter fat affect this characteristic so that customers become even more aware of this fundamental drawback (Amer et al., 1985). This drawback has been overcome by blending a butter stearin with refractionated olein, or, in other words, by eliminating the mid-fraction (Bradland, 1983). Blending also has the advantage that it eliminates the seasonal variability of the starting material. This is reflected in the yield of the various fractions but not in their properties (Dimick et al., 1996).

The product with the longest history of dry fractionation is edible beef tallow (Bussey et al., 1981). Since this product originates from a ruminant, it has been partially hydrogenated. Consequently, it contains *trans*-isomers and has such a high melting point that a poor mouth feel results. Fractionation overcomes this drawback by producing a tallow olein with acceptable mouth feel that in Europe for instance is sold as a deep-frying fat. The tallow stearin by-product can usually be sold for nonedible or feed applications. In the USA, the use of a tallow mid-fraction as a confectionery fat having a certain compatibility with cocoa butter has also been studied (Kozempel et al., 1981; Luddy et al., 1973) but has not led to commercial applications for reasons of authenticity.

A field where dry fractionation is already being used and may well gain in importance is the production of confectionery fats such as cocoa butter equivalents, cocoa butter substitutes and so on (Padley, 1997). Originally, these speciality fats were produced by solvent fractionation because their high price warranted this expensive process. However, the development of the membrane filter press (Willner et al., 1989) and the introduction of the conical sieve centrifuge (Maes and Dijkstra, 1983; Dijkstra, 1998) opened alternative ways to produce these speciality fats. Accordingly, shea butter stearin, palm mid-fraction and a mid-fraction of a trans-isomer-rich and thus partially hydrogenated blend of soybean oil and palm olein are being produced industrially by the above mentioned techniques to yield high-quality speciality fats (Kellens, 2000). Since these techniques have not yet been applied to the dry fractionation of other raw materials for confectionary fats such as sal fat (Jasko and Domek, 1984), phulwara butter (Yella Reddy et al., 1994a), kokum fat (Yella Reddy et al., 1994b), mango fat (Baliga and Shitole, 1981) and so on, it can only be expected that their as-yet nonquantified advantages will be realised in due course.

The literature also mentions the dry fractionation of lard (Wang and Lin, 1995) and interesterification products having too high a melting point for use as hardstock in margarine blends, but it is not clear to what extent these processes are being commercialised.

4.4 Discussion

Both the hydrogenation process and the fractionation process are being applied worldwide to millions of tonnes per annum. Nevertheless, both processes are incompletely understood. In the case of the hydrogenation process, its kinetics have not been resolved because insufficient attention has been paid to triglyceride selectivity or to the fact that the concept of a common fatty-acid pool turned out to be invalid (Dijkstra, 1997). In the case of the fractionation process, which has been proclaimed to be 'an art' rather than science (Tirtiaux, 1990), it is still insufficiently clear how to control the crystal morphology in such a way that crystals with the required filterability are grown in a reproducible manner. As with hydrogenation (Colen et al., 1990), operating a fractionation plant more or less amounts to repeating what worked before and hoping for the best. In this context, fractionation has the advantage over hydrogenation that batches that do not live up to expectation can always be reprocessed by mixing the fractions obtained and, again, hoping for the best.

Assuming that the subjects covered by the scientific trade journals in their recent volumes are more or less representative of current research efforts, it is unlikely that the lack of insight mentioned above will soon be remedied. 'Sexier' subjects such as the use of enzymes or functional foods monopolise the table of contents of such journals. Moreover, most industrial companies that formerly prided themselves on the strength of their research and development (R&D) effort and effectively contributed to gaining insight into what happens during edible-oil processing, patented their inventions and published their achievements now pay far less attention to internal R&D aimed at understanding what they are doing in their industrial plants.

However, process and product improvements can still be achieved fairly easily from the evaluation and application of past, possibly underutilised, achievements. Patents expire or are not maintained by their assignees so that formerly protected processes can be freely used. All that is needed is an evaluation of a process that has already been invented and perhaps even developed and industrialised. Besides, since local circumstances differ, a patented process may well have specific advantages to other companies that were not relevant to the assignee.

Consequently, an awareness of the literature and especially of patents can provide a means to counteract the otherwise stifling effects of the 'reorientation' of internal R&D effort. In this context, the use of processes that have been developed outside the field of edible oils and fats may well prove to be fertile. Examples of such processes are the crystallisation column (Arkenbout, 1981; Arkenbout *et al.*, 1981; Thijssen and Arkenbout, 1981) and/or the continuous column chromatography of triglycerides. Also the use of adjuvants such as long-chain fatty-acid esters and other materials as reported in the literature (Baur, 1962) or partially esterified sucrose (Schmid and Baur, 1962) merits further investigation. . .perhaps this is the reason why the references listed below are quite numerous.

References

Anon. (1997) FDA affirms menhaden oil as GRAS. *INFORM*, **8**, 858.

Ahmad, M.M., Priestley, T.M. and Winterbottom, J.M. (1979) Palladium-catalyzed hydrogenation of soybean oil. *J. Am. Oil Chem. Soc.*, **56**, 571-577.

Albright, L.F. and Wisniak, J. (1962) Selectivity and isomerization during partial hydrogenation of cottonseed oil and methyl oleate: effect of operating variables. *J. Am. Oil Chem. Soc.*, **39**, 14-19.

Amer, M.A., Kupranycz, D.B. and Baker, B.E. (1985) Physical and chemical characteristics of butterfat fractions obtained by crystallisation from molten fat. *J. Am. Oil Chem. Soc.*, **63**, 1551-1557.

Andrikides, A.L. (1960) Séparation et purification des acides gras, des glycérides et des alcools gras, FR Patent 1.251.738, assigned to Norco Products Company.

Arkenbout, G.J. (1981) Crystallization column, US Patent 4,257,796, assigned to Nederlandse Centrale Organisatie voor Toegepast Natuurwetenschappelijk Onderzoek.

Arkenbout, G.J., van Kuijk, A., van der Meer, J. and Schneiders, L.H.J.M. (1981) Pulsed crystallization column and method of countercurrent crystallization, US Patent 4,400,189, assigned to Nederlandse Centrale Organisatie voor Toegepast Natuurwetenschappelijk Onderzoek.

Asbeck, L.S. and Segers, J.C. (1990) Dewaxing of dried oil, EU Patent 0 397 233, assigned to Unilever.

Bailey, A.E. and Fisher, G.S. (1946) Modifications of vegetable oils. V. Relative reactivities toward hydrogenation of the mono-, di- and triethenoid acids in certain oils. *Oil & Soap*, **23**, 14-18.

Baliga, B.P. and Shitole, A.D. (1981) Cocoa butter substitutes from mango fat. *J. Am. Oil Chem. Soc.*, **58**, 110-114.

Baltes, J. (1972) Process for the selective hydrogenation of fats and fatty acids, US Patent 3,687,989.

Baltes, J. (1975) Process for manufacture of edible, flowable suspensions or solid glycerides in liquid glycerides or mixtures of liquid and solid glycerides and products obtained thereby, US Patent 3,870,807, assigned to Harburger Oelwerke Brinckman & Mergell.

Baur, F.J. (1962) Crystallization process, US Patent 3,059,008, assigned to The Procter and Gamble Company.

Bernardini, E. and Bernardini, M. (1975) Palm oil fractionation and refining using the C.M.B. process. *Oléagineux*, **30**, 121-128.

Bernardini, M. (1978) Fractionnement continue par solvent de l'huile de palme. *Oléagineux*, **33**, 297-305.

Beyens, Y. and Dijkstra, A.J. (1983) Positional and triglyceride selectivity of hydrogenation of triglyceride oils, in *Fat Science 1983, Proceeding of 16th ISF congress, Budapest* (ed. J. Holló), pp. 425-432.

Bradland, A. (1983) Production of a dairy emulsion, EU Patent 0 095 001, assigned to Arthur Bradland.

Breeding, C.J. and Marshall, R.T. (1995) Crystallization of butter oil and separation by filter-centrifugation. *J. Am. Oil Chem. Soc.*, **72**, 449-453.

Bushell, W.J. and Hilditch, T.P. (1937) The course of hydrogenation in mixtures of mixed glycerides. *J. Chem. Soc.*, 1767-1774.

Bussey, D.M., Ryan, T.C., Gray, J.I. and Zabik, M.E. (1981) Fractionation and characterization of edible tallow. *J. Food Sci.*, **46**, 526-530.

Coenen, J.W.E. (1974) Fractionnement et interestérification des corps gras dans la perspective du marché mondial des matières premières et des produits finis I—Fractionnement. *Revue française des Corps Gras*, **21**, 343-349.

Coenen, J.W.E. (1976) Hydrogenation of edible oils. *J. Am. Oil Chem. Soc.*, **53**, 382-389.

Coenen, J.W.E. (1978) The rate of change in the perspective of time. *Chem. & Ind.*, 709-722.

Coenen, J.W.E. and Boerma, H. (1968) Absorption der Reaktionspartner am Katalysator bei der Fetthydrierung. *Fette Seifen Anstrichmittel*, **70**, 8-14.

Colen, G.C.M., van Duijn, G. and Keltjens, J.C.M. (1990) Hydrogenation method, UK Patent Application 2 230 020 A, assigned to Unilever.

Coppa-Zuccari, G. (1971) Le fractionnement et l'hydrogénation des graisses et acides gras. *Oléagineux*, **26**, 405-409.

Cox, R., Lebrasseur, J., Michiels, E., Buijs, H., Li, H., van de Voort, F.R., Ismail, A.A. and Sedman, J. (2000) Determination of iodine value with a Fourier transform-Near Infra Red based global calibration using disposable vials: an international collaborative study. *J. Am. Oil Chem. Soc.*, **77**, 1229-1234.

Deffense, E.M.J. (1985) Fractionation of palm oil. *J. Am. Oil Chem. Soc.*, **62**, 376-385.

Deffense, E.M.J. (1987) Multi-step butteroil fractionation and spreadable butter. *Fett Wissenschaft Technologie*, **89**, 502-507.

Deffense, E.M.J. (1993) Milk fat fractionation today: a review. *J. Am. Oil Chem. Soc.*, **70**, 1193-1201.

Deffense, E.M.J. (1995) Dry multiple fractionation: trends in products and applications. *Lipid Technology*, **7**, 34-38.

Deffense, E.M.J. (2000) Dry fractionation technology in 2000. *European Journal of Lipid Science and Technology*, **1**, 234-236.

Denise, J. (1987) Le décirage des huiles de tournesol. *Revue française des Corps Gras*, **34**, 133-138.

Deroanne, C., Wathelet, J.P. and Severin, M. (1976) Étude de la structure des triglycérides de l'huile de palme fractionnée II—Évolution de la crystallisation des triglycérides lors du refroidissement de l'huile de palme en vue de son fractionnement par le procédé Tirtiaux. *Revue française des Corps Gras*, **23**, 28-32.

Dieffenbacher, A. (1986) Fat fractionation, US Patent 4,594,194, assigned to Nestec S.A.

Dijkstra, A.J. (1997) Hydrogenation revisited. *INFORM*, **8**, 1150-1158.

Dijkstra, A.J. (1998) Alternativas a la hidrogenación. *Revista Aceites y Grasas*, **8**, 356-357.

Dijkstra, A.J., Maes, P.J., Meert, D. and Meeussen, W.L.J. (1996) Interpreting the oil stability index. *Oléagineux Corps Gras Lipides*, **3**, 378-386.

Dimick, P.S., Yella Reddy, S. and Ziegler, G.R. (1996) Chemical and thermal characteristics of milk-fat fractions isolated by a melt crystallization. *J. Am. Oil Chem. Soc.*, **73**, 1647-1652.

Duveen, R.F. and Leuteritz, G. (1982) Der BUSS-Schleifenreaktor in der Öl- und Fetthärtungsindustrie. *Fette Seifen Anstrichmittel*, **84**, 511-515.

Evans, C.D., Beal, R.E., McConnell, D.G., Black, L.T. and Cowan, J.C. (1964) Partial hydrogenation and winterization of soybean oil. *J. Am. Oil Chem. Soc.*, **41**, 260-263.

Feuge, R.O., Cousins, E.R., Fore, S.P., DuPré, E.F. and O'Connor, R.T. (1953) Modification of vegetable oils. XV. Formation of isomers during hydrogenation of methyl linoleate. *J. Am. Oil Chem. Soc.*, **30**, 454-460.

Gee, W.P. (1948) Fractional separation of fatty oil substances, US Patent 2,450,235, assigned to Texaco Development Corporation.

Gibble, W.P. and Rhee, J.S. (1976) Dewaxing of vegetable oils, US Patent 3,994,943, assigned to Hunt-Wesson Foods Inc.

Goh, E.M. and Timms, R.E. (1985) Determination of mono- and diglycerides in palm oil, olein and stearin. *J. Am. Oil Chem. Soc.*, **63**, 730-734.

Hamm, W. (1986) Fractionation: with or without solvents? *Fette Seifen Anstrichmittel*, **88**, 533-537.

Hamm, W. (1995) Trends in edible oil fractionation. *Trends in Food Science & Technology*, **6**, 121-126.

Haraldsson, G. (1979) Pretreatment and fractionation of industrial fats. *La Rivista Italiana delle Sostanze Grasse*, **56**, 325-331.

Haraldsson, G. (1981) Selective hydrogenation and subsequent fractionation of oil. *La Rivista Italiana delle Sostanze Grasse*, **58**, 491-495.

Hartel, R.W. (1996) Applications of milk-fat fractions in confectionary products. *J. Am. Oil Chem. Soc.*, **73**, 945-953.

Hashimoto, K., Muroyama, K. and Nagat, S. (1971) Kinetics of the hydrogenation of fatty oils. *J. Am. Oil Chem. Soc.*, **48**, 291-295.

Horiuti, I. and Polanyi, M. (1934) Exchange reactions of hydrogen on metallic surfaces. *Trans. Faraday Soc.*, **30**, 1164.

Iida, M., Kato, C., Ohshima, S. and Yazawa, Y. (1981) Method for fractionating an oil or fat to separate the high melting point components thereof, US Patent 4,265,826, assigned to Ajinomoto Company Inc.

Jasko, J.J. and Domek, S.M. (1984) Hard butter and process for making same, US Patent 4,465,703, assigned to SCM Corporation.

Jonker, G.H. (1999) *Hydrogenation of Edible Oils and Fats*, PhD thesis, Rijksuniversiteit Groningen, Groningen.

Kaufmann, H.P. (1939) Wilhelm Normann. *Fette und Seifen*, **46**, 259-264.

Kawada, T. and Matsui, N. (1970) Process for the fractionation of oils and fats, US Patent 3,541,123, assigned to Kao Soap Co.Ltd.

Kaylegian, K.E. (1999) Contempory issues in milk fat fractionation. *Lipid Technology*, 132-136.

Kellens, M. (2000) Oil modification processes, in *Edible Oil Processing* (eds. W. Hamm and R.J. Hamilton), Sheffield Academic Press, Sheffield, pp. 127-173.

Kheiri, M.S.A. (1985) Present and prospective development in the palm oil processing industry. *J. Am. Oil Chem. Soc.*, **62**, 210-219.

Koetsier, W.T. (1997) Hydrogenation of edible oils, technology and applications, in *Lipid Technologies and Applications* (eds. F.D. Gunstone and F.B. Padley), Marcel Dekker, New York, Basel, Hong Kong, pp. 265-303.

Kokken, M.J. (1991) Obtention et emploi des corps gras fractionnés à sec (corps gras végétaux, matières grasses laitières). *Revue française des Corps Gras*, **38**, 367-376.

Koritala, S., Butterfield, R.O. and Dutton, H.J. (1973) Kinetics of hydrogenation of conjugated triene and diene with nickel, palladium, platinum and copper-chromite catalysts. *J. Am. Oil Chem. Soc.*, **50**, 317-320.

Kozempel, M.F., Craig, J.C. Jr, Heiland, W.K., Elias, S. and Aceto, N.C. (1981) Development of a continuous process to obtain a confectionary fat from tallow: final status. *J. Am. Oil Chem. Soc.*, **58**, 921-925.

Kreulen, H.P. (1976) Fractionation and winterization of edible fats and oils. *J. Am. Oil Chem. Soc.*, **53**, 393-396.

Lacoste, F., Raoux, R. and Mordret, F. (1999) Comparison of Rancimat stability test and ambient storage of edible oil. Paper presented at the 23rd World Congress and Exhibition of the International Society for Fat Research, Brighton.

Lampi, A.-M., Hopia, A. and Piironen, V. (1997) Antioxidant activity of minor amounts of γ-tocopherol in natural triglycerides. *J. Am. Oil Chem. Soc.*, **74**, 549-555.

Linsen, B.G. (1971) Selektivität bei der Fetthydrierung, früher und heute I. *Fette Seifen Anstrichmittel*, **73**, 411-416.

Little, T.H. (1963) Procédé pour la séparation des huiles glycéridiques sans solvent, FR Patent 1.344.962, assigned to The Sharples Corporation.

Luddy, F.E. and Longhi, S. (1983) Azeotropic fat fractionation, EU Patent 0.132 506, assigned to Société des Produits Nestlé S.A.

Luddy, F.E., Hampson, J.W., Herb, S.F. and Rothbart, H.L. (1973) Development of edible tallow fractions for specialty fat uses. *J. Am. Oil Chem. Soc.*, **50**, 240-244.

Maes, P.J. and Dijkstra, A.J. (1983) Process for separating solids from oils, EU 88 949, assigned to N.V. Vandemoortele International.

Maes, P.J., Dijkstra, A.J. and Seynave, P. (1994) Method for dry fractionation of fatty substances, European Patent 0 651 046, assigned to N.V. Vandemoortele International.

Neumunz, G.M. (1978) Old and new in winterizing. *J. Am. Oil Chem. Soc.*, **55**, 396A-398A.

Normann, W. (1902) British Patent 1 515, assigned to Herforder Maschinenfett- und Ölfabrik Leprince und Siveke.

Padley, F.B. (1997) Chocolate and confectionary fats, in *Lipid Technologies and Applications* (eds. F.D. Gunstone and F.B. Padley), Marcel Dekker, New York, pp. 391-432.

Pallmar, A.B. and Sarebjörk, K.W.H. (1981) Procédé de traitement des huiles végétales en vue de leur conservation à basse température, FR Patent 2 460 996, assigned to Alfa-Laval AB.

Pike, M., Barr, I.G. and Tirtiaux, F. (1978) Procédé de fabrication d'un succédané de beurre de cacao, FR Patent 2 369 800, assigned to Charles Lester & Co.Ltd and Fractionnement Tirtiaux S.A.

Porter Lee, A. (1939) The winterizing of cottonseed oil. *Oil & Soap*, **16**, 148-150.

Reck, J.H.M. (1973) Procédé de fractionnement de triglycérides d'acides gras, FR Patent 2.173.296, assigned to Unilever.

Rivarola, G., Añón, M.C. and Calvelo, A. (1985) Crystallization of waxes during sunflowerseed oil refining. *J. Am. Oil Chem. Soc.*, **62**, 1508-1513.

Rossell, J.B. (1985) Fractionation of lauric oils. *J. Am. Oil Chem. Soc.*, **62**, 385-390.

Rutledge, D.N., El-Khaloui, M. and Ducauze, C.J. (1988) Contribution à l'étude d'une méthode de contrôle rapide de la qualité des margarines par RMN-IBR. *Revue française des Corps Gras*, **35**, 157-162.

Schilling, K. (1968) Über den Verlauf der Hydrierung von Triolein an Raney-Nickel. *Fette Seifen Anstrichmittel*, **70**, 389-393.

Schilling, K. (1978) Der Reaktionsverlauf bei der Hydrierung von Triglyceriden: Simultanhydrierung von Tri- und Monolinolenin. *Fette Seifen Anstrichmittel*, **80**, 312-314.

Schilling, K. (1981) Der Reaktionsverlauf bei der Hydrierung von Triglyceriden, Glykolester als Modellsubstrat. *Fette Seifen Anstrichmittel*, **83**, 226-228.

Schmid, D.F. and Baur, F.J. (1962) Fat crystallization process, US Patent 3,059,010, assigned to The Procter and Gamble Company.

Siew, W.-L. and Ng, W.-L. (1995) Partition coefficient of diglycerides in crystallization of palm oil. *J. Am. Oil Chem. Soc.*, **72**, 591-595.

Smith, P.R. (2000) The effects of phospholipids on crystallisation and crystal habit in triglycerides. *European Journal of Lipid Science and Technology*, 1, 122-127.

Smorenburg, J.J. (1972) Werkwijze voor het fractioneren van vetzuren en/of esters van vetzuren. Dutch Patent 7103763, assigned to Stork Amsterdam N.V.

Stenberg, O. and Schöön, N.-H. (1985) Aspects of the graphical determination of the volumetric mass-transfer coefficient (k_La) in liquid-phase hydrogenation in a slurry reactor. *Chem. Eng. Sci.*, 40, 2311-2319.

Taylor, A.M. (1976) The crystallisation and dry fractionation of Malaysian palm oil. *Oléagineux*, 31, 73-79.

Thijssen, H.A.C. and Arkenbout, G.J. (1981) Process for the continuous partial crystallization and the separation of a liquid mixture, US Patent 4,666,456, assigned to Nederlandse Centrale Organisatie voor Toegepast Natuurwetenschappelijk Onderzoek.

Thomas, A.E. and Paulicka, F.R. (1976) Solvent fractionated fats. *Chemistry & Industry*, 774-770.

Thompson, S.W. (1951) Relative hydrogenation rates of normal and conjugated linolenic and linoleic acid glycerides. *J. Am. Oil Chem. Soc.*, 28, 339-341.

Timms, R.E. (1986) Processing of palm kernel oil. *Fette Seifen Anstrichmittel*, 88, 294-300.

Timms, R.E. (1991) Crystallisation of fats. *Chemistry & Industry*, 342-345.

Timms, R.E. (1997) Fractionation, in *Lipid Technologies and Applications* (eds. F.D. Gunstone and F.B. Padley), Marcel Dekker, New York, pp. 199-222.

Tirtiaux, A. (1990) Dry fractionation: a technique and an art, in *Edible Fats and Oils Processing: Basic Principles and Modern Practices* (ed. D.R. Erickson), American Oil Chemists' Society, Champaign, IL, pp. 136-141.

Tirtiaux, F. (1976) Le fractionnement industriel des corps gras par crystallisation dirigée—procédé Tirtiaux. *Oléagineux*, 31, 279-285.

Urosevic, D. (1986) Improvement of the Buss loop reactor in the oil and fat industry, in *World Conference on Emerging Technologies in the Fats and Oils Industry* (ed. A.R. Baldwin), American Oil Chemists' Society, Champaign, IL, pp. 128-132.

van Putte, K.P.A.M. and Bakker, B.H. (1987) Crystallization Kinetics of Palm Oil. *J. Am. Oil Chem. Soc.*, 64, 1138-1143.

Veldsink, J.W., Bouma, M.J., Schöön, N.-H. and Beenackers, A.A.C.M. (1997) Heterogeneous hydrogenation of vegetable oils: a literature review. *Catal. Rev.—Sci. Eng.*, 39, 253-318.

Veldsink, J.W., Jonker, E.W. and Beenackers, A.A.C.M. (1999) Hydrogenation kinetics of vegetable oils: review and recent developments. Paper presented at the 23rd World Congress and Exhibition of the International Society for Fat Research, Brighton.

von Rappard, G. and Plonis, G.F. (1979) Fett mit einem hohen Gehalt an 1,3Di-Dipalmitoyl-2-oleyl-glycerin, Verfahren zu seiner Gewinnung und seine Verwendung, DE Patent 27 47 765, assigned to Walter Rau Lebensmittelwerke.

Wang, F.-S. and Lin, C.-W. (1995) Turbidimetry for crystalline fractionation of lard. *J. Am. Oil Chem. Soc.*, 72, 585-589.

Weise, M. and Delaney, B. (1992) Plants using new hydrogenation agitator design. *INFORM*, 3, 817-820.

Willner, T., Sitzmann, W. and Münch, E.-W. (1989) Herstellung von Kakaobutterersatz durch fraktionierte Speiseölkristallisation. *Fett Wissenschaft Technologie*, 91, 586-592.

Willner, T. and Weber, K. (1994) High-pressure dry fractionation for confectionary fat production. *Lipid Technology*, 6, 57-60.

Wilp, Ch. (2000) Dry fractionation of fats and oils by means of centrifugation. Paper presented at the 91st AOCS Annual Meeting & Expo, San Diego.

Yella Reddy, S. and Prabhakar, J.V. (1994a) Cocoa butter extenders from kokum (garcinia indica) and phulwara (Madhuca butyracea) butter. *Journal of the American Oil Chemists' Society*, 71, 217-219.

Yella Reddy, S. and Prabhakar, J.V. (1994b) Confectionery fat from Phulwara (Madhuca butyracea) butter. *Fett Wissenschaft Technologie*, 96, 387-390.

5 Fats for chocolate and sugar confectionery

Ian M. Stewart and Ralph E. Timms

5.1 Introduction

The traditional fat ingredients for chocolate and sugar confectionery are cocoa butter (CB), milk fat (MF) and certain vegetable oils, principally coconut oil (CN). Nowadays, there is a much wider range of fats available, designed mainly to replace the expensive cocoa butter and milk fat, but also coconut oil. We term these 'alternative fats' (AF). Alternative fats may be designated by their application (e.g. milk-fat replacer, cocoa-butter extender) but the more systematic approach adopted in this chapter is to designate them according to their chemical composition, which is the fundamental basis of their properties. Three types of alternative fat can then be distinguished: symmetrical/SOS, hardened/high-trans, lauric-type (Gordon *et al.*, 1979).

Symmetrical/SOS fats are composed of the same triglycerides—POP, POS, SOS (see appendix for nomenclature)—that are found in cocoa butter. Because the unsaturated oleic acid is at the 2 position and the saturated palmitic and stearic acids are at the 1 and 3 positions, the triglycerides are described as symmetrical. These triglycerides are obtained by selecting natural fats that contain them.

Hardened/high-trans fats are complex mixtures of triglycerides with fatty acids of carbon number 16 and 18 (C_{16} and C_{18}) where the unsaturated acids have their double bonds mainly in the *trans* configuration. These triglycerides are produced by hardening (hydrogenating) liquid oils under very selective, *trans*-promoting, conditions.

Lauric-type fats are complex mixtures of triglycerides based on mainly saturated fatty acids with carbon number 8 to 18 (C_8–C_{18}). Lauric acid constitutes about 50% of all the fatty acids. The triglycerides are obtained from the two main fats containing lauric acid: coconut and palm kernel oils.

In the following sections, we describe the production and properties of cocoa butter, milk fat and the various alternative fats. We then discuss the legal and regulatory aspects governing their use in chocolate before describing their various applications in chocolate and sugar confectionery. In a chapter of this length it is not possible to go into great detail. We have aimed to cover all aspects of the topic succinctly, while providing adequate references to more detailed reviews of particular topics.

5.2   Production and properties

5.2.1   Cocoa butter and milk fat

Cocoa butter is the fat extracted from the seed of the *Theobroma cacao* tree
(Beckett, 1999; Minifie, 1989). These seeds, which are referred to as cocoa
beans, consist of about 15% shell and 85% cotyledon (referred to as 'nib'), of
which about 55% is fat. To extract the fat, the nibs are ground to a paste called
cocoa liquor or mass and the fat is then extracted by hydraulic pressing, screw
expelling or solvent extraction. The highest quality cocoa butter is produced
by hydraulic pressing and is called 'pure prime pressed' cocoa butter. Solvent
extraction is used only for extraction of cake residues from the expeller process or of other, waste, residues. The cocoa butter produced is invariably of
lower quality and contains higher levels of nontriglyceride materials such as
phospholipids and pesticides.

At some stage in the processing of the beans they are roasted to produce
the desirable chocolate flavour. Whole beans, nibs or liquor may be roasted
(Beckett, 2000). Typically, 100 g of beans produces 40 g of cocoa butter, 40 g
of cocoa powder (the residue after extraction, which contains 10–24% fat, most
commonly 10–12%) and 20 g of waste materials such as shell, moisture and dirt
(Timms and Stewart, 1999).

Most of the cocoa butter added to chocolate, particularly to milk chocolate,
which consists of about 80% of total consumption, is deodorised. The main
purpose of deodorisation is to moderate the flavour by removing harsh or acidic
flavours and off-flavours. A subsidiary purpose is to sterilise the fat. Cocoa
butter is deodorised at temperatures of 130–180°C by passing steam through
the oil under a vacuum of 1–5 mbar for 10–30 min. [In older, batch, processes
operating at lower temperatures (105°C) and poorer vacuums (up to 40 mbar),
cocoa butter could be deodorised for as long as 3 h (Hanneman, 2000).] Under
these conditions, deodorisation has no effect on the physical properties (Timms
and Stewart, 1999). There may be a small (less than 0.5%) reduction in free
fatty acids.

The milk fat used in confectionery may be added as pure fat [butter oil
or anhydrous milk fat (AMF)], as whole milk powder containing 26% fat, as
skim milk powder containing less than 1% fat, or as milk crumb containing
varying levels of milk fat and cocoa butter, typically 7% and 9%, respectively
(Jackson, 1995; Minifie, 1989). In recent years other dried ingredients have
become widely available: buttermilk powder (8% fat), whey powder (91% fat)
and high-protein and high-fat powders (Haylock, 1995). Figure 5.1 shows the
schematic composition of the four main milk ingredients. Traditionally, butter
or even cream was used for the manufacture of sugar confectionery such as
toffees, caramels or fudges. Butter and especially cream require particular

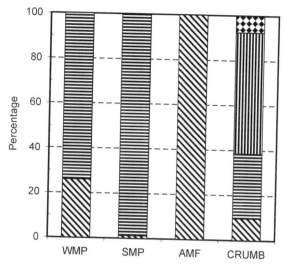

Figure 5.1 The main milk-fat-containing ingredients used in chocolate. Note: WMP, whole milk powder; SMP, skim milk powder; AMF, anhydrous milk fat; CRUMB, milk crumb. ⊠ Milk fat; ▤ Nonfat solids; ⊞ Sugar; ▣ Cocoa butter.

storage conditions and hygienic handling. Also, the water in these products must be evaporated during confectionery manufacture. It is therefore much more convenient to use AMF (Jackson, 1995).

A typical customer specification for cocoa butter is given in table 5.1. This example relates to deodorised cocoa butter, but the physical and chemical properties as opposed to the sensory properties are little affected by deodorisation. For the fat technologist, the most important properties of a fat are its melting properties, usually indicated by solid fat content (SFC), and its rate of crystallisation, in this case indicated by the Jensen cooling curve (figure 5.2). Cooling curves may also be obtained by using the Shukoff method (IUPAC standard method 2.132; IOCCC method 31-1998), by differential scanning calorimetry or by measurement of SFC over time. There are also proprietary methods such as the Barry Callebaut cooling curve (Hanneman, 2000).

Milk fat is a much softer fat than cocoa butter, as indicated in figure 5.3 in which their SFCs are compared. Both fats vary widely in their properties, depending on their origins (Cullinane et al., 1984; MacGibbon and McLennan, 1987; Padley, 1997; Shukla, 1995). Both fats are also fractionated by crystallisation into fractions of different hardness, but only milk fat fractions are readily available commercially (Versteeg et al., 1994; Weyland, 1992).

Table 5.1 Typical specification for deodorised cocoa butter (Timms and Stewart, 1999)

Parameter	Specification	Method
Appearance	Clear	Visual
Odour	Characteristic, free from rancid, smoky and foreign odours	Taste panel; particular odour and flavour required may need to be agreed with customer
Flavour	Characteristic, free from rancid, smoky and foreign flavours	Taste panel; particular odour and flavour required may need to be agreed with customer
Moisture (%)	0.05 maximum	Karl Fischer
Free fatty acid (% as oleic)	1.75 maximum	Titration (e.g. AOCS Ca 5a-40)
Diglyceride (%)	2.5 maximum	Gas–liquid chromatography after silylation
Unsaponifiable matter (%)	0.35 maximum	IOCCC method 23-1988, for example
Peroxide value (meq kg^{-1})	1.0 maximum	Titration (e.g. AOCS Cd 86-90)
Rancimat induction period (120°C)	32 h	Some companies specify different temperatures, but this should meet all requirements
Iodine value	34–38	Titration (e.g. AOCS Cc 13c-92)
Blue value	0.04 maximum	Based on IOCCC method 29-1988
E (1%) 270 nm	0.35 maximum 0.14 max after alkali wash	Based on IOCCC methods 18-1973 and 19-1973
Solid fat content (%)		
20°C	76.0 minimum	Pulsed nuclear magnetic resonance, with tempering at 26°C for 40 h (e.g. BS684: 2.22)
25°C	70.0 minimum	
30°C	42.0 minimum	
35°C	1.0 maximum	
Jensen temperature (°C)		Manual cooling curve (BS684: 1.13)
T_{max}	29.5 minimum	
T_{min}	35–55	
Rise	4.5–6.0	
Total plate count	1000 g^{-1} maximum	Standard microbiological procedure
Origin processing	As agreed with customer Deodorised to flavour agreed with customer; Filtered	Deodorised; no crude cocoa butter allowed

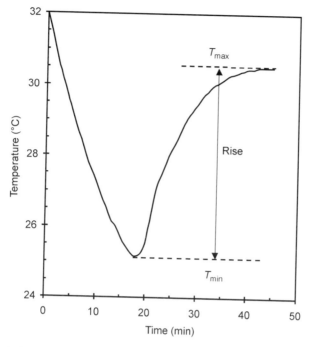

Figure 5.2 Typical Jensen cooling curve of cocoa butter T_{max}, T_{min}, maximum and minimum temperatures, respectively.

A unique property of cocoa butter is its polymorphism[1]. Six polymorphs or crystal forms exist, but only three forms—form IV (β'-2), form V (β_2-3), form VI (β_1-3)—are important in the commercial production of chocolate. Form V is the characteristic form for chocolate, produced when it is tempered (controlled crystallisation) during production. It may be considered the stable form for practical purposes. Form IV is characteristic of untempered chocolate, and form VI is a transformation of form V associated with bloom, the greyish-white discoloration sometimes found on the surface of chocolate (see section 5.4.5).

The properties of cocoa butter, as with all fats, are determined by its triglyceride composition, which is unusually simple. As shown in figure 5.4, three triglycerides compose about 80% of the total. These three triglycerides—POP, PQS and SOS—are themselves so similar that they form a single solid solution and behave almost like a pure compound, which results in the very sharp melting

---

[1] Polymorphism is the phenomenon of multiple melting points when the fat has several possible crystal packings or polymorphs. For fats, there are three basic polymorphs: α, β' and β. These can be further subdivided by adding suffixes 2 or 3 to indicate whether the unit cells or the repeat units are two or three fatty-acid chains long. Finally, subscripts are added where there is more than one polymorph of each type, with 1 indicating the highest-melting polymorph.

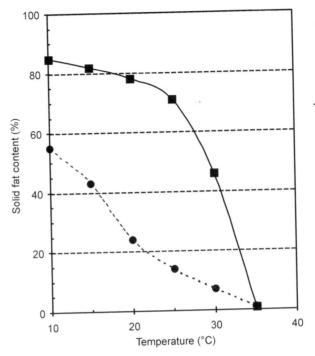

Figure 5.3 Solid fat melting curves of cocoa butter and milk fat —■— Cocoa butter; --●-- Milk fat.

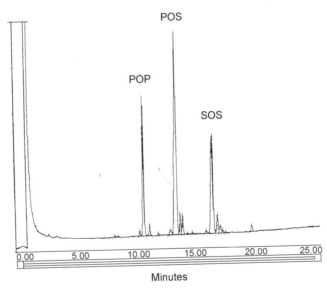

Figure 5.4 Typical triglyceride composition of cocoa butter by high-resolution gas chromatography.

curve shown in figure 5.2. In contrast, milk fat is a very complex mixture. Whereas four fatty acids—palmitic, stearic, oleic and linoleic—compose 98% of the cocoa butter fatty acids, milk fat contains at least 12 fatty acids present at more than 1%, ranging from carbon number 4 to 20 (C_4–C_{20}), including several odd carbon numbers not found in vegetable oils. The triglyceride composition is even more complex and hundreds of triglycerides have been identified. The result is a complex mixture with a relatively flat melting curve and simple polymorphism consisting mainly of one polymorph, β' (Timms, 1984).

Since, in milk chocolate, milk fat and cocoa butter are mixed together in the continuous fat phase, the final information we need to understand their functionality is the properties of their mixtures. As would be expected from the SFCs shown in figure 5.3, when increasing amounts of milk fat are added to cocoa butter the mixtures become softer (i.e. have lower SFCs), at all temperatures. Because the two fats have different stable polymorphs they cannot be expected to mix completely in the solid state (see section 5.2.5). The phase diagram of mixtures of cocoa butter and milk fat is shown in figure 5.5. The point to note is that adding milk fat to cocoa butter does not change the β polymorph of cocoa butter until about 50% is added. Since chocolate rarely contains more than 30% milk fat in the fat phase, this means that in practice the properties of the cocoa butter predominate and the effect of the milk fat can be mostly ignored except for its softening effect (Timms, 1980).

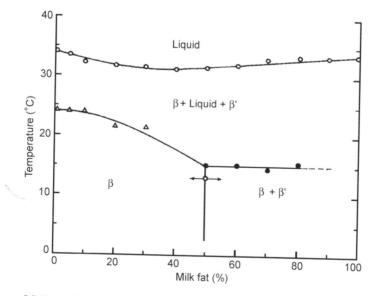

Figure 5.5 Phase diagram of mixtures of cocoa butter and milk fat. Redrawn from Timms, 1980.

5.2.2 Symmetrical/SOS-type alternative fats

The triglyceride composition of cocoa butter was known in the 1950s, which allowed Unilever to develop an alternative fat based on the assumption that if the triglyceride composition of cocoa butter could be replicated then its physical properties would be too. The subsequent patent (Unilever, 1956) discloses that the required triglycerides can be obtained from palm oil, illipé, shea and other fats from tropical countries. The patent also discloses the use of solvent fractionation to concentrate the required triglycerides, (e.g. POP from palm oil). Such fats were called cocoa butter equivalents (CBEs) as they were substantially equivalent to cocoa butter in all aspects of functionality.

The fractionation of fats by crystallisation and separation of the crystals has been reviewed elsewhere (Timms, 1997). To summarize, the fat is melted and then slowly cooled to produce crystals, which are then separated by some form of filtration. It has been shown that the fundamental efficiency of separation of the triglycerides or triglyceride phases is not affected by the solvent used, although polar solvents do affect the separation of diglycerides from triglycerides (Timms, 1983). Solvents are beneficial in diluting the crystal miscella and lowering the viscosity, thus aiding the mechanical separation of the crystals from the liquid. Partly because of the lack of fundamental effect and partly because of improved mechanical separation techniques, dry (i.e. without solvent) fractionation is now overwhelmingly the preferred technique. Nevertheless, solvent fractionation is still used for the concentration of POP, POS and SOS triglycerides because the benefits mentioned are particularly valuable for the scarce and high-cost tropical oil feedstocks. The polar solvent acetone is preferred to hexane because of its effectiveness in removing diglycerides and other polar lipids, which are known to be detrimental to the functionality of chocolate fats (Okawachi et al., 1985; Tietz and Hartel, 2000).

The raw materials and processes required to concentrate the desired triglycerides are summarised in table 5.2. The six raw materials shown are the commonest used and are now also the only ones that will be allowed for use in chocolate sold within the European Union according to recent legislation (European Parliament and Council, 2000). Except for palm oil, the raw materials are all derived from wild or semiwild trees growing in relatively poor and undeveloped tropical countries.

Although we have shown palm oil as producing a mid-fraction, it should be noted that a wide range of palm mid-fractions is now commercially available, as shown in figure 5.6, in which they are compared with cocoa butter and milk fat.

As shown in table 5.3, these various raw materials and their fractions are then blended together to produce a triglyceride composition similar to the composition of cocoa butter. The 'original' CBE consisted of approximately equal parts of a palm mid-fraction and illipé. As can be seen, although it contains about the same total symmetrical triglycerides as cocoa butter, it differs in the proportions

Table 5.2 Raw materials and processing required to produce ingredients for symmetrical/SOS alternative fat (AF) blends

Raw material	Origin	Processing required[a]	Ingredient used in AF blend	Main triglycerides
Palm oil	Indonesia, Malaysia	Two fractionations[b]	Mid-fraction[c]	POP
Shea butter	West Africa (dry tropics)	Degumming and one fractionation	Stearin[d]	SOS
Illipé butter[e]	Borneo	None	Fat	SOS, POS
Kokum butter[f]	India	None	Fat	SOS
Sal fat	India	One fractionation	Stearin	SOS, SOA
Mango kernel fat	India	One fractionation	Stearin	SOS, POS

[a]Additional to refining, bleaching and deodorising.
[b]Fractional crystallisation.
[c]Middle-melting fraction.
[d]Stearin is a high-melting fraction.
[e]Also called Borneo tallow or Tengkawang fat.
[f]Also called Kokum Gurgi fat.
Note: for triglyceride nomenclature, see appendix.

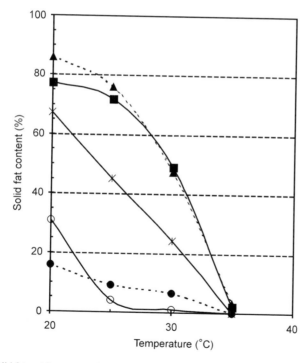

Figure 5.6 Solid fat melting curves of palm mid-fractions (PMFs) compared with those of cocoa butter and milk fat. Note: IV, iodine value; —■— Cocoa butter; --●-- Milk fat; --▲-- PMF (IV = 34); —✳— PMF (IV = 40); —⊖— PMF (IV = 50).

Table 5.3 Triglyceride compositions of symmetrical/SOS alternative fat (AF) ingredients and formulation of AFs

Ingredient fat	Typical content (%)		
	POP	POS	SOS
Cocoa butter	16	38	23
Palm mid-fraction	57	11	2
Illipé butter	9	29	42
Shea stearin	3	10	63
Original cocoa butter equivalent[a]	33	20	22
Modern cocoa butter equivalent	32	15	28

[a] 50% illipé plus 50% palm mid-fraction.

of the individual triglycerides. This is partly a question of economics—palm oil is readily available at a moderate price—and partly a question of chemistry—no natural fat contains as much POS as cocoa butter. Thus all blends will be deficient in POS relative to cocoa butter. Later developments, as illustrated by the 'modern' CBE, included shea stearin and other SOS-rich raw materials, leading to a further diminution in the amount of POS.

A large amount of information about symmetrical/SOS alternative fats and their formulation, especially using palm mid-fractions and illipé, is to be found in two books by Wong (1988, 1991).

5.2.3 High-trans-type alternative fats

These fats are produced by the hydrogenation of widely available nonlauric liquid oils such as soybean and rapeseed oils. When used in chocolate recipes they are often called cocoa butter replacers (CBRs). They are hydrogenated either singly or blended with palm olein, the liquid fraction from the fractionation of palm oil. Hydrogenation generally takes place with use of a selective nickel catalyst and with limited hydrogen availability in order to ensure a high content of trans acids (over 45% and mainly elaidic acid).

The melting profile can be altered by varying the blend proportions of the hydrogenation feedstock and/or the catalyst and conditions used. Despite this, the melting behaviour of trans-hardened fats is not generally as steep melting as cocoa butter or lauric-type alternative fats.

Although these fats are based on C_{16} and C_{18} acids, the hydrogenation process gives a different and much more complex arrangement of glyceride composition than that of cocoa butter and, consequently, they can crystallise directly in the β' form, which is their stable polymorph. This means that they are nontempering and may be used in a wide range of applications, including for compound chocolate.

The relatively poor melting profile of such products can be improved by fractionation, removing both high-melting and low-melting triglycerides to

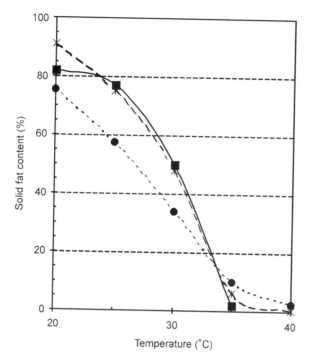

Figure 5.7 Solid fat melting curves of hydrogenated/high-trans alternative fats (AFs) in comparison with cocoa butter. Note: CB, cocoa butter; CBR, hydrogenated but unfractionated cocoa butter replacer; fractionated CBR, hydrogenated and fractionated cocoa butter replacer. —■— CB; --●-- CBR; —×— Fractionated CBR.

produce a mid-fraction that has much improved melting characteristics, as shown in figure 5.7. Because of the complex triglyceride composition of the feed oil, solvent fractionation is preferred in order to maximise this benefit (through more efficient separation of the fractions), although recent improvements in dry fractionation technology have enabled reasonable products to be produced.

When used in compound chocolate these products generally show good gloss retention, but it has been noted that fats based mainly on C_{18} acids tend to give a poorer gloss that is lost more rapidly (Padley, 1997). This defect is overcome by incorporating palm olein into the hydrogenation feedstock to introduce C_{16} acids or by the addition of products such as sorbitan tristearate.

An interesting aspect of many of these fats is the tendency for the solid fat content to increase during storage at ambient or higher temperatures. This phenomenon is known as post-hardening and is not fully understood (Padley 1997 and Wong Soon, 1991).

Recent health studies have shown that trans fatty acids can be hypercholes-terolaemic and can promote atherosclerosis. The high level of trans fatty acids in these products is a significant disadvantage at a time when consumer pressure is leading to a trend to reduce the level of trans acids in foods generally. There have been moves to seek alternatives, but the fast crystallisation and relatively steep melting characteristics are difficult to replicate.

5.2.4 Lauric-type alternative fats

Lauric-type alternative fats are widely used as confectionery fats. They are typified by the presence of large amounts of lauric (45–55%) and myristic (15–20%) acids. Because of the low melting point of these shorter-chain acids, they have a sharp melting profile and good mouth feel. When used in chocolate recipes they are often called cocoa butter substitutes (CBSs).

Both palm kernel and coconut oils are soft, with a melting point of 26°C. In order to increase the solid fat content at 20°C to an acceptable level for a confectionery fat, it is common to hydrogenate to reduce or remove the unsaturated acids. A range of different melting point materials can be produced, as shown in figure 5.8. However, hydrogenation increases the melting point above 35°C, leading to poor mouth feel and meltdown.

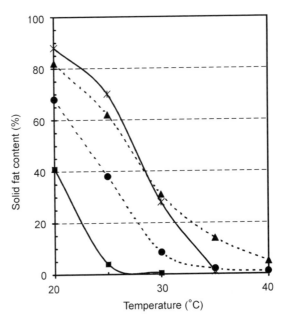

Figure 5.8 Solid fat melting curves of lauric-type alternative fats. Note: PK, palm kernel oil; hPK32, PK hardened to a melting point of 32°C; hPK41, PK hardened to a melting point of 41°C (fully hardened); PKst, palm kernel stearin with an iodine value of about 7. —■— PK; --●-- hPK32; - -▲- - hPK41 —✳— PKst.

Fractionation is an alternative and preferred method of increasing the SFC at lower temperatures to producing a steeper melting product. Palm kernel oil (rarely coconut oil) is normally dry fractionated and the crystals separated by using high-pressure presses. The stearins produced are very steep melting and have an excellent meltdown, as shown in figure 5.8. These fractions are often used as such but are sometimes hardened for specific applications, especially in hot climates.

Because of their complex triglyceride composition, lauric-type alternative fats crystallise directly in the β' form, which is their stable polymorph. This means that, like hardened/high-trans alternative fats, they do not require tempering, which is useful in many applications.

5.2.5   Comparison and compatibility

Table 5.4 shows the fatty acid composition of the various alternative fats in comparison with cocoa butter. It can be seen that the symmetrical/SOS alternative fat is similar to cocoa butter. The lauric-type alternative fat has a totally different composition, with fatty-acid carbon numbers 8–14 predominating, fatty acids that are totally absent from cocoa butter. The hydrogenated/high-trans alternative fat is composed of the same fatty acids as in cocoa butter but it also contains large amounts of *trans*-unsaturated acids.

Cocoa butter contains about two-thirds saturated acids and, by comparison with other fats of similar fatty-acid composition (e.g. tallow and palm stearin), would be expected to have a melting point much higher than 35°C. As we have seen, the reason for its low and sharp melting point is the location of the saturated acids at the 1 and 3 positions of the triglyceride molecule and the unsaturated acids at the 2 position.

Table 5.4 Typical fatty-acid compositions (percentage as methyl ester) of the three types of alternative fat (AF) in comparison with cocoa butter

Fatty acid	Cocoa butter	Symmetrical/SOS AF		Hardened/high-trans AF	Lauric-type AF
		low-PMF	high-PMF		
8:0	0	0	0	0	2
10:0	0	0	0	0	2
12:0	trace	trace	trace	0	54
14:0	0.1	0.3	0.4	trace	22
16:0	26	34	44	23	9
18:0	34	29	20	12	2
18:1cis	35	34	33	16	7
18:1trans	0	trace	trace	46	0
18:2	3	2.0	2.4	1	1
20:0	1	0.5	0.2	1	trace

Trace, < 0.1%.

Note: PMF, palm mid-fraction; for fatty-acid nomenclature, see appendix.

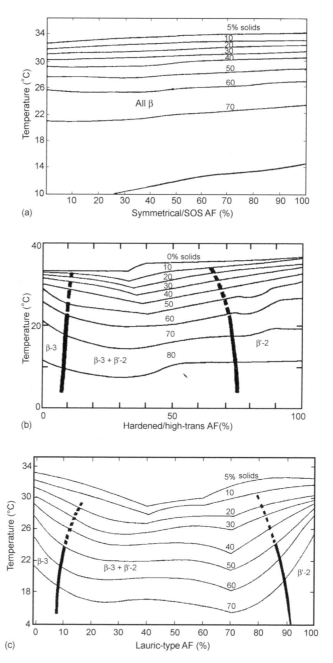

Figure 5.9 Isosolid phase diagrams of mixtures of cocoa butter with the three types of alternative fat (AF) (a) symmetrical/SOS AF compared with cocoa butter; (b) hardened/high-trans AF compared with cocoa butter; (c) lauric-type AF compared with cocoa butter. Redrawn from Gordon *et al.*, 1979.

Lauric-type alternative fats contain 85–100% saturated fatty acids. Only because the carbon numbers of the main acids are 12 and 14 is the melting point reduced to the organoleptically satisfactory level of 35–40°C.

Hydrogenated/high-trans alternative fats contain about two-thirds unsaturated acids. Only by changing the conformation of the double bonds from cis to trans is it possible to raise the melting point to the required level. Elaidic acid (*trans*-oleic acid) and trieleaidin (EEE) have melting points of 45°C and 41°C, approximately the same as the melting points of lauric acid and trilaurin (LLL) (44°C and 46°C, respectively), indicating the similarity that might be expected between the two types of alternative fat.

The isosolid phase diagrams[2] of the three types of alternative fat are shown in figures 5.9(a)–5.9(c) (Gordon *et al.*, 1979). For complete compatibility the diagram should be as in figure 5.9(a), with a single phase in the single polymorph, as in tempered cocoa butter, and more or less horizontal isosolid lines.

Figure 5.9(c) shows substantial incompatibility between cocoa butter and the lauric-type alternative fat, with only small areas on the left-hand and right-hand sides of the diagram where a single solid phase exists. The area on the left-hand side indicates how much alternative fat can mix with cocoa butter while keeping a single phase with the same properties as cocoa butter. The area on the right-hand side indicates how much cocoa butter can mix with alternative fat while keeping a single phase with the same properties as the alternative fat.

Figure 5.9(b) is similar to figure 5.9(c), but the area of single phase on the right-hand side is now significantly larger, indicating that more cocoa butter can dissolve in the alternative fat.

The reasons for these variations in compatibility can be understood from the information in table 5.5. When mixing fats in the solid state we can imagine the molecules as consisting either of cubes (β polymorph) or of spheres

Table 5.5 Compatibility of cocoa butter with the three types of alternative fat (AF)

AF	Chain length[a]	Stable polymorph	Compatibility (%)
Cocoa butter	16 and 18	β-3	N/A
Symmetrical/SOS	16 and 18	β-3	c. 100
Hardened/high-trans	16 and 18	β'-2	c. 20
Lauric-type	12 and 14	β'-2	c. 5

N/A Not applicable.
[a]of main fatty acids (carbon number).

---

[2]An isosolid phase diagram is a combination of an isosolid diagram and a conventional phase diagram. The isosolid diagram shows lines of the same SFC, which gives information about melting properties. The phase diagram gives the position of phase boundaries. In particular, we need to know the phase boundaries between different solid phases. These boundaries, separating phases usually of different polymorphs, are superimposed on the isosolid diagram as thick black lines.

(β' polymorph). (It is not to be understood that the molecules actually do have the shape of cubes or spheres.) These cubes and spheres can be either large (C_{16} and C_{18} fatty acids) or medium (C_{12} and C_{14}). Clearly, cubes of the same size should fit together easily to give a tight packing (i.e. we can say they mix easily in the solid state to give a single solid solution). This is the case for cocoa butter and the symmetrical/SOS alternative fats. Following this analogy, mixing large cubes and medium spheres is not going to be easy, and hence the compatibility of cocoa butter and a lauric-type alternative fat is poor. Finally, the hardened/high-trans alternative fats have the same chain-length fatty acids as cocoa butter but a different polymorphic form, the equivalent of mixing large cubes and large spheres. The result is still poor compatibility, but it is better than it is for lauric-type alternative fats, resulting in a larger single-phase area on the right-hand side of figure 5.9(b). The implications of these different levels of compatibility will be shown in section 5.4, on applications.

5.3   Legislation and regulatory aspects

5.3.1   Legislation

Most countries have felt it necessary to define the composition of chocolate in a prescriptive detail that is not required for most other foods. The reasons for this are many and complex. Like similar legislation for dairy products, although originally designed to protect the consumer the legislation eventually has tended to protect the producer and inhibit product development. In order to compete more effectively in the marketplace, the dairy industry in most countries has long given up such prescriptive legislation so that milk, butter and yoghurt can have widely varying compositions reflecting consumer preference and desires. In contrast, the chocolate industry has been slow to adapt to the modern philosophy that government's only job should be to ensure that food is safe and that the consumer should get all necessary compositional and nutritional information to make an informed choice.

The focus of debate about chocolate legislation has been in the European Union (EU), where the different food legislations of 15 member countries have had to be harmonised to bring chocolate into line with the EU Food Labelling Directive and the requirement for free trade between the member states. Contentious issues have concerned the minimum amounts of milk and cocoa components and the use of other vegetable fats besides cocoa butter. For example, northern European countries (the United Kingdom, Ireland and Scandinavian countries) with traditionally large dairy industries and high milk consumption have tended to prefer milk chocolate with a high milk content (and therefore necessarily lower cocoa content) than have consumers in countries further south such as France and Italy. Additionally, these northern European countries consume much more chocolate and cocoa products in total.

The history and background to the new EU Chocolate Directive (European Parliament and Council, 2000) has been lucidly explained by Eagle (1999). This new Directive must be adopted by all member states by 3 August 2003, when the current legislation, Directive 73/241/EEC, is repealed. The details are summarised in table 5.6. The biggest change for eight of the current 15 EU members will be the opportunity to produce chocolate containing up to 5% of a vegetable fat of the symmetrical/SOS type as well as cocoa butter. Nevertheless, there will also be a change for the seven countries that already permit this option. Only six component vegetable fats (the six given in table 5.3) are allowed for the 5% alternative fat; lauric fats and enzymically processed fats will no longer be permitted. Coconut oil may be used only for the manufacture of chocolate for ice cream. This automatically excludes some of the most innovative and functional ingredients developed by the European chocolate fat industry. As a result, one major company has already taken its bloom-retarding fats (which contain coconut oil) off the market (Parker, 2000).

It should also be noted that the 5% vegetable fat is in addition to the minimum values given for cocoa butter and total dry cocoa solids in the table. Also the 5% is calculated as 5% of the finished product, after deduction of the total weight of any other edible matter used. In other words, the 5%, like all the other percentage amounts given, refers to the chocolate itself, excluding any fillings, nuts and so on that may be added.

The situation in the USA is somewhat different. Chocolate is defined under the US Standards of Identity (Federal Register, Title 21, Chapter 163), which covers all chocolate products, although there is still not a standard for 'white chocolate'. Apart from prescribing minimum levels of milk and cocoa components, the US legislation also does not permit the use of vegetable fats other than cocoa butter. The majority of companies in the (US) Chocolate Manufacturers Association apparently favour keeping the standards as they are (Seguine, 2000, personal communication).

Other countries have tended to follow the legislation of US or European countries. Thus Canadian legislation is similar to that in the USA and also does not permit the use of vegetable fats other than cocoa butter.

In Australia and New Zealand the legislation has been in the process of change. The recently (end 2000) approved harmonised food standards code for Australia and New Zealand contained only temporary standards for chocolate or confectionery. Chocolate had to contain a minimum of 20% cocoa bean derivatives. This allowed the addition of vegetable fats, but it was considered unlikely to cause manufacturers to change their formulations or launch new products complying with this regulation, as a ministerial review of chocolate

Table 5.6 Summary of composition of chocolate to be permitted in the European Union as detailed in the new Chocolate Directive 2000/36/EC (European Parliament and Council, 2000)

Definition	Requirements[a]
Chocolate	
The product obtained from cocoa products and sugars	Total dry cocoa solids ≥35%
Variations are allowed for vermicelli, flakes,	cocoa butter ≥18%
Gianduja or couverture (see below) chocolate	dry nonfat cocoa solids ≥14%
Chocolate couverture	
Designates the product obtained from cocoa	Total dry cocoa solids ≥35%
products and sugars	cocoa butter ≥31%
	dry nonfat cocoa solids ≥2.5%
Milk chocolate[b]	
The product obtained from cocoa products, sugars and	Total dry cocoa solids ≥25%
milk or milk products	Dry milk solids ≥14%
Variations are allowed for vermicelli, flakes, Gianduja	Dry nonfat cocoa solids ≥2.5%
or couverture (see below) chocolate	Total fat ≥25%
	milk fat ≥3.5%
Milk chocolate couverture	
The product obtained from cocoa products, sugars	Total dry cocoa solids ≥25%
and milk or milk products	Dry milk solids ≥14%
	Dry nonfat cocoa solids ≥2.5%
	Total fat ≥31%
	milk fat ≥3.5%
Family milk chocolate[b]	
The product obtained from cocoa products, sugars	Total dry cocoa solids ≥20%
and milk or milk products	Dry milk solids ≥20%
	Dry nonfat cocoa solids ≥2.5%
	Total fat ≥25%
	milk fat ≥5%
White chocolate	
The product obtained from cocoa butter,	Cocoa butter ≥20%
milk or milk products and sugars	Dry milk solids ≥14%
	milk fat ≥3.5%

[a]In all products, vegetable fats other than cocoa butter may be used up to a maximum of 5% of the finished product, after deduction of the total weight of any other edible matter used, without reducing the minimum content of cocoa butter or total dry cocoa solids. The vegetable fats allowed are given in table 5.3.
[b]In the United Kingdom and Ireland the name 'milk chocolate' may also be used for 'family milk chocolate'.

compositional standards was to be completed by March 2001 (Webster, 2000, personal communication).

The situation is fluid and is not easily summarised. Both Australian and New Zealand manufacturers have the choice of complying with: the Australian Food Standards Code, section Q3; the New Zealand Food Regulations, section 166; or the new harmonised code, which supersedes, with some exceptions, both of the above, and which will probably come into effect in November 2002.

For plain chocolate (not milk or white) Australian standard Q3 requires a minimum of 15% nonfat cocoa solids with no vegetable fat except cocoa butter allowed. The New Zealand regulations require not less than 14% (cocoa plus nonfat milk solids) in most chocolate, but couverture chocolate for biscuits and confectionery may contain not less than 12%. No vegetable fat except cocoa butter is allowed.

Following modern food legislation philosophy, the Australia New Zealand Food Authority (ANZFA) originally preferred not to set compositional standards for chocolate, but rather to let the market decide what the acceptable quality–cost trade-off should be. However, many manufacturers, as represented by the Confectionery Manufacturers of Australasia Ltd, objected and instead favoured legislation similar to the new EU Chocolate Directive, allowing 5% of other vegetable fats as well as cocoa butter, and prescribing clear compositional standards. The latest information is that a compositional standard of this type was agreed by the Australia New Zealand Food Standards Council on 31 July 2001. Chocolate is required to have a minimum of 20% cocoa solids and up to 5% of fats. For up-to-date information it is suggested that the reader consult the ANZFA web sites at anzfa.gov.au and anzfa.govt.nz.

The situation in Japan is different again. Japan has standards for chocolate (milk and plain) and for 'quasi-chocolate' (milk and plain) (Koyano, 2000, personal communication). Plain chocolate must contain at least 35% of cocoa ingredients (cocoa liquor and cocoa butter) and at least 18% of cocoa butter. Milk chocolate must contain at least 21% of cocoa ingredients, and the total of cocoa ingredients and milk solids must be at least 35%, with at least 3% of milk fat and at least 18% of cocoa butter.

Plain quasi-chocolate must contain at least 15% of cocoa ingredients, with at least 3% of cocoa butter. Milk quasi-chocolate must contain at least 7% of cocoa ingredients, with at least 3% of cocoa butter. Also, it must contain at least 18% total fat and at least 12.5% of milk solids. For both plain and milk chocolate and quasi-chocolate, water content should be below 3%.

These standards allow a much larger amount of vegetable fat, other than cocoa butter, than is permitted under European legislation. For example, if total fat content in a chocolate is 33%, up to 15% may be a CBE or symmetrical/SOS alternative fat. This relatively liberal legislation has undoubtedly accelerated the development of alternative fat technology in Japan, putting Japanese companies at the forefront of technical developments such as the use of enzymes to synthesise new triglycerides and the development of products such as BOB (1,3-dibehenoyl-*sn*-2-oleoyl-glycerol) fat, which obviates the need for tempering and ensures that chocolate is stable under high ambient temperatures (Hachiya *et al.*, 1989a, 1989b).

In Malaysia, following the British regulations, up to 5% alternative vegetable fat is permitted in chocolate. It is likely that the changes taking place in the EU and in the Codex Alimentarius will have a bearing on the future legislation in Malaysia (Sudin, 2001, personal communication).

The Codex Alimentarius Commission is also looking at chocolate legislation. The next meeting of the Committee on Cocoa Products and Chocolate will be held in Fribourg in October 2001, when it will further consider the draft standard for chocolate and cocoa products. Within the draft there is a proposal for the use of 5% vegetable other than cocoa butter as in the new EU Chocolate Directive.

5.3.2 Adulteration and its detection

A frequent rationale for objecting to the use of vegetable fats other than cocoa butter in chocolate is that because symmetrical/SOS alternative fats are so similar to cocoa butter they are difficult to detect, allowing the possibility of fraud. There is some justice in this claim, although the objection does often seem like special pleading, because the other argument brought against symmetrical/SOS alternative fats is that they lower the quality of the chocolate. If this were really so, of course, there would be no problem in detecting adulteration, either in the laboratory or in the marketplace.

In reality, because the triglycerides in cocoa butter and in symmetrical/SOS alternative fats are identical, differing only in their relative proportions, and because only 5% of alternative fat is to be allowed in a total fat content of about 30%, we need to detect a relatively small change in fatty acid or triglyceride composition. There are some other differences between cocoa butter and the alternative fats and their components, principally sterols and their derivatives, but although these may allow qualitative results they are not capable of quantification. The analytical approaches for identification and determination of alternative fats in chocolate have recently been reviewed (Lipp and Anklam, 1998). A particularly useful method for the detection, but again not quantification, of an alternative fat in chocolate is to determine sterol degradation products (sterenes) (Crews et al., 1997). Sterenes are produced when an oil is bleached, a process that all alternative fats but not cocoa butter undergo.

A definitive method for the determination of CBEs (symmetrical/SOS alternative fats) in chocolate, including milk chocolate, has been developed (Padley and Timms, 1978, 1980). By using gas chromatography (GC), one can analyse the intact triglycerides on the basis of their carbon number. The method depends on the fact, alluded to above, that no natural fat contains as much POS (carbon number 52) as cocoa butter and, further, that a plot of carbon number 50 (POP) against carbon number 54 (SOS) shows all pure cocoa butters to be distributed along a straight line. As a result, precise quantitative predictions about the amount of alternative fat present can be made, including statistical estimates of errors. The method is robust, accurate and rapid. Although there have been some further developments of the method (Fincke, 1982; Young, 1984) it remains essentially the same as when it was first developed. With the advent of modern capillary GC columns capable of resolving the individual triglycerides POP, POS and SOS rather than just the triglycerides grouped by carbon number, there is scope for further improvement in the method. This approach is currently being evaluated by the European Commission's Joint Research Centre at Ispra, Italy.

The other alternative fats—hardened/high-trans and Lauric-type—are easy to detect simply on the basis of their fatty-acid compositions. Since cocoa butter contains neither *trans*-acids nor lauric acid and the alternative fats typically contain at least 45% of either acid, even 1% of either of these alternative fats in cocoa butter would lead to an increase of 0.45% in lauric or *trans*-(elaidic) acid. Such an amount is easily detectable; in fact, even 0.1% of lauric acid is detectable with ordinary laboratory GC equipment, so that very low levels of lauric fats can be detected.

5.4 Applications

5.4.1 Real chocolate

Chocolate is a suspension of cocoa solids and sugar in a continuous fat phase. Where the fat phase consists only of cocoa butter or of cocoa butter and up to 5% symmetrical/SOS alternative fats, where legislation permits, the resulting product may be labelled 'chocolate'. We designate this product as 'real chocolate'. Where the fat phase contains other alternative fats the product is designated as 'compound chocolate'. Milk chocolate, either 'real' or 'compound', contains a milk component such as milk powder. So-called 'white' chocolate is basically milk chocolate without any cocoa solids (i.e. no cocoa liquor or powder is used, only cocoa butter or other vegetable fat).

It is the fat phase that gives chocolate its desirable texture and mouth feel. With milk chocolate, milk fat is added to the basic ingredients and, where legislation allows, an alternative fat can be added to improve functionality or reduce cost. As discussed in section 5.3.1, the allowed ingredients and composition of chocolate vary between countries, and the reader is advised to study the appropriate legislation.

Milk is never added in liquid form, but only as a dry ingredient as described in section 5.2.1. Milk crumb imparts a characteristic 'caramel' flavour to the chocolate and is widely used in the United Kingdom, Australia and the USA. Typical recipes for plain and milk chocolates are given in table 5.7.

Chocolate production involves mixing cocoa liquor together with some fat. This mixture or paste is then ground between rollers to a particle size of typically 20–30 μm microns (Beckett, 1999). This process is known as refining. If the

Table 5.7 Real chocolate recipes including use of symmetrical/SOS alternative fats

Ingredient	Chocolate type		
	Basic plain	Basic milk, with SMP[a]	Basic milk, with milk crumb
Cocoa mass	40.0	15.0	
Added cocoa butter	7.0	11.5	15.0
Cocoa butter equivalent	5.0	5.0	5.0
Whole milk powder	0.0	13.5	0.0
Skim milk powder	0.0	7.0	0.0
Milk crumb	0.0	0.0	63.0
Anhydrous milk fat	0.0	0.0	0.0
Sugar	47.5	47.5	16.5
Lecithin	0.5	0.5	0.5
Total	100.0	100.0	100.0
Fat as percentage of chocolate:			
cocoa butter	28.2	19.8	19.6
cocoa butter equivalent	5.0	5.0	5.0
milk fat	0.0	3.6	5.8
total fat	33.2	28.4	30.4
Fat as percentage of fat phase:			
cocoa butter	85.0	69.5	64.5
cocoa butter equivalent	15.0	17.6	16.5
milk fat	0.0	12.8	19.1
total	100	100	100

[a] Skim milk powder.
Note: percentages are relative to weight (wt%).

particle size is much lower than this, the chocolate will have a greasy mouth feel; if much higher, it will taste gritty.

Extra fat is then added and the mixture is heated with continuous mixing for a long period (several hours to several days). This process is known as conching and improves the flavour of the product along with its texture and viscosity. Both batch and continuous conches are available.

Because of the polymorphic nature of cocoa butter it must be tempered in order to promote crystallisation in form V. The process of tempering involves initial cooling to promote crystallisation followed by heating to melt the unstable crystals. This used to be a batch process, but is now usually carried out in continuous tempering equipment.

It is possible to replace all the added cocoa butter in a chocolate recipe with a symmetrical/SOS alternative fat. Such a product is commonly known as a Supercoating (Talbot, 1999). The Supercoating will require processing in the same manner as chocolate and will be suitable to replace it.

One advantage of Supercoatings is to reduce cost, although this will depend on the prevailing price of cocoa butter. Further advantages include increasing or decreasing hardness and viscosity for particular applications. Some Supercoating recipes are given in table 5.8.

5.4.2 Compound chocolate

In some cases the use of real chocolate in confectionery products is not desirable, because of either cost or functionality (e.g. real chocolate is inconveniently brittle for use in ice cream and cakes). Compound or substitute chocolate is then used. This can be formulated with either lauric or high-trans alternative fats and is most frequently used as a coating to enrobe biscuits, centres or other fillings for snack bars.

5.4.2.1 Compound chocolate with lauric-type alternative fats
Palm kernel stearins have a very steep melting curve not unlike that of cocoa butter and thus have good snap, gloss and eating characteristics. A major disadvantage of lauric fats is that they form strong eutectics with cocoa butter [figure 5.9(c)]. This means that, in the manufacture of compound chocolate, low-fat (10–12% fat) cocoa powders must be used, making it difficult to achieve a strong cocoa flavour. Even so, if more than about 5% cocoa butter is present

Table 5.8 Supercoating recipes using symmetrical/SOS alternative fats (Talbot, 1999)

Ingredient	High-milk	Low-milk	Plain
Cocoa mass	10	10	40
Added cocoa butter equivalent	22	24	12
Sugar	46	46	48
Whole milk powder	22	10	0
Skim milk powder	0	10	0
Total	100	100	100
Fat as percentage of chocolate:			
cocoa butter	5.3	5.3	21.2
cocoa butter equivalent	22.0	24.0	12.0
milk fat	5.9	2.7	0.0
total	33.2	32.0	33.2
As percentage of fat phase:			
cocoa butter	16.0	16.6	63.9
cocoa butter equivalent	66.3	75.0	36.1
milk fat	17.7	8.4	0.0
total	100.0	100.0	100.0

Note: percentages are relative to weight (wt%).

Table 5.9 Compound chocolate recipes using lauric-type alternative fats

		Milk		
Ingredient	Dark	1	2	White
Cocoa powder[a]	14	5	7	0
Whole milk powder	0	10	0	0
Skim milk powder	6	8	19	20
Lauric-type alternative fat	32	32	29	32
Sugar	48	45	45	48
Total	100	100	100	100
Added lecithin (%)	0.2–0.4	0.2–0.4	0.2–0.4	0.2–0.4
Total fat content	33.5	35.2	29.9	32.2
Cocoa butter as percentage of total fat	4.2	1.4	2.3	0
Total cocoa solids	14	5	7	0

[a]Low-fat cocoa powder (10–12% fat).
Note: percentages are relative to weight (wt%).

in the fat phase, excessive softening and bloom may occur. A recent paper describes the use of low-temperature solvent extraction with propane to produce a very low-fat powder with excellent flavour. The resulting powder produced no eutectic softening and a chocolate judged to have a flavour as good as a premium quality plain real chocolate (Trout, 2000).

Typical recipes are shown in table 5.9. The method of production is similar to that of real chocolate, involving mixing, followed by refining and conching, although the conching time is less. However, the nontempering nature of these fats means that processing is easier and the chocolate can be used instantly for enrobing or moulding. This is advantageous in many ways and the lower viscosity can be useful in enrobing (because unlike tempered real chocolate, no crystals are present).

Because the finished chocolate is hygroscopic, products must be well wrapped to avoid ingress of water, which could result in rancidity (see section 5.4.5). For the same reason, care must be taken to select milk and cocoa ingredients and to reduce the moisture content of the chocolate.

5.4.2.2 Compound chocolate with hydrogenated/high-trans alternative fats

Hydrogenated/high-trans alternative fats are often used in place of lauric fats because they are more compatible with cocoa butter. Consequently, cocoa liquor may be used instead of cocoa powder, and this means that products can have a more rounded chocolate flavour. Additionally, because of their compatibility with cocoa butter, the same production line can be used for both real chocolate and for the compound chocolate with only minimal cleaning between runs,

Table 5.10 Compound chocolate recipes using hardened/high-trans alternative fats (AFs) (adapted from Talbot, 1999)

Ingredient	Dark 1	Dark 2	Milk 1	Milk 2	White 1	White 2
Cocoa liquor	10	0	10	0	0	0
Cocoa powder[a]	15	20	0	5	0	0
Hardened/high-trans AF	28	33	28	34	30	35
Sugar	47	47	44	44	45	45
Whole milk powder	0	0	6	0	20	0
Skim milk powder	0	0	12	17	5	20
Total	100	100	100	100	100	100
Total fat content	35.0	35.0	35.2	34.7	35.4	35.2
Cocoa butter as percentage of total fat	20.0	5.7	15.6	1.4	0	0
Total cocoa solids	25	20	10	5	0	0

[a] Low-fat cocoa powder (10–12% fat).
Note: in all recipes, 0.4% lecithin is added; percentages are relative to weight (wt%).

without bloom occurring in either the real or the compound chocolate products. Typical recipes are shown in table 5.10. The method of production is as described for lauric-type compound chocolate.

5.4.2.3 Centre filling creams

Centre filling creams are widely used in confectionery products and biscuits. Applications vary widely from fillings for chocolates such as pralines to the cream fillings used in sandwich-type biscuits. Compositions and the fats used vary correspondingly. Frequently, lauric-type or hardened/high-trans alternative fats are used for economy and for ease of processing (i.e. they do not require tempering). Hardened/high-trans alternative fats have the disadvantage of a high trans level. A second disadvantage, especially if the product is enrobed with real chocolate, is that, because they are incompatible with cocoa butter, migration from the centre can cause problems with softening and bloom. This problem is even more acute with centres based on lauric-type alternative fats.

Palm fractions or CBEs (see section 5.2.2) can be used in fillings and have none of these disadvantages but, as with cocoa butter, they usually require at least minimal tempering and tend to be more expensive than other alternative fats. Recent developments have produced centre filling fats based on palm fractions that do not require tempering, and these are likely to become more extensively used (Duurland and Smith, 1995).

Biscuit cream fillings are usually made from hardened palm kernel oil products or from a nonlauric equivalent. Again, these can be usefully replaced with palm fractions to avoid bloom problems.

Table 5.11 Some typical centre filling cream recipes

Ingredient	Chocolate	Custard	Praline	Biscuit
Sugar	45	60	40	54
Cocoa powder[a]	7	0	0	0
Cocoa Liquor	0	0	10	0
Skim milk powder	10	0	10	9
Whole milk powder	0	4	0	0
Starch	0	0	0	5
Hazelnut paste	0	0	10	0
Fat	38	36	30	32
Total	100	100	100	100
Added lecithin	0.2–0.5	0	0.5	0

[a]Low-fat cocoa powder (10–12% fat).
Note: flavouring, colour, vanillin and salt are added as required.

Some recipes are shown in table 5.11. Production consists of little more than stirring all the ingredients together with the liquid fat, followed by cooling to the depositing temperature, or possibly tempering.

Sometimes the eutectic interaction between cocoa butter and lauric-type alternative fats can be used to advantage. A mixture of about 65% cocoa butter and 35% lauric-type alternative fat was found to give a filling with very good mouth feel and 'melt-away' characteristics. When the filling was enrobed with real chocolate, bloom occurred within one month, but was still absent after three months when enrobed with a lauric-type alternative fat compound chocolate (Rahim *et al.*, 1998).

5.4.3 Toffees and other sugar confectionery

The main use of fats in sugar confectionery is in toffees, caramel, fudge and nougat. Toffee is an emulsion of fat in a complex aqueous system (Stansell, 1995). There is no definite distinction between caramel and toffee, and the two names may be regarded as synonymous. The main constituents of these products are fat and sugar, and the basic process consists of boiling the ingredients together to dispel water. By heating the ingredients, characteristic flavours develop as a result of the reaction between reducing sugars and milk proteins. This is known as the Maillard reaction.

Traditionally, the fat ingredient used is either AMF or butter. However, butter is preferred because of the contribution made by the nonfat solid components (i.e. milk proteins) to unique flavour development through the Maillard reaction. AMF and butter are frequently replaced by vegetable fat, although there is some loss of flavour because of the absence of milk proteins and the natural milk/butter flavours derived from fatty acids and flavour precursors. The hardness of toffee or caramel is purely dependent on the water content, which depends on the final boiling temperature. The type of vegetable fat is not particularly important. It should, however, be solid at room temperature and be relatively sharp melting as it must be fully liquid at body temperature to avoid an unpleasant mouth feel (Stansell, 1995). Hardened palm kernel oil was used for many years but is now often replaced by other fats, typically hardened/high-trans alternative fats of the simple, unfractionated type. Further information on toffees, caramel, fudge and nougat may be found in the comprehensive book by Jackson (1995). Some recipes are given in table 5.12.

5.4.4 Truffles

There are three main types of truffle—American, European and Swiss (Minifie 1989). The American truffle is usually a mixture of dark and milk chocolate, together with milk fat and hardened coconut oil. It has a good shelf-life because it contains virtually no moisture.

The European truffle contains syrup combined with cocoa powder, milk powder, fat, sugar, glucose syrup and invert sugar. The final truffle is an oil-in-water emulsion adjusted to give a water activity of 0.7 or greater and a syrup phase concentration of greater than 75%. Provided the truffle is manufactured correctly, the shelf-life can be good.

The Swiss truffle is made from dairy cream, dark chocolate and butter. It is made by adding melted chocolate to the boiling mixture of cream and butter. The approximate proportions are 60% chocolate, 10% butter and 30% cream. These truffles have desirable eating qualities but have a shelf-life of only a few days. They tend therefore to be used only by specialist confectioners selling fresh products. The shelf-life of Swiss truffles can be improved by adding alcohol, usually in the form of a liqueur, and by using sweetened condensed milk instead of cream.

In all truffle types, the chocolate can, and often does, contain a symmetrical/SOS alternative fat (CBE quality) up to total replacement of all the cocoa

Table 5.12 Typical toffee and caramel recipes (Kempas Edible Oil Sdn. Bhd., 81700 Pasir Gudang, Johar, Malaysia)

Ingredient	European	Tropical
Granulated sugar	23.5	15
Glucose syrup	35	43.5
Skimmed sweetened condensed milk	29.5	29.5
Fat	12	12
Total	100	100
Added salt	2	2

Note: flavouring and water are added as required; percentages are relative to weight (wt%).

butter (see section 5.4.1 and table 5.9, relating to Supercoatings). It can also be replaced with a mixture of vegetable fat and cocoa liquor or powder. If a CBE is used instead of cocoa butter, the texture and eating characteristics of the product are the same. In fact, by using harder or softer vegetable fats the consistency of the truffle can be tailored to suit specific applications, thus improving the eating characteristics.

5.4.5 Bloom and rancidity

Fat bloom is a common occurrence in the confectionery and biscuit industries. It is caused by a change in crystal morphology after the product is made and is manifested by the appearance of either a white 'frosting' on the surface of the product or a loss of gloss. There are two main causes of bloom: (a) polymorphic change in the fat, from form V to form VI, (b) migration of fat from fillings or centres to their chocolate coatings. Both types of bloom are frequently seen and, if not controlled, can be a major source of losses and rejection. Bloom problems generally occur because of incorrect processing, use of incompatible fats or poor storage conditions.

An example of bloomed chocolate is shown in figure 5.10. This is classic form VI bloom, usually caused by poor processing or poor storage. An illustration of bloom caused by migration is shown in figure 5.11. The filling based on lauric-type alternative fat, has migrated into the chocolate, causing bloom as well as softening and cracking of the chocolate.

Food products are not static entities but are dynamic, which gives rise to the phenomenon of shelf-life (Hammond and Gedney, 2000). Bloom problems will thus occur eventually but can be retarded, thus increasing shelf-life, by choosing appropriate temperatures for the operation of cooling tunnels and storage. Bloom can be exacerbated by nonoptimised temperature control or by choosing incompatible fats for centres on multipart products such as enrobed

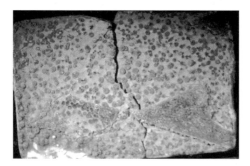

Figure 5.10 Severe bloom caused by form VI crystal growths on the surface of chocolate. Reproduced from Hammond and Gedney, 2000.

— Cream filling

— Chocolate coating
in form VI

Figure 5.11 Bloom caused by migration, resulting in major structural change. Reproduced from Hammond and Gedney, 2000.

biscuits. It is particularly important to avoid temperature cycling, as heating and then cooling can lead to uncontrolled recrystallisation.

In a comprehensive series of papers, Ziegleder and co-workers investigated and explained the formation of bloom, especially the bloom in pralines caused by migration of the nut oil from the praline centre into the chocolate shell (Ziegleder and Mikle, 1995a, 1995b, 1995c; Ziegleder and Schwingshandl, 1998; Ziegleder *et al.*, 1996a, 1996b). The development of fat bloom in pralines depends particularly on the type of chocolate and the product storage temperature. For milk chocolate, bloom is greatest between 18°C and 22°C; for plain chocolate it is greatest between 18°C and 26°C, with a maximum at 20°C. This maximum at 20°C may be considered surprising since the tendency of plain chocolate to bloom is known to increase over this temperature range. The difference results because migration is the mechanism of bloom formation in a praline. As the temperature rises there is a balance between an increasing diffusion rate (positive for bloom) and a decreasing crystallisation rate (negative for bloom).

Another method of retarding bloom is the addition of antibloom agents. Traditionally, milk fat has been used. This softens the chocolate and inhibits the transformation from form V to form VI. Bloom-inhibiting fats that mimic this behaviour have also been developed (Talbot, 1995) but, as noted in section 5.3.1 will shortly become illegal in real chocolate to be sold in the EU.

Another approach is to add the symmetrical/SOS triglyceride BOB, which crystallises directly into the β polymorph (Hachiya *et al.*, 1989a, 1989b). Essentially, chocolate containing BOB does not require tempering, so even complete melting does not result in bloom. Real chocolate containing BOB will also become illegal in the EU under the new European legislation.

Rancidity arising from oxidation is not usually a problem with chocolate products. Cocoa butter is very stable to oxidation, both because of its low content of polyunsaturated acids and because of its high content of natural antioxidants, which can also protect the other ingredients from oxidation. Similarly, both lauric-type and hydrogenated/high-trans alternative fats are stable to oxidation, although the fractionated products are inferior in this respect because the natural antioxidants present are concentrated into the olein during fractionation. Padley

(1994) has reviewed the control of rancidity in confectionery products. A general review of the stability of oils and fats in foods has been published by Kristott (2000).

White chocolate is particularly sensitive to oxidation on exposure to light. If white chocolate is stored in transparent foil, light causes a rancid off-flavour to develop and a loss of the typical yellowish colour. The loss of quality may be reduced by using a nitrogen atmosphere and a high-oxygen-barrier foil, or by using a CBE instead of cocoa butter. CBEs are less sensitive to light because they may contain less or no chlorophyll, the photo-sensitiser for the oxidation (Krug and Ziegleder, 1998a, 1998b).

Hydrolytic rancidity, leading to the production of free fatty acids, is only a problem with products containing lauric-type alternative fats. They are prone to lipase-catalysed rancidity in the presence of low levels of moisture; this is normally described as 'soapy' rancidity because of the characteristic flavour development. It occurs because the low flavour threshold of the short-chain fatty acids (6:0 to 12:0) means that a low level of these are organoleptically detectable. As little as 0.2% of total free fatty acids will cause a detectable flavour. Development of soapy rancidity can be avoided by reducing moisture and by avoiding microbiological contamination. Lecithin is often added to lauric-type alternative fat products in order to bind any free water and make it unavailable for hydrolysis (Andersen and Roslund, 1987).

Microbiological contamination, usually from poor-quality cocoa powder, can also lead to ketonic rancidity where the flavour compounds are ketones. It is now thought that off-flavours in lauric-type alternative fat products are caused by a combination of ketonic and soapy rancidity (Padley, 1994).

5.5   Conclusions

We have seen that a wide variety of fats is available for use in the confectionery industry. The traditional fats—cocoa butter, milk fat and coconut oil—have long since been augmented with a range of alternative fats. Initially, the aim was merely to simulate the properties of the traditional fats, but in recent years the aim has been to improve on them. Improvements that have been achieved include:

- lower cost and improved availability
- better melting properties
- resistance to bloom
- convenience of use (e.g. no need to temper)
- resistance to both oxidative and lipolytic rancidity

Fat can be seen as the most functional ingredient in both chocolate and sugar confectionery and can be tailor-made for each particular application.

References

Andersen, B. and Roslund, T. (1987) Low temperature hydrolysis of triglycerides, paper presented at the Lipid Forum conference, Mora, Sweden, June.

Beckett, S.T. (ed.) (1999) *Industrial Chocolate Manufacture and Use*, 3rd edn, Basil Blackwell, Oxford.

Beckett, S.T. (2000) *The Science of Chocolate*, RSC paperbacks, Royal Society of Chemistry, Cambridge.

Crews, C., Calvet-Sarrett, R. and Brereton, P. (1997) The analysis of sterol degradation products to detect vegetable fats in chocolate. *J. Am. Oil Chem. Soc.*, **74**, 1273-1280.

Cullinane, N., Aherne, S., Connolly, J.F. and Phelan, J.A. (1984) Seasonal variation in the triglyceride and fatty acid composition of Irish butter. *Irish J. Food Sci. Tech.*, **8**, 1-12.

Duurland, F. and Smith, K. (1995) Novel non-temper confectionery filling fats using asymmetric triglycerides. *Lipid Tech.* (January) 6-9.

Eagle, R. (1999) The Chocolate Directive. Bob Eagle Associates and Britannia Food Ingredients Ltd, www.britanniafood.com.

European Parliament and Council (2000) Directive 2000/36/EC of the European Parliament and of the Council of 23 June 2000 relating to cocoa and chocolate products intended for human consumption, *Off. J. Eur. Com.*, **L197**, 19-25.

Fincke, A. (1982) Möglichkeiten und Grenzen einfacher gaschromatografischer Triglyceridanalysen zum Nachweis fremder Fette in Kakaobutter und Schokoladenfette: 4. Auswertung gaschromatografischer Triglyceridanalysen von Milchschokoladenfetten. *Deutsche Lebensmittel Rundschau*, **78**, 389-396.

Gordon, M.H., Padley, F.B. and Timms, R.E. (1979) Factors influencing the use of vegetable fats in chocolate. *Fette Seifen Anstrichmittel*, **81**, 116-121.

Hachiya, I., Koyano, T. and Sato, K. (1989a) Seeding effects of solidification behavior of cocoa butter and dark chocolate: I. Kinetics of solidification. *J. Am. Chem. Soc.*, **66**, 1757-1762.

Hachiya, I., Koyano, T. and Sato, K. (1989b) Seeding effects of solidification behavior of cocoa butter and dark chocolate: II. Physical properties of dark chocolate. *J. Am. Oil Chem. Soc.*, **66**, 1763-1770.

Hammond, E. and Gedney, S. (2000) Fat bloom. United Biscuits (UK) Ltd and Britannia Food Ingredients Ltd, www.britanniafood.com.

Hanneman, E. (2000) The complex world of cocoa butter. *The Manufacturing Confectioner*, **80**, 107-112.

Haylock, S.J. (1995) Dried ingredients for confectionery. *49th PMCA Production Conference*, 21-29.

Jackson, E.B. (ed) (1995) *Sugar Confectionery Manufacture*, 2nd edn, Blackie Academic & Professional, Glasgow.

Koyano, T. (2000) Personal Communication, *Meiji Seika Kaisha Ltd.*, Saitama 350-0289, Japan (www.meiji.co.jp).

Kristott, J. (2000) Fats and oils, in *The Stability and Shelf-life of Food* (eds. D. Kilcast and P. Subramaniam), Woodhead Publishing, Cambridge, Chapter 12, 279-309.

Krug, M. and Ziegleder, G. (1998a) Lichtanfälligkeit weisser Schokolade (part I–Problem und Ursachen). *Zucker u. Süsswarenwirtschaft*, **51**, 24-27.

Krug, M. and Ziegleder, G. (1998b) Lichtanfälligkeit weisser Schokolade (part II–Erste Versuche zur Problemlösung). *Zucker u. Süsswarenwirtschaft*, **51**, 102-104.

Lipp, M. and Anklam, E. (1998) Review of cocoa butter and alternative fats for use in chocolate: part B. Analytical approaches for identification and determination. *Food Chem.*, **62**, 99-108.

MacGibbon, A.K.H. and McLennan, W.D. (1987) Hardness of New Zealand patted butter: seasonal and regional variations. *NZ J. Dairy Sci. Tech.*, **22**, 143-156.

Minifie, B.W. (1989) *Chocolate, Cocoa, and Confectionery: Science and Technology*, 3rd edn, Chapman & Hall, New York and London.

Okawachi, T., Sagi, N. and Mori, H. (1985) Confectionery fats from palm oil. *J. Am. Oil Chem. Soc.*, **62**, 421-425.

Padley, F.B. (1994) The control of rancidity in confectionery products, in *Rancidity in Foods* (eds. J.C. Allen and R.J. Hamilton), 3rd edn, Blackie Academic & Professional, Glasgow, Chapter 14, 230-255.

Padley, F.B. (1997) Chocolate and confectionery fats, in *Lipid Technologies and Applications* (eds. F.D. Gunstone and F.B. Padley), Marcel Dekker, New York, pp. 391-432.

Padley, F.B. and Timms, R.E. (1978) Determination of cocoa butter equivalents in chocolate. *Chemistry and Industry*, 918-919.

Padley, F.B. and Timms, R.E. (1980) The determination of cocoa butter equivalents in chocolate. *J. Am. Oil Chem. Soc.*, **57**, 286-293.

Parker, R. (2000) Chewing the fat. *Confection*, 18-20.

Rahim, M.A.A., Kuen, L.P., Fisal, A., Nazaruddin, R. and Sabariah, S. (1998) Fat interactions and physical changes in 'melt-away alike' chocolate during storage, in *Proceedings of Conference on Oilseed and Edible Oils Processing, Istanbul 1996, Volume 2* (eds. S.S. Koseoglu, K.C. Rhee and R.F. Wilson). AOCS Press, Champaign, Illinois, pp. 141-143.

Seguine, E. (2000) Personal Communication, *Guittard Chocolate Company*, Burlingame, CA 94010 (www.guittard.com).

Shukla, V.J.S. (1995) Cocoa butter properties and quality. *Lipid Tech.*, **7**, 54-57.

Stansell, D. (1995) Caramel, toffee and fudge, in *Sugar Confectionery Manufacture* (ed. E.B. Jackson), 2nd edn, Blackie Academic & Professional, Glasgow, Chapter 8, 170-188.

Sudin, N. (2001) Personal Communication, Technical Advisory Service, Malaysian Palm Oil Board PO Box 10620, 50720 Kuala Lumpur, Malaysia.

Talbot, G. (1995) Chocolate fat bloom: the cause and the cure. *International Food Ingredients* (January/February), 40-45.

Talbot, G. (1999) Vegetable fats, in *Industrial Chocolate Manufacture and Use* (ed. S.T. Beckett), 3rd edn, Basil Blackwell, Oxford, pp. 242-257.

Tietz, R.A. and Hartel, R.W. (2000) Effects of minor lipids on crystallization of milk fat–cocoa butter blends and bloom formation in chocolate. *J. Am. Oil Chem. Soc.*, **77**, 763-771.

Timms, R.E. (1980) The phase behaviour of mixtures of cocoa butter and milk fat. *Lebensmittel Wissenschaft Technologie*, **13**, 61-65.

Timms, R.E. (1983) Choice of solvent for fractional crystallisation of palm oil, in *Palm Oil Product Technology in the Eighties* (eds. Pushparajah and Rajadurai), Palm Oil Research Institute, Kuala Lumpur, pp. 1-17.

Timms, R.E. (1984) Phase behaviour of fats and their mixtures. *Prog. Lipid Res.*, **23**, 1-38.

Timms, R.E. (1997) Fractionation, in *Lipid Technologies and Applications* (eds. F.D. Gunstone and F.B. Padley), Marcel Dekker, New York, pp. 199-222.

Timms, R.E. and Stewart, I.M. (1999) Cocoa butter, a unique vegetable fat. *Lipid Technology Newsletter*, **5**, 101-107.

Trout, R. (2000) Manufacturing low-fat cocoa. *The Manufacturing Confectioner*, **80**, 75-82.

Unilever (1956) Improvements in or relating to cocoa-butter substitutes. Patent specification GB 827,172, The Patent Office, London.

Versteeg, C., Thomas, L.N., Yep, Y.L. and Papalois, M. (1994) New fractionated milkfat products. *Australian J. Dairy Tech.*, **49**, 57-61.

Webster, G. (2000) Personal Communication, *FoodChem Associates Ltd.*, Auckland (www. foodchem.co.nz).

Weyland, M. (1992) Cocoa butter fractions: a novel way of optimizing chocolate performance. *The Manufacturing Confectioner*, **72**, 53-57.

Wong Soon (1988) *The Chocolaty Fat from the Borneo Illipe Trees*, Atlanto Sdn. Bhd., Petaling Jaya, Malaysia.

Wong Soon (1991) *Speciality Fats versus Cocoa Butter*, Atlanto Sdn. Bhd, Petaling Jaya, Malaysia.

Young, C.C. (1984) The interpretation of GLC triglyceride data for the determination of cocoa butter equivalents in chocolate: a new approach, *J. Am. Oil Chem. Soc.*, **61**, 576-581.

Ziegleder, G. and Mikle, H. (1995a) Fettreif: Probleme, Ursachen, neure Ergebnisse (part I), *Süsswaren Tech. Wirtschaft*, **39**(9) 28-32.

Ziegleder, G. and Mikle, H. (1995b) Fettreif: Probleme, Ursachen, neure Ergebnisse (part II–Probleme, Ursachen, neuere Ergebnisse). *Süsswaren Tech. Wirtschaft*, **39**(10) 23-25.

Ziegleder, G. and Mikle, H. (1995c) Fettreif: Probleme, Ursachen, neure Ergebnisse (part III–Probleme, Ursachen, neuere Ergebnisse). *Süsswaren Tech. Wirtschaft*, **39**(11) 26-28.

Ziegleder, G. and Schwingshandl, I. (1998) Kinetik der Fettmigration in Schokoladenprodukten (part 3). *Fett/Lipid*, **100**, 411-415.

Ziegleder, G., Moser, C. and Geier-Greguska, J. (1996a) Kinetik der Fettmigration in Schokoladenprodukten (part 1–Grundlagen und Analytik). *Fett/Lipid*, **98**, 196-199.

Ziegleder, G., Moser, C. and Geier-Greguska, J. (1996b) Kinetik der Fettmigration in Schokoladenprodukten (part 2–Enfluss von Lager temperatur, Diffusions, Koeffizient, Festfettgehalt). *Fett/Lipid*, **98**, 253-256.

6 Spreadable products

David J. Robinson and Kanes K. Rajah

6.1 Introduction

6.1.1 Definition of spreads: margarine, low(er) fat spreads and butter

The term 'yellow fat spreads', often just 'spreads', refers to all products that are described as butter, margarine and their low-fat alternatives, the vast majority of which are made mainly from vegetable oils. However, in recent years two other product groups have emerged—sweet and savoury spreads—and their use in place of yellow fats is so significant that it is appropriate to consider these, albeit briefly, within this chapter.

In Asia and the Middle East ghee is used instead of butter as a spread on local flour-based hosts such as nan bread in Pakistan and Afghanistan, chapati and so on in Northern-India and pancakes (e.g. thosai) in South India and Sri Lanka. Rice is often also an important host. The vegetable oil alternative is known as vanaspati; both products are discussed in the chapter on culinary fats.

6.1.1.1 Competition between butter and margarine

Butter, which was developed over 5000 years ago, remained unaltered and without serious competition until the invention of margarine by Mège Mouriès, a French chemist, in 1869. He utilised a mix of beef tallow oleine, with about 10% milk, some water and 0.4% udder tissue as flavour. This patented mix was agitated to produce a cream that was then processed like a butter to obtain margarine, or butterine, as it was known in the UK and the USA respectively in the early days.

Industrialists rapidly adopted his invention and world consumption grew, overtaking butter in the late 1970s and reaching its zenith, of 9.5 million tons, in the early 1990s. Since then it has levelled off as other spreads and snacking, and probably the lack of innovation in yellow fats, has taken its toll (figure 6.1). For instance spreads such as soft cream cheese and chocolate spreads have eroded the yellow fat market. Latterly, the growth of perceived soft 'butter' spreads has further reduced the traditional (minimum 80% fat) 'margarine' market.

Today margarine is fast becoming a generic term for vegetable yellow fat spreads containing less than 80% fat. By 2000 in the USA the percentage of fat in vegetable yellow fat spreads had dropped to 53% and in Unilever's products in Europe the average fat level is 63%. In Unilever Europe only 38% of these products are now called 'margarine', having over 80% fat.

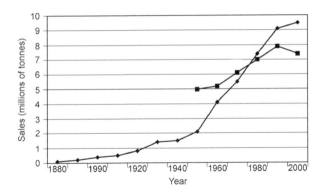

Figure 6.1 World margarine and butter sales (*Nahrungs fette und-öle*, Michael Bockisch. Updated by *Oil World* statistics). ◆ Margarine, ■ Butter.

Numerous on-pack descriptors have replaced the term 'margarine' over the past 20 years. Butter, with over 80% fat, is also fast disappearing as a descriptor, with fat levels falling and butter being replaced by vegetable oils, making them more spreadable and affordable. So we enter the twenty-first century with the descriptor 'margarine', invented in the nineteenth century, almost extinct and another, 'butter', with its origins stretching back over 5000 years following a similar fate.

6.1.1.2 Market and usage considerations

All these spreads are unique in that they are used in conjunction with a 'host', such as bread, biscuits or vegetables. Alternatively, they are used as an ingredient, or aid, in the cooking of meat, cakes, pastry, creams, sauces and other food products. No spread is eaten alone and as such is totally dependent on the market and the properties of the 'host' or destination product. Having said this, the functionality of the spread can significantly improve the attractiveness of the 'host' or destination product and both benefit from mutual innovation.

For example, a spread, such as a margarine or spreadable butter sold in a soft plastic tub will suit softer breads, as often sold in the UK, whereas a harder spread such as butter or margarine sold in a wrapper will be more acceptable on harder breads, such as those made in Germany or France. Also, in markets where the 'host', such as bread, is practically nonexistent, as in China, the market for yellow fats is very small. In the USA it is traditional for vegetables to be coated with a yellow fat and therefore the 'host' is not always confined to a baked product.

In designing the 'perfect' yellow fat spread it is necessary to consider the multifunctional nature of the product, as often the single pack in the household is used in conjunction with many food products. Any consumer market analysis will show that a single yellow fat is used not only for spreading but typically will be used in the shallow frying of meat, for coating vegetables and for making cakes, pastry and sauces.

Additionally, the taste, colour and texture of the spread will tend to follow the characteristics of the traditional butter sold in that region. Hence, in Sweden and the UK a high-salted product is preferred, whereas in Holland and Germany lightly salted products are the norm.

Lastly, health and nutrition have become an important factor in the yellow fat market and products, with cholesterol-lowering properties, added essential nutrients, such as vitamins and low-calorie products making up a significant proportion of the market.

When designing a product all these various factors need to be taken into consideration. Some functions will conflict whereas others will be mutually exclusive. A trade-off needs to be made between the essential USP (ultimate selling property) and the other desirable, but not essential, requirements.

6.1.2 Summary of product development

One way to show the history of margarine is on a time line indicating the years when various product types were introduced together with the enabling process technologies (figure 6.2). From this it is clear that the first 90 years were characterised by innovation in margarine processes, as well as by considerable oil processing changes, which will not be covered here. The only major product innovation during this period was in 1927 with the adding of vitamins to margarine. This is taken for granted today but without this vital nutritional supplement it is likely that this industry would be much smaller today. In the UK a government committee in 1926 ruled that margarine should not be considered as an effective substitute for butter because of its vitamin deficiency and should certainly not be used by children. It is interesting to note that in the UK today 30% of the daily intake of vitamin D and 15% of vitamin A comes from margarines.

The past 40 years show a significant number of product innovations, with few fundamental process advances (figure 6.2(b)). It was only in the early 1960s that spreadable products sold in tubs were introduced, closely followed by liquid margarine, primarily in the USA. The mid 1960s saw the introduction of tubs of 'heart-health' high-PUFA (polyunsaturated fatty acids) as well as low-calorie halvarine products, with an empty water phase. These products received considerable publicity in the 1980s with the publication of the COMA (Committee on the Medical Aspects of Food in the UK) report which promoted the reduction of total fat and saturated fatty acid in the diet.

Although milk fat and vegetable fat mixtures, known as melanges, have been used over the years, usually with butter levels below 10%, it was not until the 1970s that an 80% melange, Bregott, with around 75% butter, was marketed in Sweden. Also in Sweden a similarly constituted fat phase is used for a half-fat product, Latt au Lagom. With the benefit of hindsight it is a mystery that similar products took so long to become established in other countries. In the late 1970s

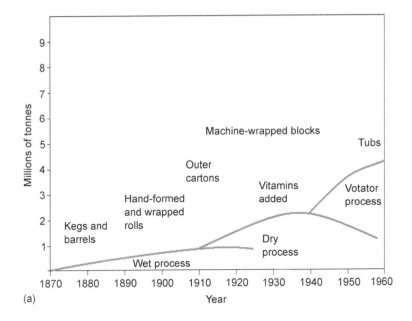

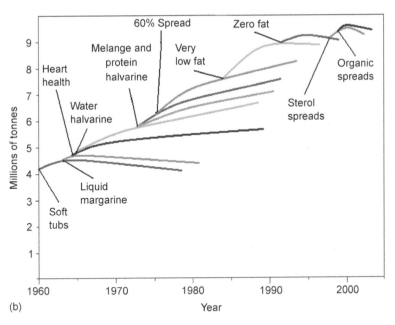

Figure 6.2 Margarine products and processes: (a) 1870–1960; (b) 1960–2000.

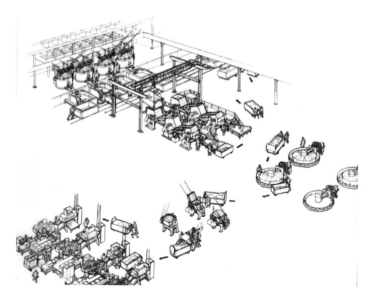

Figure 6.3 Margarine manufacturing in the late nineteenth century: plant layout.

the low-protein and high-protein halvarines were introduced with, for the first time in the UK, the use of preservatives.

By the 1980s the fat levels in most traditional brands started to fall, and the ubiquitous 'reduced 60% fat spread' entered our vocabulary. The 20%, very-low-fat, and 'almost-zero-fat' spreads were introduced in the late 1980s and early 1990s.

In 1998 the first significant cholesterol-lowering yellow fat, Benecol, was introduced, followed in 2000 by Unilever's Proactiv (both Proactiv and Benecol are registered). Perhaps after 10 years of market decline in volume the market will increase in terms of value, as these products sell at four times the price of equivalent products because of the high price of plant sterols.

Last, organic spreads are now starting to appear on the supermarket shelves in the UK and Sweden. Both the lack of confidence with governments in dealing with the BSE (bovine spongiform encephalopathy) crisis and the introduction of GMOs (genetically modified organisms), as well as the green agenda of the 1990s, has provided the catalyst for these products to become accepted despite their price premium.

6.1.3   Summary of process development

As mentioned above, the first 90 years of the industry was characterised by process, rather than product development. Although not covered here, developments in oil processing—particularly hardening, deodorisation and interesterification—greatly influenced the raw materials used and the quality and range of the products manufactured.

6.1.3.1   Unit processes

Mège Mouriès and the early margarine manufacturers used the same basic unit processes that are in use today—mixing, emulsification, chilling, working and resting—but there the comparison must end, for today's manufacturing process bears little relationship to the open, wet and time-and labour-consuming process at the end of the nineteenth century (figure 6.3).

6.1.3.2   Wet chilling

For a long period of time the mechanical churn drum (figure 6.4) was used to mix the oil and aqueous ingredients and to create an oil-in-water emulsion. The churn cooled the product through the cooling jacket but also ice was used as an ingredient to aid this process. During this process the emulsion partially inverts to a water-in-oil emulsion.

When the process was complete the cooled emulsion was flushed out of the churn, sprayed with iced water and poured into open troughs through which cold water flowed. The water and emulsion mix was held in wagons to continue fat crystallisation and separation. The crystallised emulsion was separated from the free water and kneaded in multiplex rollers to start the working process.

The margarine was then stored in wagons for about 24 h to initiate postcrystallisation, before being transferred to the French rolls. These rolls worked

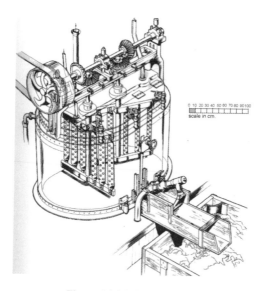

scale in cm.

Figure 6.4 Mechanical churn.

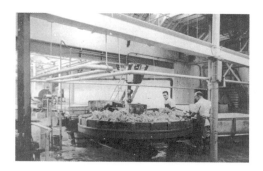

Figure 6.5 French rolls.

and softened the product further to make it less brittle and easier to spread (figure 6.5).

Salt and colour were sometimes added in the French rolls or in a kneading blender. Salt and colour were added towards the end of the process to prevent losses and protein precipitation during the wet chilling process. It also allowed small batches of product to be salted and coloured to suit regional tastes. For example, in Wales it was said they liked to 'chew their salt'; whereas in Kent a very dark colour known as 'double Kent' was required; in Glasgow it was believed the people would eat only pale margarine.

More resting in wagons was required before being packed into kegs, rolls or packets. In total, some 60 h was taken from mixing of the ingredients to the packing of the final product. Estimates suggest that one person was required to produce about 40 tonnes of margarine per year.

6.1.3.3 Open dry chilling

It must have appeared a significant advantage in the 1920s for the wet process to be superseded by the open dry chilling drum (figure 6.6). The churned emulsion,

Figure 6.6 Chill drum.

now of the water-in-oil type, including salt and colour, was passed directly to the feed roll, or trough, on the chilling drum. This was initially chilled internally by brine, but later by ammonia.

The flakes could be held in wagons to continue crystallisation or be fed direct to a complector, which worked the product as well as applying vacuum to remove air. The blocks of product were then stored in wagons to await filling into hopper-fed wrapping machines, developed in the 1920s.

The complete process cut the cycle time in half, to about 30 h, with productivity at about 200 tonnes per person per year.

6.1.3.4 Votator process

In the 1940s a scraped surface heat exchanger was developed to chill margarine. This 'A unit', as it was known, in conjunction with enclosed crystallisers (known as 'C units') and resting tubes (or 'B units') were used to feed product direct to a packing machine (figure 6.7). For the first time the complete process could be totally enclosed. Surprisingly, this continuous process is still in use today despite a major drawback in that it requires a rework system to accommodate packing machine stoppages. Direct-fed packing machines were in operation packing at speeds up to 220 packs per minute; now the figure is 360 tubs per minute.

With little significant change this led the way to today's totally enclosed process, with a total time from mixing to packing of 10 min. Productivity has risen, in larger plants, to 800 tonnes per person per year.

6.1.4 Summary of ingredient development

The ingredients used in traditional margarines can be subdivided into oil-phase ingredients, oil-soluble ingredients and aqueous-phase ingredients.

6.1.4.1 Oil-phase ingredients

Looking behind the products at the raw materials used shows interesting trends in history (figure 6.8). The first oil to be used was beef tallow, used as the

Figure 6.7 A Votator.

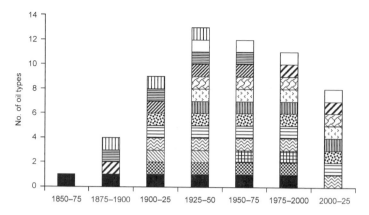

Figure 6.8 Major oil types used in margarine. ■ Tallow, ▨ Lard, ⊞ Fish, ▨ Whale, ▨ Palm,
⊟ Coconut, ▨ Palm kernel, ▥ Sunflower, ▫ Soybean, ▨ Rapeseed, ▨ Groundnut, ▨ Olive,
▤ Cottonseed, □ Maize, ▥ Seasame.

soft fraction, oleine. Tallow was first softened by olive oil, a practice that was
quickly abandoned because of the high price of olive oil, in favour of groundnut
and cottonseed oil. It has taken over a century for olive oil to reappear under a
Mediterranean life-style platform.

Early on, sesame oil was used as a marker in margarine to distinguish it from
butter, with the aid of the long-established phytosteryl-acetate colour test. Palm
and the lauric oils, coconut and palm kernal were introduced in the early 1900s
and still enjoy high usage today. Most of the liquid oils in use today, such as rapeseed, sunflower, maize and soybean oils, were used in the early twentieth century.

The use of whale oil needed to wait for the hardening process to be developed
by Normann in 1902, but it was not until the second quarter of the twentieth
century that sizeable quantities were used in margarine. By the 1960s public
concern and catch restrictions stopped its use.

Hardened fish oil was used from the 1950s and in some countries was the
major oil ingredient until 1993, when worries arising from Willett's trans fatty-
acid studies and later the sustainable fishing issue cut their use dramatically
(Willett *et al.*, 1993). This reduction of the raw material supply was closely
followed by the BSE crisis, first in the UK and now in mainland Europe, which
has directly limited the use of tallow. Indirectly, the public questioned the use
of all animal fats, so lard has also become unacceptable. Genetically modified
oils came under significant public scrutiny in the 1990s, which has limited the
use of soybean and maize oil, particularly in Europe.

Fat levels and triglyceride types. Coincident with these changes the reduction
of fat levels and trans fatty acids combined with an increase in use of polyunsaturated fats has led to a change in dietary intake from margarine (figure 6.9). This

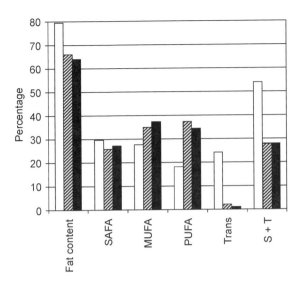

Figure 6.9 Fat content of yellow fats, 1984–2000, as produced by Unilever, UK (in the products Stork, Flora, Olivio and 'I can't believe it's not butter'). Note: SAFA, saturated fatty acid; MUFA, monounsaturated fatty acid; PUFA, polyunsaturated fatty acid; Trans, trans fatty acid; S + T, saturated plus trans fatty acids; the values for SAFA, MUFA, PUFA and Trans are expressed as a percentage of the fat phase only. □ 1984, ▨ 1996, ■ 2000.

figure shows the UK situation for Unilever's products over the past 15 years, with significantly lower fat and trans fatty acid levels and higher PUFA and MUFA (monounsaturated fatty acid) levels. Surprisingly SAFA levels have not fallen much, as a percentage of the fat phase, but in total have fallen due to lower fat levels in the products. This position will be similar for most Western Countries.

6.1.4.2 Oil-soluble ingredients
The oil-soluble ingredients typically involve the emulsifiers, flavours, colorants and oil-soluble vitamins. An emulsion is created in the margarine by mechanical shearing, originally in the churn but now in the votator, and is in part stabilised by the crystallised fats. However, on storing or temperature cycling the emulsion will start to break down unless stabilised by an emulsifier. Also, during hot use, such as in shallow frying, a margarine without an emulsifier will spatter. Before the 1900s egg yolk was used as an emulsifier and antispattering agent but was replaced in the 1920s with vegetable lecithin, mainly from soybeans. At about the same time monoglycerides and diglycerides were introduced to stabilise the margarine emulsion at ambient temperatures. To this date these emulsifiers are still in general use.

Until the 1950s flavouring came from the cultured milk incorporated in the product. Since the 1950s butter flavours have been developed, in particular delta lactones and diacetyl.

Owing to consumer expectations, colorants have always been an important additive in the manufacture of margarine. A variety of food colours have been available for many years, but the main source has been from the carotenoids in red palm oil and bixin (obtained from the seeds of the annatto tree).

Vitamins were developed in the 1920s and first introduced in margarine in 1927. Vitamins A and D are still used in almost all margarine. Additionally, many margarines contain vitamin E, an antioxidant, which is added to supplement the vitamin E that naturally occurs in crude vegetable oils but is partially lost during deodorisation.

6.1.4.3 Aqueous-phase ingredients

The only major change in the aqueous phase of margarine over the years has been the change from the use of cultured milk to skim milk powder, or whey. This change happened in conjunction with improvements in flavours in the 1950s. Ripening of milk continued up to the 1980s although some plants continued with this process until the end of the twentieth century. The process was both time- and space-consuming, requiring the use of large souring tanks needing up to 30 h to culture the milk.

More interesting changes have taken place in the aqueous phase of low-fat products (figure 6.10). Initially, a protein empty aqueous phase was used in the first halvarines in the 1960s. In the 1970s high protein levels of dairy-based low-fat spreads such as Latt au Lagom appeared. With the price of milk protein equating to that of milk fat it was not long before medium-level protein-containing spreads entered the market. Protein levels continued to decline as thickener systems changed from gelatine, to alginate, then to starch during the 1980s and 1990s.

6.1.5 Summary of packaging developments

Margarine was originally packed into kegs or barrels (in the late 1800s). At about the turn of the century it was packed largely in hand-wrapped rolls or bricks. Machine wrapping was introduced in the 1920s. Initially, vegetable parchment was used as the wrapping material, but by the middle of the twentieth century some premium products were upgraded to aluminium foil wrappers. The folded carton was introduced in the USA in 1907 to protect the softer wrapped product usually sold as 'fingers'. This type of pack was, and still is, sold on both sides of the Atlantic for premium softer products.

Tubs were introduced in the late 1950s and the materials and forming processes have changed from waxed card, to thermoformed PVC (polyvinyl chloride), ABS (acrylonitrile butadiene styrene) or polypropylene, to injection-moulded high-density polyethylene and polypropylene, and, latterly, to in-mould labelled polypropylene.

For the best part of the 130-year history of margarine standard primary pack sizes have been prescribed by law with, depending on the measurement used

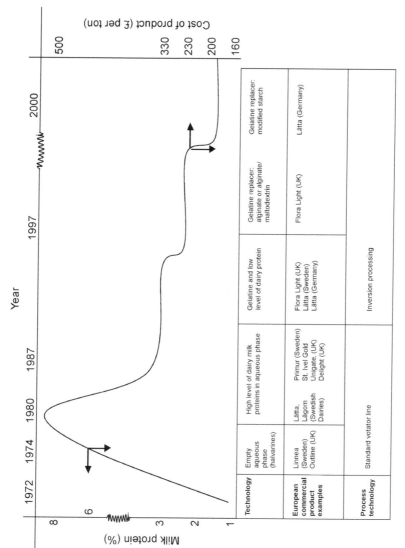

Figure 6.10 Development of Unilever low-fat spreads (40% fat content).

within a country, packs of 250 g ($\frac{1}{2}$ lb), 500 g (1 lb) or 1 kg (2 lb) being the most common. In recent years unit pricing (price per kilogram or pound stated on shelf, or pack) has been introduced, which has overcome the main reason for standard ranges; namely, consumer price confusion. As such, a growing number of countries no longer prescribe a standard range of packs.

Primary packs were hand packed into wooden boxes until replaced by cardboard, around the 1940s. Secondary packaging machines were introduced in the middle of the 1990s and the American box as well as the simple cardboard tray, usually followed by shrink or stretch wrapping, were being mechanically packed on the larger lines by the 1970s. In the 1990s more sophisticated secondary packs consisting of trays with reinforced corners were being formed on-line by robotic packing machinery.

Throughout the second half of the twentieth century the pallet was used as tertiary packaging, in common with most other industries. For about 30 years the display pallet has also been used, particularly in countries where ambient distribution is still used, such as Germany. In Sweden the display pallet, or cage, has been in use in the chilled cabinet for many years. This form of tertiary packaging enables the primary packs to be merchandised within the shop without the need to unbox the product.

6.2   Legislation

Almost from the start, the margarine industry was subject to what can be viewed as discriminatory regulations brought about by strong political activity from the dairy lobby. Margarine was seen to be produced by 'big business', whereas butter was made predominately on a large number of small farms.

Past regulations have played a significant part in the development of the product and its packaging. Regulations proposed that margarine must

- be colourless (with colour capsules provided for the consumer to add at home)
- be coloured (pink, brown or blue!)
- be sold only in cubic packs
- be sold in a separate area from butter
- have a red band printed on the pack
- have sesame and starch tracers
- have a lead seal closure
- be made in quantities set by legal production quotas

Apart from setting adverse taxes, governments have been obsessed with preventing the product being mixed with butter or from being coloured, or packed, as butter. Although blue margarine was never sold, a cubic pack with the warning red band was.

In the USA capsules of colorant were sold with the colourless margarine to allow consumers to 'mix their own'. In fact, even in the 1980s in Wisconsin, a dairy state, colour capsules were still sold. Unbelievably, in 1885, the clash between the butter and margarine lobbies led to the dissolution of the Danish parliament.

The very regulations that made butter so secure also restricted butter to such a narrow definition that the dairy industry locked itself into a single product. The flexibility of the margarine regulations with regard to ingredients and processes—provided milk fat was not used—enabled margarine to thrive at the expense of the dairy industry, at least until the latter part of the twentieth century.

In the 1970s developments in Sweden with Breggott and Latt au Lagom started to push butter into the melange area. But it was the introduction of the single unifying EU Yellow Fats Spreads regulation in 1994 that legally freed up the use of milk and edible fats, at varying fat levels. The dairy industry at last had the legal flexibility to exploit the advantages of butter, this time not only as a raw material but with brand names perceived by the consumer to be 'butter'.

6.2.1   EU regulations

The initial European regulations for spreadable fats were laid down in 1994 (Council Regulations [EC] No. 2991/94) but were subsequently modified by five regulations between 1997 and 1999 (EC Regulations No. 577/97, No. 1278/97, No. 2181/97, No. 623/98, No. 568/99). These regulations apply to products made from three fat groups; namely, 'milk fats', 'fats', and 'fats composed of plant and/or animal products'. Hence butter, margarine and melanges are all covered within one set of regulations.

In principle the regulation applies to yellow fats with a fat content between 10% and 90% by weight. The fat content must be at least two-thirds of the dry matter, excluding salt. The products must remain solid at room temperature and be suitable for use as spreads and be intended for human consumption. The sales descriptions of the products are set out in table 6.1.

6.2.2   US regulations

The US Federal regulations for butter and margarine are contained in separate documents. Additionally, there are laws governing margarine in many of the individual States. The standards for margarine-type products apply to all substances, mixtures and compounds known as margarine and to all substances, mixtures and compounds that have a consistency similar to that of butter. These products may contain any edible oils and fats other than milk if made in imitation or semblance of butter.

The term 'spread' is the commonly used term for lower-fat margarine, but there is no Federal standard for spread. 'Diet margarine' is a name applied to

Table 6.1 Fat groups: description and product category

Sales description	Product category[a]
Milk fats[b]:	
Butter	Milk-fat content not less than 80% but less than 90%; maximum water content of 16%; maximum dry nonfat milk-material content of 2%
Three-quarter-fat butter	Milk-fat content of not less than 60% but not more than 62%
Half-fat butter	Milk-fat content not less than 39% but not more than 41%
Dairy spread (x % fat)	Milk-fat content: less than 39% more than 41% but less than 60% more than 62% but less than 80%
Fats[c]:	
Margarine	Obtained from vegetable and/or animal fats with a fat content of not less than 80% but less than 90%
Three-quarter-fat margarine	Obtained from vegetable and/or animal fats with a fat content of not less than 60% but not more than 62%
Half-fat margarine	Obtained from vegetable and/or animal fats with a fat content of not less than 39% but not more than 41%
Fat spreads (x % fat)	Obtained from vegetable and/or animal fats with fat content: less than 39% more than 41% but less than 60% more than 62% but less than 80%
Fats composed of plant and/or animal products[d]:	
Blend	Obtained from a mixture of vegetable and/or animal fats with a fat content of not less than 80% but less than 90%
Three-quarter-fat blend	Obtained from a mixture of vegetable and/or animal fats with a fat content of not less than 60% but not more than 62%
Half-fat blend	Obtained from a mixture of vegetable and/or animal fats with a fat content of not less than 39% but not more than 41%.
Blended spread (x % fat)	Obtained from a mixture of vegetable and/or animal fats with fat content: less than 39% more than 41% but less than 60% more than 62% but less than 80%

[a]Including a description of the category with an indication of the percentage fat content by weight.
[b]Products in the form of a solid, malleable emulsion, principally of the water-in-oil type, derived exclusively from milk and/or certain milk products, for which the fat is the essential constituent of value. However, other substances necessary for their manufacture may be added, provided those substances are not used for the purpose of replacing, either in whole or in part, any milk constituents.
[c]Products in the form of a solid, malleable emulsion, principally of the water-in-oil type, derived from solid and/or liquid vegetable and/or animal fats suitable for human consumption, with a milk-fat content of not more than 3% of the fat content.
[d]Products in the form of a solid, malleable emulsion, principally of the water-in-oil type, derived from solid and/or liquid vegetable and/or animal fats suitable for human consumption, with a milk-fat content of between 10% and 80% of the total fat content.

a product with 40% fat made like a margarine except for the lower fat content. Again there is no Federal standard or definition.

The composition of margarine is defined by two Federal standards, one for vegetable products, administered by the Food and Drug Administration (FDA), and one for products containing animal fats, administered by the US Department of Agriculture (USDA). The FDA standard, which applies to the majority of margarine, defines margarine as 'food in plastic form or liquid emulsion containing not less than 80% fat'. There is no maximum to the milk fat that can be used. It is obligatory to add vitamin A, but not vitamin D. However, if vitamin D is added it must be at a prescribed minimum level. Addition of vitamin E is not permitted.

6.2.3   Codex standards

There are three separate standards covering margarine (Codex Standard 32-1981 [Rev. 1-1989]), minarine (Codex Standard 135-1981 [Rev. 1-1989]) and butter (Codex Standard A-1-1971 [Rev. 1-1999]). Essentially the margarine standard applies to 'a food in the form of a plastic or fluid emulsion, which is mainly of the type water/oil, produced principally from edible fats and oils, which are not derived from milk. The margarine must contain a minimum fat content of 80% and a maximum water content of 16%. Vitamins A, D and E are permitted as well as other ingredients including milk or milk products, salt, sugars, suitable edible proteins, colours, flavours, emulsifiers, preservatives, antioxidants, antioxidant synergists, acidity regulators and antifoaming agents.

Minarine has a similar product description to margarine but requires a fat content between 39% to 41%. Also, the water content must not be less than 50%. It has a similar ingredient list to margarine, although gelatine, natural starches and thickening/stabilising agents can be used.

Butter is described as a fatty product derived exclusively from milk and/or products obtained from milk, principally in the form of an emulsion of the type water-in-oil. It must contain a minimum of 80% milk fat and a maximum of 16% water and a maximum of nonfat milk solids of 2%. The only additives that can be used are salt, starter cultures, potable water, prescribed colorants and acidity regulators.

At the time of writing a single Codex standard for 'margarine', 'minarine' and 'blended spreads'(which can contain milk fats) is being drafted. Butter will remain covered by a separate standard.

6.3   Emulsion technology

6.3.1   Properties of emulsions

6.3.1.1   Texture, plasticity and consistency
The crystallisation of fat in emulsions is discussed in chapter 1. As a result of such processing, spreads take the form of viscoelastic materials, the physical

properties typically being measurable by elasticity, viscosity and work softening. These properties are also discussed in terms of the physical quality of the emulsion (i.e. texture, plasticity and consistency; De Man *et al.*, 1976). Texture, plasticity and consistency are equally important because they encapsulate those aspects of product benefit sought by the consumer in food applications. For instance, texture is primarily used to describe the state of the emulsion structure, and can range from smooth to floury, grainy, granular, or sandy to coarse and lumpy.

Very close to this is the aspect of consistency, a temperature-dependent property. It describes a smooth, even plastic, state varying from soft, medium, firm and tough to hard and brittle. Since it is possible to encounter smooth as well as grainy and coarse plastic products, plasticity is associated with both these terms.

Plasticity remains distinct from the other two terms because it describes the ability of a fat or emulsion to retain its shape under slight pressure, such as that encountered during rolling, mixing or spreading. Three conditions must apply for a fat system to be considered plastic.

- Two phases must be present, one of which must be a solid and the other preferably a liquid or clearly behaving as a liquid.
- The solid phase must be finely dispersed to enable the entire mass to be effectively held together by internal cohesive forces. The particle size of the solids must be small enough for the force of gravity on each to be negligible in relation to the adhesion of the particle to the mass. Equally, the interstitial spaces must also be small enough to prevent the liquid phase seeping from the material.
- Mass flow must be facilitated. Flow behaviour is usually described with reference to Newtonian, Bingham and pseudo-plastic materials. Liquid oils are typically Newtonian but crystallised emulsions behave as pseudo-plastic materials. In accordance with Bingham's (1922) concept of plasticity, flow is achieved by ensuring the proportion of the phases present are such that the solid particles remain small enough to prevent obstruction or to engage in the formation of a rigidly interlocking crystalline structure.

Melting and solidification or solid–liquid and liquid–solid phase changes are amongst the most important of physical properties of all fats, including milk fat (Patton and Jensen, 1975; Sonntag, 1979). They are determined by the glyceride composition, that is by the amount and type of fatty acids present, the triacylglycerols in which they occur, including their position and rate of occurrence, and polymorphic form (table 6.2). Larsson (1966) reported that complex triacylglycerol mixtures such as margarine exhibit four polymorphic crystal forms—α, β', β_2, and β'_1 in increasing order of stability.

In rheological terms, plastic fats possess a yield stress and hence a yield value. This measure can be used to distinguish between the consistency of butter, margarine and low-fat spreads by studying the alterations in yield value

Table 6.2 Plastic behaviour of spreads

Crystal property	Behaviour or performance
Amount	Solid:liquid ratio
Melting point of crystals	Melting point of fat
Geometry	Melting behaviour, consistency, etc
Mixed crystals	Fat stability
Flocculation to form networks	Firmness

following work softening (mechanical working). After normal work softening, those substances with a high degree of secondary bonds display comparatively higher structural hardness relative to those with a mainly primary bonded structure. Recovery for the primary bonded group is generally limited to achieving a degree of hardness, but not the original values, unless of course standard procedures are adopted for recrystallisation.

Haighton (1965) found that there is a correlation between lower degrees of work softening with plastic fats and high degrees with brittle fats. He also reported on a number of values for different fats (Haighton, 1959).

6.3.1.2 Spreadability

The term 'spreadability' is often used in connection with the solid state of table spreads. It is connected with consistency, texture and shear but is also influenced by the material on which the fat is spread (i.e. the host). These properties are not only important for their physical effect but also for their physiological and psychological effect on flavour and general palatability (table 6.3). These effects are primarily determined by glyceride composition, state of emulsion and type of crystallisation of the fat.

Spreadability can be measured by testing the strength of the structure of spreads. A number of common methods are available. Cone penetrometry is a routine test based on the determination of the yield value. Mortenson and Danmark (1982) have reported on the use of the yield strength to determine the spreadability of butter. Moran (1994) has reviewed a number of techniques: extrusion method, cutting or sectility method, oil exudation, microscopy, sizing of emulsions and electrical conductance.

Table 6.3 Performance expected of spreads at specified temperatures

Temperature (°C)	Performance
0–10	Spreadability from the cold or from the fridge
10–20	Efficient creaming, rolling or sheeting out for baking; ease of packing
15–25	Stability at ambient temperatures in product preparation areas (e.g. in a warm kitchen or bakery)
20–30	Butter, margarine or pastry fats for baking (e.g. pastry products) must retain their plastic character during the early stages of baking
30–35	Organoleptic property or eating quality (i.e. complete melt is required for sandwich spreads but a higher melting property is expected in fats for meat pies)

6.3.1.3   Emulsion quality

A stable emulsion state is important to the quality of the spread. Fundamental to this is the water droplet size. Typically, an 80% fat margarine would contain 95% water droplets in the diameter range 1–5 μm, 4% between 5–10 μm and 1% around 20 μm. It is estimated that when the water droplet diameter is 3 μm, and given that 1 g of margarine contains 16% water, there will be 10–20 million droplets. It is known that bacteria will grow to a length of 15 μm and so it is imperative to keep the diameter of the water droplets below this. However, with the development of lower-fat spreads there is a consequent increase in water content. The water droplet size can increase to 80 μm and therefore the shelf-life of such products is affected. To overcome this, preservatives are often added, particularly to very low-fat spreads, while factory hygiene standards and microbiological quality of raw materials have also to be monitored closely and maintained at an optimum level.

The functional benefits of a fine emulsion are as follows (Andersen and William, 1965):

- it contributes towards the plasticity of the product
- it improves the shelf-life of the product by making it more resistant to bacterial attack (in a small droplet, a bacterium cannot multiply and soon dies)
- added flavours are best perceived when in the aqueous phase because it is held that only water can wet the taste buds; nevertheless, some good oil-soluble flavours are also available (figure 6.11).

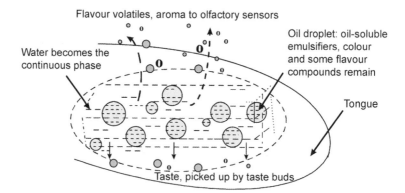

Figure 6.11 Phase inversion of a water-in-oil emulsion and organoleptic quality. Phase inversion is caused by the combined effect of an increase in moisture level from saliva, an increase in temperature to that of the mouth (typically 37.8°C) and mechanical action from mastication. These factors weaken the fat crystal structure, allowing water to become the continuous phase, thereby releasing the flavour compounds held within that phase.

However, it should be noted that too fine a dispersion can also mask the flavour of the margarine. Equally, such an emulsion could also be absorbed without being broken on the palate and therefore never really be tasted.

An early important book covering margarine manufacture is that by Andersen and Williams (1965). Recent reviews and updates are available from Hoffman (1989) and Rajah (1992). Moran (1994) has reported on the preparation of lower-fat emulsions. Details on buttermaking were covered by McDowall (1953). Recent advances have been reviewed by Munro (1986). Jebsen (1994), and Ranjith and Rajah (2001) have also presented an update on the manufacture of cream.

The measurement of butter quality still remains largely a manual process, undertaken by butter graders. The following factors are taken into consideration when grading and assessing the suitability of the butter for retailing as a spread:

- flavour and aroma
- body and texture
- colour, appearance, finish and salt content
- absence of free moisture

Those that fall outside the specification for free moisture may be processed into anhydrous milk fat (butter oil). However, if the moisture level is either much lower than the legal maximum allowed of 16% (e.g. below 15%) or exceeds it, the batch is likely to be blended in with other batches of butter to reach a level as close to 16% as possible. Similar action could also be taken when the salt level falls outside the required limit. However, if the product fails in terms of free moisture, poor texture or flavour it is more than likely to be processed into butter oil. If there are slight flavour defects it could still satisfy the specification for ghee. For instance, a well-developed butter flavour is considered a rancid product in many Western countries. However, it is a much desired product in the Middle East and Far East as a spread and fat base both for sweet confections and for savoury preparations.

6.3.2 Emulsifiers and hydrophilic–lipophilic balance values

The crystallisation of fat contributes a solid crystal network to stabilise the emulsion. However, this alone is insufficient to reduce the interfacial tension at the interface of the two fluids (i.e. water and oil). Emulsifiers are necessary to overcome this problem. The most widely used products are the monoacylglycerols and diacylglycerols, both fulfilling this function by providing the base materials for emulsifiers that are used in water-in-oil emulsions (e.g. margarines). Monoacylglycerols and diacylglycerols of low HLB (hydrophilic–lipophilic balance) numbers help to stabilise the water-in-oil emulsions in margarine. Distilled monoglycerides are used for reduced-fat spreads.

Spreadable butter that is manufactured by emulsifying vegetable oils into the cream prior to churning requires additional emulsifiers because the natural emulsifiers, the phospholipids from the milk-fat globule membrane, are reduced as a result of the blending. A detailed treatment of emulsifiers is available in chapter 7.

6.3.3 Stabilisers

Oil-in-water emulsions such as milk and mayonnaise are stabilised by milk proteins such as casein and lactalbumin, vegetable protein derived from, for example, soybeans, and proteins and lecithin from egg yolk (Hoffman, 1989). Equally, when low-fat water-in-oil spreads, typically below 40% fat, are to be stabilised similar ingredients are used, including starches and gelatine.

6.3.4 Preservatives and microbiological stability

Reduced shelf-life of spreads can be caused either by microbiological or by chemical processes.

6.3.4.1 Microbiological rancidity

Yeast, bacteria or mould cause microbiological rancidity. Their activity leads to the hydrolytic decomposition of fats and sometimes the splitting of the proteins in the water phase, with either event resulting in unpleasant taste and odour. Hence prevention of microbiological activity is important. Since microorganisms cannot usually grow in fat, it follows that it takes place in the water droplets and on the surface of yellow fat spreads. A number of additives and procedures can be used to address the problem.

Preservatives. Benzoates or sorbates are added to the water phase, and benzoic acid or sorbic acid are added to the fat phase, to prevent the growth of microorganisms. Usually, the decomposition of fats as a result of microbiological activity will result in an increase in its acid value, from below 0.1% in a neutral oil to 0.3%–1.0%, which is considered too high. It should be noted that addition of sorbic acid or benzoic acid may cause the acid value to increase by 0.5–1.0% without any decomposition of the fats having taken place.

Dosage rates depend on the composition of the water phase and the degree of contamination. Typical rates are 0.1–0.2%, calculated on the finished margarine. Preservatives are more effective at a pH of 4.0–4.5 than at a pH of 5.5–6.0.

Salt. Sufficient addition of salt to the water phase should stop the growth of microorganisms, but this is dependent on the type of microorganism. In normal margarine recipes, a 1% addition of salt will prevent the growth of many microorganisms, but the addition of 2% will stop most. However, equally, in some instances small quantities of salt, (i.e. 0.1–0.2%), may even support the growth of microorganisms.

It is maintained that it is the actual percentage of salt within the water phase that is important and not the salt content of the overall margarine. A 1% addition in a margarine containing 16% water should result in 6% in the water phase with 2% addition naturally leading to 12% in the water phase.

pH in the water phase. A low pH, of 4.0–4.5, retards the growth of microorganisms whereas a high pH, 5.5–6.0, furthers growth.

Starch. Microorganisms can thrive in the conditions created by the presences of starches but they can also be adversely affected.

Pasteurisation. The process described for cream pasteurisation (see section 6.3.5) applies here. Pasteurisation of the aqueous phase or the liquid emulsion will improve shelf-life stability.

Fat blend. Some fats are more prone to hydrolytic activity by microorganisms than others (e.g. the lauric oils, mainly coconut oil and palm kernal oil, which develop a soapy taste).

Temperature. Low storage temperatures, of 5–10°C for butter and margarine is recommended. Most microorganisms thrive at 20–30°C.

6.3.5 Emulsion preparation

A stable emulsion is a fine dispersion of one immiscible liquid in another. For buttermaking the following procedures are carried out (Jebsen, 1994).

- Separation: cream (an oil-in-water emulsion) is concentrated to an optimal fat level of 38–42% by expelling a considerable amount of water (as skimmed milk) through centrifugal separation. With use of modern separators, the fat level in the skimmed milk may be as low as 0.05%. Milk-fat globule diameter varies from 1 μm to 10 μm. Those below 1 μm are lost in the skimmed milk, and 10% of those between 2–3 μm microns are also lost in this way.
- Pasterisation: heat treatment destroys many microorganisms and enzymes. Hence, a typical pasteurisation treatment of 95°C (203°F) for 15 s is often used.
- Deodorisation: dairy herds feeding on fresh pasture (e.g. in New Zealand) sometimes graze on weeds and shrubs. Some varieties of these weeds and shrubs impart such strong flavours that can carry through to the milk fat during biosysnthesis. In other instances, oils and high-fat fluids

readily pick up aromas from the environment. Deodorisation removes these unwanted aromas and flavour taints. A vacreator is used for this purpose. Cream is heated to 95–98°C (203–208°F) through direct steam injection, which strips out and removes the aromas in the outgoing vapours. This is a severe treatment which leads to more fat loss in the buttermilk. The cream is then cooled to 5–8°C (41–46°F) to crystallise the fat and to inhibit the growth of microorganisms.

- Cream treatment: this takes place in cream ripening tanks. Crystallisation technology for fat in cream differs from that of milk fat. Milk fat has a tendency to become supercooled (i.e. cooling to below its melting point without crystallisation or solidification having taken place). At this point, further subcooling cannot follow, but once crystallisation is triggered it does not abate until crystallisation is evident in the entire mass. With cream, however, crystallisation must be triggered for every single fat globule. Consequently, crystallisation of cream is time-dependent, and typically takes a minimum of 2 h at 8°C (46°F) (Berntsen, 2001). Sufficient crystallisation will minimise fat loss in the buttermilk and lead to obtain good consistency in butter. However, cream churned directly from the cooling section of the pasteuriser will result in fat loss between 50 to 100% higher than that which had been held and 'ripened' over time. If fat loss is the only criteria, then deep cooling alone would suffice but this will inevitably also contribute towards a very hard butter.

- Cream temperature treatment: the temperature treatment varies with the type of butter to be produced (i.e. whether it is lactic or sweet cream butter). For instance, lactic batters are preferred in Scandinavia and in most Western European countries, whereas sweet cream butter is more popular in the United Kingdom, the Republic of Ireland and New Zealand.

  - Lactic butter: butter from cultured cream (pH 4.7–4.8) is held in a jacketed cream-ripening tank. Considerations of volume of cream to heat-transfer surface (heating/cooling area) suggest that tank volumes should not exceed 30 000 l. Tight control of temperature is also required. This allows the culture added to promote fermentation of the cream to form lactic acid and the desired flavours. Equally, the type of fat in the cream will influence the choice of cooling profile to be used. Hence the final selection is a compromise between the two. For fermented winter cream [i.e. for low-iodine-value fats (more saturated fatty acids)] the temperature cycle often used is 8°C to 19°C to 16°C. For fermented summer cream [i.e. high-iodine-value fats (relatively higher levels of unsaturated fatty acids)] the profile is 19°C to 16°C to 8°C. There may be local variations in the temperature profile used (Berntsen, 2001)

- Sweet cream butter: although the temperature treatment of sweet cream can be carried out in the same tanks as for lactic cream, it is often in much larger silos. For winter sweet cream (low-iodine-value fats), the temperature profile is typically 18–21°C then 14–16°C followed by 8–10°C before churning.

The important aspects of process change that take place during churning are (Jebsen, 1994; Ranjith and Rajah, 2001):

- The emulsifiers, particularly the milk-fat globule membrane, which stabilise the original oil-in-water emulsions, are inactivated by aeration.
- Phase inversion results, and the oil becomes the continuous phase (i.e. the water-in-oil emulsion found in butter). However, the butter emulsion is complex. Unlike margarine emulsions it will contain some oil-in-water droplets, arising from still-intact milk-fat globules. Hence, butter is a mixed emulsion. The free fat (continuous phase) acts as a lubricant. If the amount of lubricant is insufficient, the consistency of the butter is likely to be 'short' or 'brittle'. Rapid cooling of cream results in fast crystallisation and this leads to the creation of many small crystals with a large total surface area. These crystals can bind much of the liquid (noncrystallised) fat; that is, the fat is retained within the globule and thereby the amount of fat available to form the continuous fat phase is reduced, with the consequence that the butter will be hard. Conversely, with slow cooling of the cream, large crystals, with total individual surface area far exceeding that of the small crystals obtained from rapid crystallisation, will form, thereby squeezing out a larger portion of noncrystallised liquid fat from the globules. This forms a more distinct and larger continuous fat phase and hence a relatively softer butter. Therefore, for a more plastic and spreadable butter there needs to be a distinct and dominant continuous fat phase.
- Butter consistency is affected by the melting point of the milk fat and the size and composition of the fat crystals.

For the manufacture of margarine and low-fat spreads the following stages are of primary importance:

- preparation of the water phase
- preparation of the fat phase
- emulsion preparation
- chilling, crystallisation and kneading
- packing or filling

Table 6.4 Spreads in the market: yellow fats

Percentage of fat				
80–82[a]	55–75[b]	39–41[c]	20–25[c]	0–5[c]
Butter and recombined butter	Butter with 60% fat	Butter with 40% fat	Mainly vegetable oil blends	Vegetable oils Nonfat varieties (e.g. mimetics such as globulised egg or whey protein, sucrose polyester)
lactic	Clotted cream	Dairy		
sweet cream (unsalted,	Dairy blends	blends		
slightly salted and	Vegetable	Vegetable		
salted varieties)	oil blends	oil blends		
Other butters				
savoury additives				
(e.g. garlic, chives)				
sweet additives				
(e.g. brandy butter)				
Margarine	Mainly slightly	Only slightly	Only slightly	Only slightly
unsalted	salted and	salted and	salted and	salted and
slightly salted	salted varieties	salted	salted varieties	salted varieties
salted		varieties		

[a]Both groups of products are available in spreadable or packet (stick) form. Pourable margarine is also available.
[b]All groups are available mainly in spreadable form.
[c]All groups are available in spreadable form only.

6.4   Process technology

6.4.1   Current yellow fat range

The current market for yellow fat spreads is summarised in table 6.4. Higher-fat-content spreads such as those from Denmark based on omega-3-rich nonhydrogenated fish oil have been prepared with up to 95% fat. Equally, concentrated butter (up to 97% fat) is available in Europe as beurre cuisine (cooking butter), at a part-subsidised price, strictly for domestic consumption. If required, this butter can be emulsified further, with addition of water and mechanical working with a domestic blender, into a spread. In Italy, bread is eaten dipped into fresh virgin olive oil (100% liquid oil). Ghee (99.8%) and vanaspati (100%) are also used on appropriate hosts which, in some cultures, include bread.

6.4.2   Scraped-surface cooling

6.4.2.1   Drum

This is no longer used as a commercial route for producing spreads and so will not be considered here. The most well-known system is the Diacooler, manufactured by Gersternberg and Agger (Denmark).

6.4.2.2 Tubular action

The schemes available for producing butter or butter–vegetable oil spreads with use of scraped-surface cooling and mechanical working are as follows.

- Phase inversion of high-fat cream. Three pathways are available:

  - creams of with greater than 75% fat undergo phase inversion and are then cooled and texturised mechanically, following one of two routes: in the Alfa Laval, Alfa, New Way and Meleshin process the cream is destabilised during cooling and mechanical working; in contrast, in the Creamery Package and Cherry Burrell (Gold 'n' Flow) process phase inversion takes place before the cooling and working stage
  - 60% fat, phase inversion to 60% butter
  - 40% fat, phase inversion to 40% butter (although this is claimed to be feasible, it is necessary to ensure that, without additives, the water droplet size will not be too large and cause either microbiological or emulsion breakdown)

- Ammix process: cream is added to anhydrous milk fat (AMF) to produce 80% butter. The Ammix process, developed in New Zealand, claims to improve on earlier processes. In principle, a proportion of the cream from the same batch is converted to AMF. The two phases are brought together and scraped-surface cooling and pin working is used to convert an emulsion of the AMF, cream and brine into butter (Rajah, 1992).

- Pacillac process: a mixture of butter and vegetable oils produces 80%-fat or lower-fat spreads.

- Pacillac process: butter and milk protein serum are mixed to produce 60%-fat and 40%-fat butter.

- AMF [or a substitute with milk-fat fractions for spreadable butter (Deffense, 1987, 1993)] is added to milk solids to produce 80%-fat or lower-fat-content butter by using a butter recombination process (Munro, 1982).

- AMF, vegetable oils/fats and milk solids are mixed to produce 80%-fat butter blends.

Today, some of the most well-known units are manufactured by Gerstenberg & Agger (Perfector), Schröder (Kombinator), Cherry Burrell (Votator), Chemtech (Chemetator), and Chris James Consultants (Phrocessor).

A typical plant manufacturing margarine, shortening and yellow fat spreads will feature the following.

- Holding or mixing tanks: to enable minimum holding times, small batch sizes are used.

- Automated load cell batch control: this ensures maximum weight accuracy.

- Digitalised control system interfaced with computer-driven programmes: this facilitates repeatability, record tracking and advanced stock or plant performance monitoring.
- Pumps: typically variable-speed triplex piston pumps fitted with a hygienic pulsation damper in order to achieve a constant flow, free of pressure variation, are used. This feature is particularly important in 100% fat products where compressed gas injection and entrainment is required (e.g. aerated shortenings). Pressure pulsations result in coagulated bubbles appearing as streaks in the final product.
- A pasteurisation unit for emulsions: scraped-surface heat exchangers (SSHXs) are also being used for pasteurisation of low-fat emulsions containing large amounts of emulsifiers and stabilisers. This treatment is carried out at about 80°C. The use of SSHXs prevent 'burn' of the milk proteins (i.e. caseinates) and of starches used in the formulations; consequently fouling of the pasteurisation surfaces are avoided. Cooling to 40°C is then completed in tubular coolers.
- Aerating unit: this is a gas injection system capable of sparging compressed gas (usually nitrogen) into the fat and is required for the purpose of creaming or whitening fats or to add body (volume) and/or soften the fat. The injection probe is typically manufactured from sintered stainless steel in order to ensure a very fine dispersion of gas, a significant technological improvement over the earlier crude orifice-type injection system. Typically, the injection system is located before entry into the first SSHX.
- Inverter: this to phase inverts an oil-in-water emulsion into a water-in-oil emulsion.
- Scraped surface units (also referred to as A units): the development of crystals and crystallisation takes place in these units.
- Working section units (also referred to as B units): the fat passes into a worker unit and undergoes further crystal development and growth while at the same also being physically broken and worked to aid texturisation. The worker unit is designed as a tubular structure with a central rotor shaft containing an arrangement (in some systems a semihelical arrangement) of pins that intermesh with static pins on the tube surface. The pins on the rotor shaft are removable in some systems (e.g. Phrotex), therefore offering the opportunity to reduce shear when all other conditions of temperature and pressure are adjusted to aid crystal development.
- Resting section or tube (also referred to as C units): this unit is required for packet or stick-type spreads and for higher melting pastry margarines and fats.
- A range of stainless steel piping and holding and melting facilities for reworking leftover blends and fats.

The principle underlying the operation of an SSHX remains unchanged, but the technology has been revised to account for market needs. For instance, in the 1960s SSHXs were rated up to about 40 bar g pressure, and these coped adequately with the margarine and shortening manufacture of the day. Many were also driven by gear pumps. However, with increased use of higher melting fats, units capable of 100–160 bar g pressure are now standard. To overcome the inherent problems of slip in gear pumps, triplex piston pumps have been introduced.

Most SSHX units are constructed of high-yield carbon steel tube and hard chromium plated on a nickel base giving maximum heat exchange capability within a high-pressure cylinder. A number of blade arrangements are available. Typically, the Phrotators are fitted with a balanced 2 mm cut-away 'floating' blades of a design that promotes turbulence, and hence further heat exchange, at the rear of the blade tip. The Perfectors may be fitted with a 'sword' knife system consisting of long heavy stainless steel holders with inserted replaceable plastic blades. The knives are mounted on the knife rotors so that they are movable at their fixing points. At high rotation speed, the blades or knives are pressed against the tube surface mainly by centrifugal force. Towards the end of the crystallisation process, a low rotation speed may be used, the blades or knives now being pressed against the tube mainly by product pressure. The blades or knives are fitted onto temperature-controlled rotor shafts. Close control of the rotor temperature is achieved through the flow of tempered water (close to the temperature of the fat) in the annular space between the rotor shaft and an inner tube.

Manufacturers offer two-bladed or three-bladed shafts at the periphery. It is often argued that three-bladed shafts offer 'a balanced pressure stress reaction at the shaft and so resists the promotion of stress related shaft vibration, always a potential with high viscosity products such as puff pastry fats'. However, the high level of engineering evident in both systems appears to mitigate against this argument.

Eccentric shafts (BP 842,310) were developed by the Chemetron Corporation (Votator Division). They are claimed to provide more intensive cooling for high melting bakery margarines as well as a certain amount of kneading and a compressive action similar to that provided by the roller, table and helical worm treatment in the older type of plant. Most of the SSHX units available today have the rotor in a central position.

SSHXs also contain an annular space. The annular space is the area between the centre shaft and the inner wall of the SSHX. The distance between the shaft and wall is important for crystallisation. A small annular space contributes to a high degree of crystallisation; important, for example, in puff pastry production, whereas larger dimensions cause less shear and therefore may be important for low-fat (high-moisture) products. Typically, the annular space between the rotor shaft and the tube surface is 8–20 mm.

The rotor speed is also of importance. The frequency at which surface scraping (or cutting speed) takes place determines the types of crystals being formed and worked. The conditions will therefore be set based on the product being processed. Two factors contribute towards achieving the best results: the rotor speed (typical range, 300–700 rpm) and the number of blades fitted at the periphery. To aid in the control of rotor speed manufacturers offer variable speed or fixed-speed drives (most offer two-speed drives).

6.4.3 Churning technology

6.4.3.1 Batch process
The development of the butter churn pre-dates the 1900s. They consisted primarily of a wooden barrel that rotates on its horizontal axis. These barrels were fitted with internal side baffles that lift and drop the cream to create the necessary working action.

6.4.3.2 Continuous process
Significant change took place in the 1930s, when continuous butter-making techniques and processes were developed (McDowall, 1953). Details have been described by Jebsen (1994). Several process techniques are now available to produce spreadable butter:

Butter is subject to seasonality effects. In many Western countries cattle feeding practices are dictated by the season (i.e. cows graze during the milder spring and summer months and are on solid feed during winter). Hence, selection of milk (and therefore cream) from cows fed only on fresh pasture, which leads to higher levels of unsaturated fatty acids in the milk fat, will result in a relatively softer butter. Typically, in the UK, summer butter churned from Friesian cow's milk has N_{20} values of 13–18% solids [solid fat content (SFC)] compared with 18–22% for winter butter. Irish butter is typically 13–16% solids. Supermarkets sell these butters packed in tubs, adding further to the perception of a more spreadable butter relative to what is perceived to be normal butter.

Monobutter is produced from the milk of lactating cows fed a diet rich in full-fat rapeseed (in excess of 20% of diet) or soybean in a concentrated mixture. The milk produced has an altered fatty-acid content in the milk fat (Charteris, 1991). This change improves the physical and nutritional properties of butter such that the high stearic acid (C18 : 0) content from the oilseeds is desaturated in the gut wall and mammary gland to oleic acid (C18 : 1). This is subsequently secreted into the milk. Murphy et al. (1990) found that, typically, the palmitic acid (C16 : 0) level is reduced and oleic acid is increased. Consequently, the SFC at 10°C is reduced from 40–45% (characteristic of Irish butter) to 30–32%, making the butter spreadable from the fridge.

Other techniques consist of:

- whipping: this consists of aerating with nitrogen
- work softening: mechanical working of the butter
- temperature treatment (Alnarp treatment): this involves temperature cycling of cream prior to churning
- addition of fractionated milk fats to the cream
- addition of vegetable oils (e.g. Bregott development; see figure 6.12)
- addition of vegetable-oils-based cream (artificial cream)

Moran (1991) elegantly encapsulated the variations in the manner of adding vegetable oil by identifying three inventions that best describe them:

- British Patent (1968): this involves addition of liquid oil to dairy cream, emulsification prior to ripening and churning, up to 30% of the fat phase being nondairy
- European Patent (1984): this involves addition of liquid oil and partially hydrogenated bean oil to dairy cream, emulsification prior to pasteurisation, ripening and churning, the final fat phase containing at least 35% nondairy fat
- European Patent (1985): crystallisable vegetable cream is prepared separately and blended in with a dairy cream; the mixture is churned with liquid vegetable oil; the final fat phase contains 50% or more nondairy fat

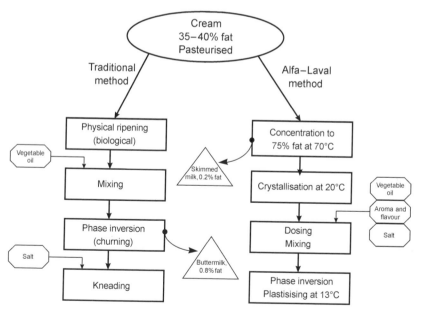

Figure 6.12 Bregott manufacture; percentages are the solid fat content.

A drawback of these processes is that the buttermilk will contain vegetable oils. Since authentic buttermilk or buttermilk powder is a premium-priced food ingredient in its own right this may restrict its application. To overcome this, another route is now available (Studer, 1990): butter is produced from the churn. Fresh or stored butter is then softened and re-emulsified with a serum rich in dairy protein (i.e. caseinates) and vegetable oils to produce a spreadable 80% fat product. Equally, with the absence of vegetable oils 60% and 40% fat butter can be produced. The re-emulsified blend is crystallised in a closed tubular chiller (scraped-surface cooling) and is mechanically worked before packing. These lower-fat butters are also more spreadable compared with the freshly churned 80% butter.

6.4.4 Storage conditions

It is good practice to store butter in tightly sealed, polymer-lined cartons. If the product is partially used, then the container should be resealed without removing the balance of product from its original polymer lining and should be stored in a humidity-controlled (80–85%) room. Typically, good-quality bulk butter can be stored up to four months when refrigerated (0–3°C; 32–38°F), or up to one year when frozen (−23°C to −29°C; −10°F to −20°F). In Europe, butter stored frozen in EU intervention stores have been found to be of satisfactory quality after two years.

When thawing, or 'conditioning' as it is sometimes referred to by butter traders in Europe, it is important to remove butter cartons (boxes) from tightly stacked pallets and to distribute them around a temperature-controlled room with good ventilation and good air circulation. The cartons should also be moved occasionally, to ensure that all sides of each carton is thawed slowly and evenly. Humidity should be controlled and not exceed 20%; room temperature should not exceed 21°C and preferably be at 16–18°C (60–65°F). Under these conditions, typically, a 25 kg block (Europe) or 68 lb block (USA) will take 4–5 days to thaw to 0–3°C (32–38°F).

6.5 Yellow fat blends

6.5.1 Trans-fatty-acid-free oil blends

Margarines with low or no trans fatty acids are now an important part of the shopping basket in many developed economies. Consumer preference for low-trans-fatty-acid products has focused attention on the C18 : 1 and C18 : 2 cis–trans isomers. These are normally absent in vegetable oils while milk fat has about 5% occuring naturally. However, during the hydrogenation reaction the cis–cis isomers convert to the cis–trans form. This is dependent on the process conditions and catalyst used and can result in large amounts of the latter being formed.

Munro, D.S. (1982) *Bulletin Int. Dairy Fed.*, **142**, 137.

Munro, D.S. (1986) *Bulletin Int. Dairy Fed.*, **204**, 17-19.

Murphy, J.J., *et al.* (1990) *J. Dairy Res.*, **57**, 295-306.

Patton, S. and Jensen, R.G. (1975) in *Progress in the Chemistry of Fats and Other Lipids* (ed. R.T. Holman), Pergamon Press, Oxford, p. 163.

Rajah, K.K. (1992) *Lipid Technology* (November/December), 129-137.

Ranjith, H.M.P. and Rajah, K.K. (2001) in *Mechanisation and Automation in Dairy Technology*, (eds. A.Y. Tamime and B.A. Law), Sheffield Academic Press, Sheffield, pp. 119-151.

Sonntag, N.O.V. (1979) in *Bailey's Industrial Oil and Fat Products, Volume 1*, 4th edn (ed. D. Swern), John Wiley, Chichester, Sussex, p. 292.

Studer, F. (1990) EPA 0368805.

Teah, Y.K., Sudin, N.A. and Hamirin, K. (1994) Interesterification: a useful means of processing palm oil products for use in table margarine. PORIM information series, PORIM, Malaysia.

Willett, W.C., Stampfer, M.J., Manson, J.E. *et al.* (1993) *Lancet*, **341**, 581-585.

7 Emulsifiers and stabilizers

Clyde E. Stauffer

7.1 Introduction

Fats in food systems interact with other components, usually at an interface between two phases. Three specific kinds of interfaces are of particular importance in foods:

- liquid–liquid (emulsions)
- air–liquid (foams)
- solid–liquid (dispersions)

Controlling the physical nature of interfaces is often crucial to making a high-quality food product. This is frequently done by including emulsifiers, stabilizers, and surface-active agents among the ingredients used to prepare the food. These may be as components of an ingredient (e.g. egg yolk) or as an additive (e.g. monoglyceride).

The US Food and Drug Administration has defined these ingredients as follows:

- emulsifiers and emulsifier salts: substances that modify surface tension in the component phase of an emulsion to establish a uniform dispersion or emulsion
- stabilizers and thickeners: substances used to produce viscous solutions or dispersions, to impart body, improve consistency, or stabilize emulsions, including suspending and bodying agents, setting agents, jellying agents, bulking agents
- surface-active agents: substances used to modify surface properties of liquid food components for a variety of effects, other than emulsifiers, but including solubilizing agents, dispersants, detergents, wetting agents, rehydration enhancers, whipping agents, foaming agents, and defoaming agents

Generally, emulsifiers and surface-active agents are relatively small molecules (molecular weight less than 1000 Da), whereas stabilizers and thickeners are polymers such as gums and proteins. In the following discussion the general term 'surfactant' (a coined word meaning surface active agent) will be used to include all the above categories.

Numerous books present a more complete discussion on the subject of surface activity than is appropriate for this chapter. The orientation covers the range from physical chemistry research (Adamson, 1967) to food research (Friberg and Larsson, 1997) and food technology (Stauffer, 1999a,b); this listing is by no means exhaustive.

7.2 Surface activity

7.2.1 Amphiphiles

A surfactant is a molecule that migrates to interfaces between two physical phases and is more concentrated in the interfacial region than in the bulk solution phase. The key molecular characteristic of a surfactant is that it is amphiphilic in nature, with the lipophilic (or hydrophobic) part of the molecule preferring to be in a lipid (nonpolar) environment and the hydrophilic part preferring to be in an aqueous (polar) environment (structures **2–5** illustrate some surfactant types; the R group is a lipophilic hydrocarbon, typified by stearic acid, **1**). The term 'preferring' is used in the sense that the thermodynamic free energy of the system is at a minimum when the lipophilic part is in an

$$HOOC - (CH_2)_{16}CH_3$$

Stearic acid $\left[R = -(CH_2)_{16}CH_3 \right]$

1

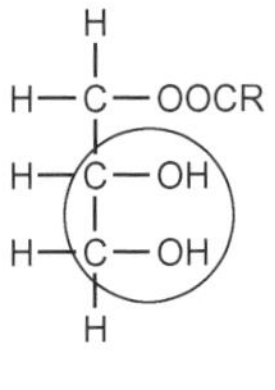

Nonionic 1-monoglyceride

2

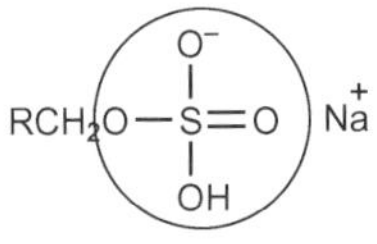

Anionic lauryl sulfate

3

Amphoteric lecithin

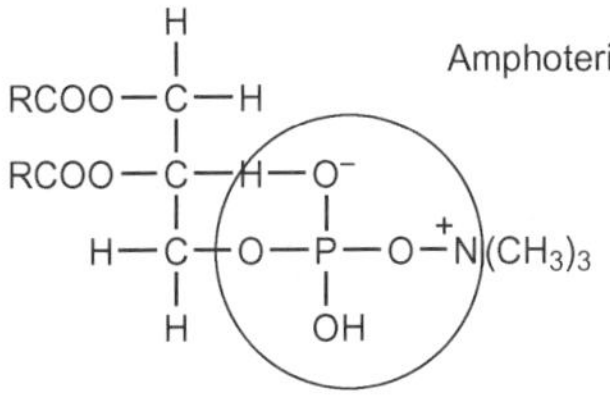

4

$$RCH_2O - \overset{+}{N}(CH_3)_3Br^-$$

Cationic cetyl trimethyl ammonium bromide

5

oil (or air) phase and the hydrophilic part is in water. If a surfactant is dissolved in one phase of an ordinary mixture of oil and water, some portion of the surfactant will concentrate at the oil–water interface, and at equilibrium the free energy of the interface (called the interfacial or surface tension, γ) will be lower than it would be in the absence of the surfactant. Putting mechanical energy into the system (e.g. by mixing) in a way that subdivides one phase will increase the total amount of interfacial area and energy; the lower the amount of interfacial free energy per unit area, the larger the amount of new interfacial area that can be created for a given amount of energy input. The subdivided phase is called the discontinuous phase, and the other phase is called the continuous phase.

The lipophilic part of emulsifiers (food surfactants) is usually a long-chain fatty acid obtained from a food-grade fat or oil. The hydrophilic portion is either nonionic (e.g. glycerol), anionic (negatively charged; e.g. sulfate) or amphoteric, carrying both positive and negative charges (e.g. phosphatidyl choline). Cationic [positively charged; e.g. cetyl trimethyl ammoniun bromide (CTAB)] surfactants are usually bactericidal, somewhat toxic and are not used as food additives. Examples of food surfactants are monoglyceride (nonionic), lauryl sulfate (anionic) and phosphatidylcholine (amphoteric). The nonionic surfactants are relatively insensitive to pH and salt concentration in the aqueous phase, but the functionality of the ionic types may be markedly influenced by pH and ionic strength.

7.2.2   Surface and interfacial tension

Increasing the amount of interface in a system requires work (energy) input. In practice this energy is supplied by shaking or mixing. If a stoppered bottle containing oil and water is gently inverted, only a small additional amount of interface (a few large oil drops) is generated. More vigorous shaking subdivides the oil droplets further, that is, it creates more interface. The system has a higher energy content or, in physical chemical terms, a higher total free energy. This energy is attributable to the presence of the interface and is denoted by the symbol γ. The usual unit for γ is milli-Newtons per meter (mN m^{-1}), numerically equal to the older (non-SI) unit of dynes per centimeter or ergs per square centimeter. The energy resides at the surface between two condensed phases (e.g. water and oil) or between a condensed phase and a gas (e.g. water and air) and is called interfacial (or surface) tension.

The terms surface (interfacial) free energy and surface (interfacial) tension are synonymous.

If the surface is curved, the radius of curvature plays a role. Given an air bubble of radius r the total surface energy is $4 \pi r^2 \gamma$. Decreasing the radius by the amount dr decreases the total surface energy by $8 \pi r \gamma\, dr = 4 \pi r^2 \gamma\, dr$. This change must be balanced by a pressure increase, ΔP, otherwise the bubble will be compressed to 'nothingness'. This pressure difference multiplied by the change in surface area equals the change in total surface energy:

$$\Delta P 4 \pi r^2 dr = 8 \pi r \gamma\, dr \tag{7.1}$$

and

$$\Delta P = \frac{2\gamma}{r} \tag{7.2}$$

Equation (7.2) indicates that the internal pressure of a small bubble is greater than that of a large bubble. This has practical consequences in aerated food systems. In a cake batter, for example, time-lapse photography shows that small bubbles (containing carbon dioxide) disappear and large bubbles increase in size. The carbon dioxide in the small bubbles dissolves into the aqueous phase because of a higher pressure then enters the larger bubbles, which represent regions of lower internal pressure. Similar effects are expected in other foods where the continuous phase may act as a conduit for gases dissolved from within the bubble. The same pressure differential applies to a dispersed condensed phase (e.g. oil droplets in water) but it has no practical consequences for food systems.

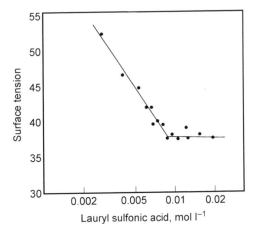

Figure 7.1 Surface tension of lauryl sulfonic acid solutions.

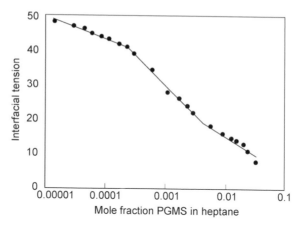

Figure 7.2 Interfacial tension at the interface of water and solutions of propylene glycol monostearate (PGMS) in heptane.

The surface tension of a solution of a surfactant is lower than that of the pure solvent. Surface tension is, roughly, a linear function of ln (surfactant concentration) up to the critical micelle concentration (CMC) (figure 7.1). Above the CMC the thermodynamic activity of the surfactant does not increase with the addition of more surfactant, and the surface tension remains constant. Interfacial tension also decreases with the concentration of an emulsifier dissolved in one of the phases. In figure 7.2 the decrease in γ does not level off, because the emulsifier [propylene glycol monostearate (PGMS)] does not form micelles in the organic solvent phase (heptane). The changes in the slope of the plot are attributed to changes in orientation of emulsifier molecules at the interface (Stauffer, 1968).

7.3   Interface formation

7.3.1   Division of internal phase

Simply adding oil to water does not result in emulsion formation. Input of mechanical energy subdivides the droplets of internal phase until they reach the final average droplet diameter, in the range 1–100 μm. A cylinder of liquid whose length is more than 1.5 times its circumference is unstable and tends to break up into droplets. Mechanical stirring of an oil–water mixture forms drops of liquid that are then distorted into cylinders (along the lines of flow) and that break up into smaller droplets (figure 7.3). The process is repeated until the droplets are so small they cannot be further distorted, and further subdivision ceases.

Shear

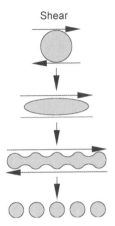

Figure 7.3 Breakage of cylinders of liquid into smaller droplets, under shear.

A suspended liquid drop forms a sphere, because this shape has minimum surface area (hence minimum interfacial free energy) for a given volume; area is related to the cube of droplet radius. Distortion is a flow shear effect, depending on droplet cross-section, related to the square of the radius. At large diameters, shear forces are greater than interfacial tension forces, droplets are distorted into cylinders and subdivision occurs. Droplet radius decreases until the interfacial tension forces balance (or exceed) shear forces, and further division stops. In emulsification experiments in which the amount of mixing energy is constant and γ is changed by adding emulsifier it is found that the average oil droplet diameter parallels γ (i.e. as more emulsifier is added, γ decreases and so does average droplet size). If γ is unchanged but mixing energy is increased, droplet size is also decreased. This is because of the change in the balance of shear and interfacial forces, allowing cylindrical distortion of smaller droplets. Equipment design that enhances shear is more effective at dividing droplets.

7.3.2 Emulsions

If oil and water are vigorously shaken, they form a dispersion of water droplets in oil and of oil droplets in water. When shaking is stopped the phases start to separate; small water drops fall toward the container bottom and oil drops rise. When they come into contact they coalesce and the 'emulsion' quickly breaks. Adding an emulsifier to the system changes the outcome; after standing, one phase becomes continuous, whereas the other (the discontinous phase) remains dispersed. The nature of the emulsion is determined by the emulsifier. As a general rule, the continuous phase is the one in which the emulsifier is soluble. Thus sodium stearate promotes an oil-in-water (O/W) emulsion, whereas zinc distearate promotes a water-in-oil (W/O) emulsion.

7.3.3 Foams

Food foams are usually made by whipping an aqueous solution of a foamer. Foaming agents are quite diverse, including proteins (e.g. egg white), fat crystals (e.g. whipping cream) and emulsifiers (e.g. one of the polysorbates). Air is first entrained by the action of the mechanical element (paddle, whip, mixer blades), then air bubbles are elongated and subdivided into smaller bubbles, just as described in section 7.3.1 for liquid internal phases.

Air is a nonpolar medium; surfactants concentrate at the air–water interface with the hydrophobic portion extending into the gas phase. In proteins (common foam formers in foods) some amino acid side-chains are hydrophilic whereas others are lipophilic. In their natural configurations the protein chain is usually folded so that the lipophilic residues are in the interior, whereas the hydrophilic residues are on the surface. (Protein molecules have been described as 'an oil droplet surrounded by a hydrated shell', a characterization with some merit as applied to relatively small albumin and globulin proteins.) When a protein such as egg albumen is exposed at an air–water interface it tends to unfold, with the hydrophobic side-chains entering the air phase and the hydrophilic chains remaining in the water phase. If oil is present it will spread at the air–water interface, displacing the protein and destabilizing the foam. Traces of oil (or egg yolk) in egg white, for instance, make it difficult if not impossible to whip the egg white into a foam.

7.3.4 Wetting

Some surfactants are good wetting agents. This property is useful in many circumstances: enhancing dispersion of dry mixes in liquid; improving spread-ability of chocolate and cocoa-based coatings; and incorporation of dietary fiber materials in dressings. Qualitatively, a drop of water is placed on the solid surface. If the contact angle θ is greater than 90° (figure 7.4) the water does

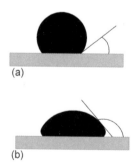

(a)

(b)

Figure 7.4 Contact angle θ at the liquid–solid–air juncture of (a) nonspreading ($\theta < 90°$) and (b) spreading ($\theta > 90°$) liquid drops.

not spread; it is said that the solid is not wetted. If θ is less than $90°$, the water spreads, and the solid is wetted.

The angle θ is determined by the surface tension at the three interfaces involved:

$$\cos\theta = \frac{(\gamma_{SV} - \gamma_{SL})}{\gamma_{LV}} \qquad (7.3)$$

The spreading coefficient is defined as:

$$S_{L/S} = \gamma_{SV} - \gamma_{SL} - \gamma_{LV} \qquad (7.4)$$

in which S denotes solid, L denotes liquid and V denotes vapor (air). When $S_{L/S}$ is greater than 0, wetting occurs and the liquid spreads. An efficient wetting agent is one that minimizes the surface tension of the air–water and solid–water interfaces while leaving the air–solid surface tension unchanged. This is the situation, for example, when dry beverage powder is added to water. Sodium lauryl sulfate lowers the air–water and solid–water interfacial tensions and enhances dispersibility. In the absence of the surfactant, the contact angle at many of the (irregular) solid surfaces is greater than $90°$ and the powder, with its entrapped air, floats on the top of the water.

In chocolate coating the liquid (continuous) phase is an oil (cocoa butter), and the solid (dispersed) phase consists of finely ground cocoa particles and sugar crystals. The addition of lecithin or some other appropriate emulsifier aids the wetting of the solid particles by oil, most probably by lowering γ_{SL}. Unwetted, the solid particles tend to adhere to each other (flocculate), forming a matrix that decreases the flowability of the coating. Wetting by oil disrupts this matrix, lowering the apparent viscosity of the heterogeneous mass as well as giving a smoother mouth feel to the final product.

7.4 Stabilization

Surfactants have two primary functions: promoting dispersion of the discontinuous phase and stabilizing the resulting emulsion or foam. The promotion function has been addressed in section 7.3. Stabilization is effected by numerous mechanisms; in some instances more than one of these may be present at the same time in food systems. The intent is to prevent coalescence—that is, many small oil droplets coming together to form one large drop (emulsion breakdown) or many small gas cells being converted into one large cell (foam breakage).

7.4.1 Creaming

Creaming occurs when the dispersed phase of an emulsion is lighter than the continuous phase and the dispersion remains quiescent for a period of time. The rate at which the particles rise is given by Stokes's law:

$$v = \frac{2gr^2(\rho_1 - \rho_2)}{9\eta} \tag{7.5}$$

where v is the rate of creaming (or sedimentation, if the droplets are heavier than water), g is the gravitational constant, r is the droplet radius, ρ_1 and ρ_2 are the densities of the oil and water phase, respectively, and η is the viscosity of the water phase.

Oil droplets in an O/W emulsion float to the top because the density of vegetable oil is about $0.91\,\text{g ml}^{-1}$, $0.08\,\text{g ml}^{-1}$ less than that of water. The rate at which they rise depends on particle diameter. A drop having a $1\,\mu\text{m}$ diameter rises at a rate of $4\,\text{cm day}^{-1}$, and one with a $10\,\mu\text{m}$ diameter rises $4\,\text{m day}^{-1}$. Obviously, reducing the average droplet size reduces the rate of creaming. Fat globules in raw milk have an average diameter of $3\,\mu\text{m}$; after homogenization the average diameter is $0.5\,\mu\text{m}$. In raw milk, the average flotation rate is $36\,\text{cm day}^{-1}$, and in homogenized milk the rate is $1\,\text{cm day}^{-1}$. Creaming brings the oil droplets closer together, and if contact is not prevented (e.g. by ionic repulsion) coalescence occurs. A simple creamed layer of stabilized oil droplets is readily redispersed by inverting the container a few times.

Sedimentation occurs when the dispersed phase is more dense than the continuous phase. The most common food example of this is in salad dressings containing solid particles of spices and vegetables. Gums or other stabilizers are added to increase water viscosity; after dispersing the particles throughout the dressing (by shaking) the particles remain uniformly suspended for as long as needed to apply the mixture to the salad.

7.4.2 Flocculation

In some emulsions, particularly those made with high-molecular-weight emulsifying agents (proteins, gums), the oil drops rise to the top and form a layer that is rather resistant to redispersion. The size (and number) of the oil droplets does not change; coalescence does not occur. The material adsorbed at the interface interacts, probably through hydrogen bonds and perhaps some ionic bonds, to hold the droplets together. The phenomenon is termed flocculation or clumping. Positive use of it is made in water treatment plants, for example, where certain surface active materials adsorb to solid impurities and then promote flocculation, allowing easy removal by filtration. Flocculation is much less significant in food applications.

Flocculation is differentiated from creaming by the amount of energy needed to redisperse the collected phase. A creamed material is dispersible by gentle hand agitation of the whole system, whereas a flocculated material requires rather more energetic treatment.

7.4.3 Ionic stabilization

When two surfaces approach each other, two forces exist: one repulsive and one attractive. Whether or not the surfaces touch and coalesce depends on the relative sizes of the two forces. This is equally true for liquids (e.g. oil droplets in an emulsion), solids (e.g. finely divided $CaCO_3$) and films (air bubbles in a foam).

Two surfaces carrying like charges repel each other. For example, if an O/W emulsion is stabilized by an anionic surfactant the oil droplets have a negative charge on their surface. Electrical repulsion then tends to keep the droplets from making contact. Suspensions of solids (cellulose fibers, finely divided $CaCO_3$, etc.) can be stabilized in the same way. Ionic surfactants are used that selectively adsorb to the solid surface, generating a surface charge and stabilizing the suspension. The strength of the repulsion depends on the ionic strength of the aqueous continuous phase.

Attractive forces, collectively called van der Waal's forces, exist between two oil droplets. Simplistically, these forces may be thought of as the attraction between oil molecules at the O/W interfaces, which have lower energy when in contact with each other than when in contact with water. Several phenomena are involved; hydrophobic interactions and London dispersion forces are most commonly considered. These are unaffected by ionic strength.

At the oil (or solid) surface the electrical potential (or charge) is denoted by ψ_0. Cations are attracted into the region, partially neutralizing the surface negative charge (assuming the surfactant is anionic). The value of ψ decreases as the distance from the oil-drop surface increases and at some point becomes essentially zero. The rate of decrease of ψ is directly related to the ionic strength of the aqueous phase. Ionic strength, μ, is related to the concentration (c_i) of individual salt ions and the square of the ionic charge (z) of each ion:

$$\mu = \frac{1}{2} \sum c_i z_i^2 \qquad (7.6)$$

Divalent ions are four times as effective as monovalent ions in decreasing ψ. Thus 0.25 M zinc sulfate, for example, is as effective as 1 M sodium chloride in decreasing the thickness of the electrical double layer (the region where $\psi > 0$), thus promoting emulsion flocculation or coalescence.

As long as ionic strength is low, electrical repulsion is greater than van der Waal's attraction, and the droplets remain suspended. If gravity (creaming) is the only force bringing the droplets together, they will approach to a distance where repulsion due to ψ is just balanced by gravitational effects, and the emulsion will then be stable. If, by addition of salt (particularly divalent or trivalent salts) the ψ potential is markedly suppressed, the surfaces can approach so closely that van der Waal's forces override repulsion, and the droplets can touch and coalesce. At some intermediate ionic strength, the two forces are approximately equal, and the droplets will remain separated by about one droplet diameter. The practical conclusion to be drawn from this is as follows. If the emulsifier used is ionic in nature, the salt concentration of the aqueous phase markedly affects emulsion stability; low salt concentration enhances stability, whereas high salt concentration increases coalescence and/or flocculation.

7.4.4   Steric hindrance

Another form of stabilization is relatively independent of ionic strength: the oil droplets are prevented from making contact by simple steric hindrance. This may take two forms, either an immobilized water layer at the interface or a solid interfacial film. Emulsion stabilization by proteins, gums and polyoxyethylene derivatives occurs by the first mechanism. Hydrophobic parts of the stabilizers adsorb at the oil surface, but adjacent large hydrophilic segments are hydrated and form an immobilized layer on the order of 10–100 nm thick. As mentioned, these hydrated segments frequently interact to cause flocculation, while coalescence of the oil drops themselves is prevented. Such emulsions are frequently used as carriers for oil-soluble flavors, essences and colorants.

Gum-stabilized emulsions may be spray-dried, yielding free-flowing powders in which the oil-soluble flavor is encased in a shell of dried gum. Such powders are significantly more shelf-stable than the liquid emulsion. When dispersed in water the gum rehydrates and the emulsion is reconstituted. Spray-dried flavors and essences find wide use in the food industry.

The α-tending emulsifiers such as propylene glycol monostearate are oil soluble. The emulsifier adsorbs at the oil–water interface, but under certain conditions (low temperature, presence of a free fatty acid) the emulsifier forms a solid interfacial film (figure 7.5). Although the oil droplets may make contact, the film prevents coalescence. The interfacial layer actually appears to be crystalline, with a well-defined melting point.

7.4.5   Foam drainage and film breakage

With respect to the forces involved a foam is highly similar to an O/W emulsion. The terminology is somewhat different, but the results are the same: either a foam is stable or the gas bubbles coalesce and the foam breaks. Rather than referring to 'creaming', a foam is said to 'drain'; the effect is the same, with the water phase concentrating at the bottom and the dispersed phase concentrating towards the top of the container. The volume fraction (ϕ) of gas in a foam is usually much higher than that of oil in an emulsion. Whipping egg white, for example, may easily give a 10-fold to 15-fold expansion ($\phi = 0.9$–0.93), and mayonnaise ($\phi = 0.7$–0.8) has the highest oil content of any food emulsion.

made either by direct esterification of propylene glycol with fatty acids or by interesterification of fat (triglycerides) with propylene glycol. The direct esterification product typically contains about 55–60% monoester, with the remainder being diester. Although a product containing more than 90% monoester is made by molecular distillation, for most commercial uses (cake-mix production) the extra cost is not warranted. The interesterified product is more complex, containing not only monoesters and diesters of propylene glycol, but also some 10–15% monoglyceride and a small amount of diglyceride. Again, as with DATEM, because such a wide range of product compositions is possible from different manufacturing processes, it is advisable to have rather stringent raw material specifications for this ingredient.

Lactic acid esters of monoglyceride are usually made by reacting lactic acid with a distilled monoglyceride. The complication here is that lactic acid contains a hydroxyl group, and the fatty acid moiety may migrate. As an example, suppose lactic acid is heated with 1-monostearin. The main initial product would be 3-lactoyl-1-stearoyl glycerol. However, during the reaction some portion of the stearic acid may migrate to the lactyl hydroxyl, giving glyceryl 3-(stearoyl)-lactylate. In addition, lactic acid can polymerize (to form lactoyl lactic esters), and lactoyl dimers and trimers may also be present. In sum, the reaction product mixture from heating lactic acid with a monoglyceride is a complex mixture containing as many as 10 identifiable molecular species. Production parameters must be tightly controlled to obtain a product with consistent functionalities.

7.5.3 Sorbitan derivatives

When the sugar alcohol sorbitol is heated with stearic acid in the presence of a catalyst, two reactions occur: sorbitol cyclizes to form the five-membered sorbitan ring and the remaining primary hydroxyl group is esterified by the acid. The resulting sorbitan monostearate (13) is oil soluble, with a rather low HLB value, and is an approved additive for food use in many countries. Other sorbitan esters of importance are the monooleate and the tristearate (14). The tristearate is not an emulsifier but finds use as a modifier of crystal polymorphs in many foods.

Sorbitan monostearate

13

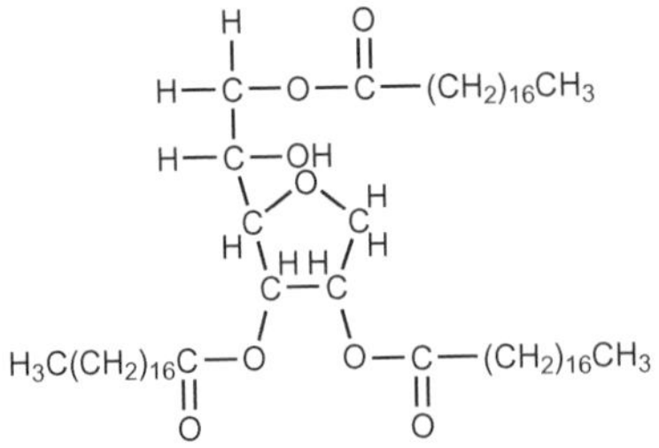

$$H-\overset{\overset{\displaystyle H}{|}}{C}-O-\overset{\overset{\displaystyle O}{\|}}{C}-(CH_2)_{16}CH_3$$

Sorbitan tristearate

14

Any of the three esters may be reacted with ethylene oxide to give polyoxyethylene derivatives (**15** and **16**; the polyoxyethylene side-chain is hydrophilic). The monostearate derivative **15** is known as polysorbate 60, the tristearate **16** is polysorbate 65; the monooleate is polysorbate 80. The remarks made in connection with EMG regarding the length and location of the polyoxyethylene chains apply to these compounds. The average number of oxyethylene monomers is 20 ($n = 20$) and, in the case of the monoesters, chains may be located on more than one hydroxyl group of the sorbitan ring. (With the triester, of course, only one hydroxyl group is available for derivatization.)

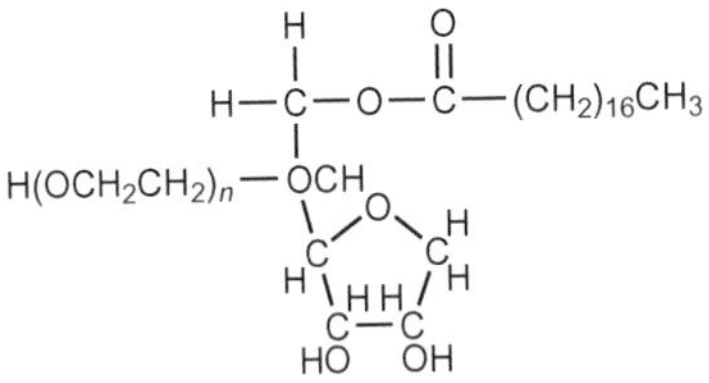

Polyoxyethylene (20) sorbitan monostearate (polysorbate 60)

15

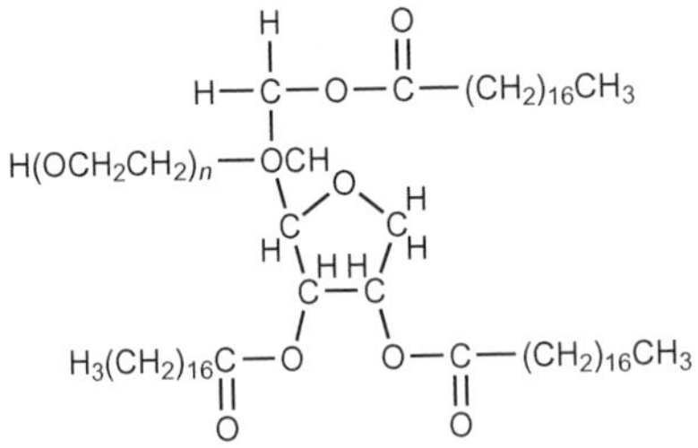

Polyoxyethylene (20) sorbitan tristearate (polysorbate 65)

16

Sorbitan monostearate (**13**) finds use as a good emulsifier for making icings that have superior aeration, gloss and stability characteristics. It is also used as part of the emulsifier system in whipped toppings and in coffee whiteners. The polyoxyethylene derivatives have found more acceptance, with the monostearate polysorbate 60 (**15**) being the most widely used of the group. At a level of 0.25% (flour basis) the ability of polysorbate 60 to strengthen dough against mechanical shock is greater than that of SMG (**8**) and about equal to those of EMG (**7**) and SSL. Polysorbate 60 has also been used in fluid oil cake shortening systems, generally in combination with GMS and PGMS.

7.5.4 Polyhydric emulsifiers

Polyglycerol esters (**17**) have a variety of applications as emulsifiers in the food industry. The polyglycerol portion is synthesized by heating glycerol in the presence of an alkaline catalyst; ether linkages are formed between the primary hydroxyls of glycerol. In **17** n may take any value, but for food emulsifiers the most common ones are $n = 3$ (triglycerol), $n = 6$ (hexaglycerol), $n = 8$ (octaglycerol) and $n = 10$ (decaglycerol), remembering that in all cases n is an average value for the molecules present in the commercial preparation. The polyglycerol backbone is then esterified to varying extents, either by direct reaction with a fatty acid or by interesterification with a triglyceride fat. Again, the number of acid groups esterified to a polyglycerol molecule varies around some central value, so an octaglycerol octaoleate really should be understood as an (approximately octa)-glycerol (approximately octa)-oleate ester. By good control of feedstocks and reaction conditions, manufacturers manage to keep the properties of their various products relatively constant from batch to batch.

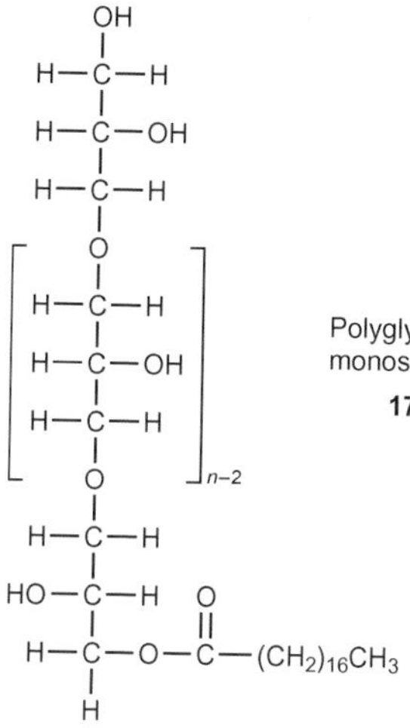

Polyglycerol monostearate

17

The HLB balance of these esters depends on the length of the polyglycerol chain (the number of hydrophilic hydroxyl groups present) and the degree of esterification. For example, decaglycerol monostearate has an HLB of 14.5, and triglycerol tristearate has an HLB of 3.6. Intermediate species have intermediate HLB values, and any desired value may be obtained by appropriate blending, as described below. The wide range of possible compositions and HLBs makes these materials versatile emulsifiers for food applications.

The most recent addition to the stable of food emulsifiers is polyglycerol polyricinoleate (PGPR). Ricinoleic acid (12-hydroxy oleic acid) forms about 90% of the fatty-acid content of castor oil. Heating the fatty acid under dehydrating conditions leads to an interesterified polyricinoleic acid. This is then esterified with polyglycerol yielding an emulsifier with a moderately low HLB. PGPR is used mainly in the confectionery coating (chocolate) industry, for modifying the flow properties of the coating.

Sucrose has eight free hydroxyl groups, which are potential sites for esterification to fatty acids. Compounds containing six or more fatty acids per sucrose molecule are marketed as noncaloric fat substitutes under the name Olestra: this material acts like a triglyceride fat and has no surfactant properties. Compounds containing one to three fatty acid esters (for a possible positional isomer of a sucrose diester, see **18**), however, do act as emulsifiers and are approved for food use in that capacity. They are manufactured by the following steps:

1. an emulsion is made of fatty-acid methyl ester in a concentrated aqueous sucrose solution
2. the water is removed under vacuum at elevated temperature
3. alkaline catalyst is added, and the temperature of the dispersion is raised slowly to 150°C under vacuum, distilling off methanol formed on transesterification
4. the reaction mixture is cooled and purified

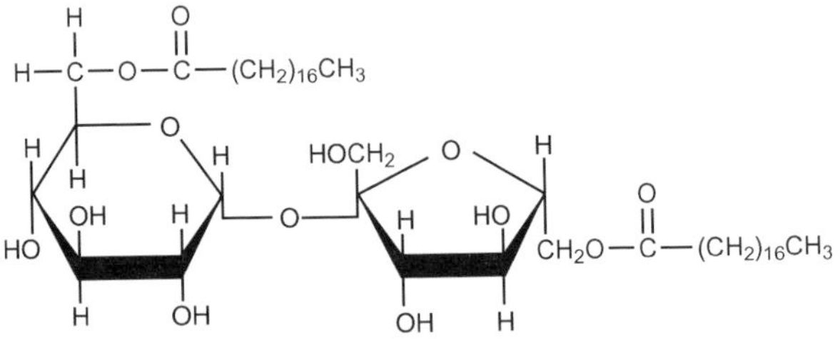

Sucrose diester

18

The degree of esterification is controlled by the reaction conditions, especially the sucrose: methyl ester ratio, and the final product is a mixture of esters (table 7.1). The HLB value of a particular product is smaller (more lipophilic) as the degree of esterification increases, as would be expected.

Table 7.1 Ester distribution in sucrose ester emulsifiers and hydrophilic–lipophilic balance (HLB) values

	Percentage				
Designation	mono-	di-	tri-	tetra-	HLB value
F-160	71	24	5	0	15
F-140	61	30	8	1	13
F-110	50	36	12	2	11
F-90	46	39	13	2	9.5
F-70	42	42	14	2	8
F-50	33	49	16	2	6

7.5.5 Anionic emulsifiers

In addition to SMG and DATEM, some other anionic surfactants that have been tried as dough strengtheners are SSL (**19**), sodium stearyl fumarate (**20**) and sodium dodecyl sulfate (**21**). SSL is currently the surfactant most widely used in the USA; sodium stearyl fumarate did not find acceptance, and sodium lauryl sulfate is used mainly as a whipping agent with egg whites.

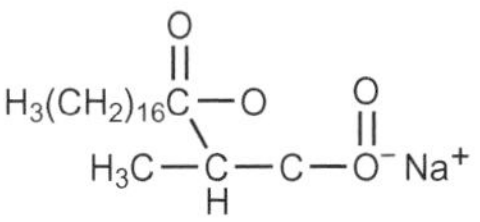

$$H_3(CH_2)_{16}\overset{\overset{O}{\|}}{C}-O \qquad \overset{\overset{O}{\|}}{C}$$
$$H_3C-\overset{|}{\underset{H}{C}}-C-O^-\,Na^+$$

Sodium stearoyl lactylate (SSL)
19

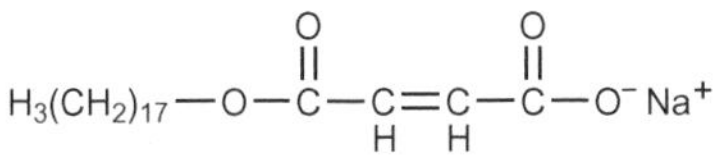

$$H_3(CH_2)_{17}-O-\overset{\overset{O}{\|}}{C}-\overset{|}{\underset{H}{C}}{=}\overset{|}{\underset{H}{C}}-\overset{\overset{O}{\|}}{C}-O^-\,Na^+$$

Sodium stearyl fumarate
20

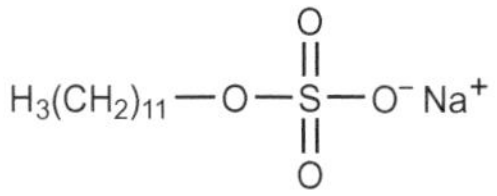

$$H_3(CH_2)_{11}-O-\overset{\overset{O}{\|}}{\underset{\underset{O}{\|}}{S}}-O^-\,Na^+$$

Sodium dodecyl sulfate (SDS)
21

Lactic acid, having both a carboxylic acid and a hydroxyl function on the same molecule, readily forms an ester with itself. In commercial concentrated solutions almost all the acid is present in this polylactic form, and to get free lactic acid it must be diluted with water and refluxed for a period of time. When stearic acid is heated with polylactic acid under the proper reaction conditions and then neutralized with sodium hydroxide, a product having the structure shown in **19** is obtained. The monomer lactylic acid shown represents the predominant product; the dilactylic dimer is also present as well as lactylic trimers and tetramers. As with all compounds based on commercial stearic acid derived from hydrogenated fats, some percentage of the fatty acid is palmitic, with small portions of myristic and arachidic acids also being present. Sodium stearoyl lactylate is readily water soluble, but the calcium salt is practically insoluble. In this respect, it mimics a soap (e.g. sodium stearate is water soluble but calcium stearate is oil soluble). Either form may be used, depending on the details of the intended application, but as a dough strengthener the sodium salt is more commonly used. As a stabilizer for hydrated monoglyceride the sodium form is used, because ionization in the water layer is necessary.

Sodium dodecyl sulfate (SDS), **21**, is a sulfate ester of the C_{12} alcohol dodecanol. Commercially, this alcohol is produced by reduction of coconut oil, and the resultant mixture is called lauryl alcohol (from lauric acid, the predominant fatty acid in coconut oil). The alcohol portion of sodium lauryl sulfate is a mixture of chain lengths, the approximate composition being 8% C_8, 7% C_{10}, 48% C_{12}, 20% C_{14}, 10% C_{16} and small amounts of longer chains. The most common food use of sodium lauryl sulfate is as a whipping aid. The compound is added to liquid egg whites at a maximum concentration of 0.0125%, or to egg white solids at a level of 0.1%. It promotes the unfolding of egg albumin at the air–water interface and the stabilization of the foam.

7.5.6 Lecithin

The lecithin generally used by food processors is a by-product of the processing of crude soybean oil; it is the 'gum' that is removed during the degumming step of oil refining. The crude gum is treated and purified to give the various commercial lecithin products that are available today. Crude soybean oil contains about 2% lecithin. Crude corn and cottonseed oils contain about 1% lecithin, but because smaller amounts of these oils are processed in the USA (compared with soybean oil) the amount of 'gum' obtained is usually too little for economical processing for human food uses. Instead, it is added back to animal feed formulations as a valuable source of energy. Egg yolk contains about 20% phospholipid, which accounts for its excellent emulsifying functionality, for example, in mayonnaise. However, isolated lecithin from egg yolk is too expensive to be used for food manufacture.

The crude gum is dehydrated (removing water used in degumming), then insoluble fines are removed by filtration. The crude material is brown to dark brown (depending on the amount of heat applied during processing) and contains some pigments extracted from the original soybean. It is bleached to give a more acceptable light brown color. Treatment with up to 1.5% hydrogen peroxide gives the product known as single-bleached lecithin, and further addition of up to 0.5% benzoyl peroxide yields double-bleached lecithin. Reaction with hydrogen peroxide at even higher levels, plus lactic acid, hydroxylates unsaturated fatty-acid side-chains at the double bond, yielding, for example, dihydroxystearic acid from oleic acid. Hydroxylated lecithin is formed, which is more dispersible in cold water than are the other types and is more effective as an emulsifier for O/W emulsions.

Phospholipids are insoluble in acetone, and the phospholipid content of lecithin is specified by the acetone insolubles (AI). The standard lecithin of commerce will have a minimum AI of 65%. Crude bleached lecithin is quite viscous. The addition of vegetable oil fluidizes lecithin, and commercial fluid lecithin products are standardized to have a viscosity in the range of 7500–10 000 centipoise at 25°C. On the other side of the coin, a fully de-oiled lecithin is produced that is a granular, free-flowing product with a typical AI content of 95%–98%.

The main surface-active components of soy lecithin are the phosphatidyl (22), ethanolamine (23), choline (24) and inositol (25) groups.[3] The phosphatidyl group (22) is a phosphate ester of diglyceride. The fatty-acid composition of the diglyceride is similar to that of the basic oil so a number of different fatty acids are found, not just the stearic and oleic acids depicted. The three species are found in approximately equal amounts. Phosphatidylethanolamine (PE) and phosphatidylcholine (PC) are amphoteric surfactants, whereas phosphatidyli-nositol (PI) is anionic. Other surface active materials are found at somewhat lower concentrations. These include: phosphatidic acid (the phosphatidyl moiety plus a hydrogen atom); lysophosphatides (the above species but with one fatty acid removed); and glycolipids (a sugar residue, either galactose or digalactose, attached to the free hydroxyl of a diglyceride).

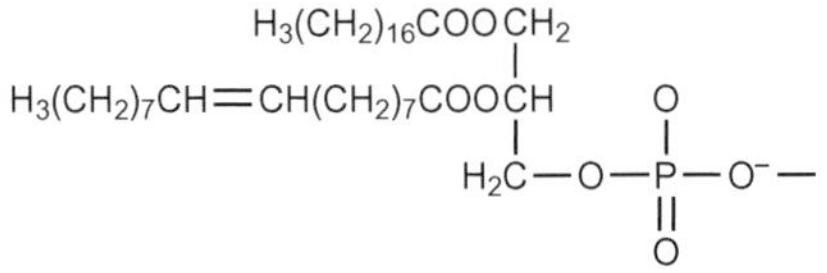

$$H_3(CH_2)_{16}COO\,CH_2$$
$$H_3(CH_2)_7CH = CH(CH_2)_7COO\,CH$$
$$H_2C-O-P-O^--$$

Phosphatidyl group

22

$$-CH_2CH_2\overset{+}{N}H_3$$

Ethanolamine group

23

$$-CH_2CH_2\overset{+}{N}(CH_3)_3$$

Choline group

24

---

[3] Serine is another possible substituent, but phosphatidylserine is found mainly in animal phospholipids such as those obtained from egg yolk. Removing one fatty acid from the phosphatidyl moiety gives a lysophospholipid.

Inositol group

25

The HLB values of the three types are varied, with PC having a high HLB, PE an intermediate HLB and PI having a low value. The HLB of the natural blend is in the range of 9–10, and emulsifier mixtures having values in this range will tend to form either O/W or W/O emulsions, although neither type is highly stable. Emulsifiers with intermediate HLB values are excellent wetting agents (a major application for lecithin).

The emulsifying properties of lecithin can be improved by ethanol fractionation. Phosphatidylcholine is soluble in ethanol, PI is rather insoluble and PE is partially soluble. Adding de-oiled lecithin to ethanol gives a soluble and an insoluble fraction. The phosphatide compositions of the two are: (1) ethanol soluble, 60% PC, 30% PE and 2% PI plus glycolipids; and (2) ethanol insoluble, 4% PC, 29% PE and 55% PI plus glycolipids (the remaining few percent in both cases include oil, free fatty acids and lysophosphatides). The soluble fraction is effective in promoting and stabilizing O/W emulsions, and the insoluble portion promotes and stabilizes W/O emulsions. The ethanol-soluble fraction can be chromatographed to give a material containing 90% PC. This is used to make an egg-yolk replacer. Part of the emulsifying ability of egg yolk is a result of its high HLB value (70% PC, 15% PE, of the total phospholipid), plus the protein. At present, several European companies are using the fractionation process to produce industrial food-grade emulsifiers having special functionalities.

7.6 Hydrophilic–lipophilic balance

7.6.1 Basic principal of the concept

Emulsifiers consist of a hydrophilic portion (a wide variety of structures are used) and a lipophilic part (usually a fatty acid, occasionally a fatty alcohol). The balance between these two tendencies governs the functionality of the emulsifier at interfaces, and hence its utility in foods. This is called the hydrophilic–lipophilic balance (HLB) and attaches a number to emulsifiers that guides the food technologist in choosing an emulsifier for a particular application.

The initial proposal was to calculate HLB as follows

$$\text{HLB value} = \frac{20\,L}{T} \tag{7.7}$$

where L is the molecular weight of the hydrophilic part of the molecule and T is the total molecular weight. Thus, a pure hydrocarbon would have an HLB value of 0, and a pure hydrophile (say, sugar) would have an HLB value of 20.

Another method of calculating the HLB value is to add up contributions by various functional groups. The functional groups and their associated group numbers as listed in table 7.2. The group values for the hydrophilic and lipophilic functions in the emulsifier are summed and inserted in the equation:

$$\text{HLB value} = \sum (\text{hydrophilic values}) - \sum (\text{lipophilic values}) + 7 \quad (7.8)$$

The agreement between calculated and experimentally determined HLB values is generally within a few tenths. Considering that neither method is highly precise, one cannot expect much better, and the calculated value will give a good indication of HLB for a new emulsifier under investigation.

HLB values give a guide to the functionality of the emulsifier system. The following guidelines are based on experience:

- HLB of 2–6 characterizes a good W/O emulsifier
- HLB of 7–9 characterizes a good wetting agent
- HLB of 10–18 characterizes a good O/W emulsifier

Table 7.2 Hydrophilic–lipophilic balance (HLB) functional group numbers

	Groups number
Hydrophilic groups	
—SO$_4$Na	38.7
—COOK	21.1
—COONa	19.1
Sulfonate	c.11
—N(CH$_3$)$_3$	9.4
Ester	
sorbitan ring	6.8
other	2.4
—COOH	1.9
—OH:	
sorbitan ring	0.5
other	1.9
—(CH$_2$—CH$_2$—O)—	0.33
Lipophilic groups	
—CH—	0.475
—CH$_2$—	0.475
—CH$_3$	0.475
=CH—	0.475

7.6.2   Experimental determination of hydrophilic–lipophilic balance

Although theoretical calculations are fine, emulsifiers are used in food systems that for the most part are anything but theoretically simple. For practical purposes experimental determination of optimum HLB for a particular system is best done experimentally. This approach starts with the concept that the HLB of a blend of emulsifiers is the algebraic sum of their contribution. For example, suppose a blend of sorbitan monostearate (HLB = 4.7) and polysorbate 60 (HLB = 14.9) is used to determine the optimum HLB for a salad dressing formulation. A 50/50 mixture of the two has an HLB value of 9.8 $[(0.5 \times 4.7) + (0.5 \times 14.9)]$, a 25/75 blend has an HLB value of 12.35 $[(0.25 \times 4.7) + (0.75 \times 14.9)]$, etc. This gives a straightline relationship, but it should be noted that in practice the prediction is not precisely true.

A series of emulsifier blends at intervals of 0.5 HLB units are prepared. The range should be selected according to the desired product characteristics (W/O or O/W emulsion), according to the guidelines above. An excess of emulsifier (approximately 10% of the weight of the oil phase) is used, to ensure emulsion formation. The emulsifier is dissolved in the oil, the aqueous components are added and emulsions are made for each trial, using a standardized agitation technique. Then the emulsions are allowed to stand and are assessed for stability, say, by measuring the thickness of the oil (or water) layer formed at various times. (If all the emulsions are too stable to break down in a reasonable length of time, the experiment is repeated with use of less emulsifier.) This should give an idea of the approximate optimum HLB range for the system.

The next step is to choose the best chemical type of emulsifier to use. Experience shows that a blend of two emulsifiers (one lipophilic, one hydrophilic) generally gives the most stable emulsions. Numerous such pairs are available, for example:

- sorbitan monostearate plus a polysorbate
- monoglyceride plus a fatty-acid salt (sodium or potassium)
- triglycerol tristearate plus decaglycerol monostearate
- sucrose ester (e.g. F-160 plus F-50)

In addition, the nature of the fatty-acid chain (chain length, degree of unsaturation) can sometimes make a difference, readily seen by comparing some of the entries in table 7.3. The melting point can be a factor in some processing systems; solid monostearin may be difficult to add whereas liquid mono-olein (with the same HLB value) may be more convenient to use. If the system is acidic (e.g. a salad dressing) acid-stable emulsifiers such as the sorbitans or polyglycerols may be the best choice, whereas if the pH is in the neutral range the monoglyceride plus fatty acid salt pair may be the most effective.

The various combinations are then subjected to emulsifying trials, as described above, but using blends that are in the middle of the apparent optimum

Table 7.3 Experimental hydrophilic–lipophilic balance (HLB) values for food emulsifiers

Emulsifier	HLB value
Sodium lauryl sulfate	40
Sodium stearoyl lactylate	22
Potassium oleate	20
Sucrose monoester	20
Sodium oleate	18
Polysorbate 60	15
Polysorbate 80	15
Decaglycerol monooleate	14
Decaglycerol monostearate	13
Ethoxylated monoglyceride	13
Decaglycerol dioleate	12
Polysorbate 65	11
Hexaglycerol dioleate	9
Decaglyerol hexaoleate	7
Triglycerol monostearate	7
Glycerol monolaurate	7
Sorbitan monostearate	5.9
Sucrose triester	5
Propylene glycol monolaurate	4.5
Propylene glycol monostearate	3.4
Glycerol monostearate	3.8
Sorbitan tristearate	2.1

HLB range. These trials should include relatively small HLB variations (i.e. optimum and optimum ± 0.5 HLB units), and also a limited variation in concentration. The goal is to choose the emulsifier system that gives the best stability at the lowest use level.

After selecting the type of emulsifiers to use, the system can be fine-tuned by running a series of trials at intervals of 0.1 HLB units and at various concentrations to find the minimum amount that yields the desired emulsion stability. At this stage of testing, use of a response surface methodology maximizes the information gained while minimizing the amount of work needed.

Two final points need to be emphasized. First, the HLB is an empirical system for characterizing emulsifiers, and the values assigned to emulsifiers (as in table 7.3) are necessarily somewhat imprecise. Thus, if a particular emulsifier combination gives best results at an HLB value of 11.4, a different combination might function best at an HLB value of (say) 11.7. Any substitutions must be checked out, although the use of the HLB system greatly decreases the amount of work needed. Second, emulsifiers interact with other food ingredients, and if formula changes are made (additions or subtractions) the assessment of optimum HLB needs to be repeated. This is especially so if the changes involve proteins or gums.

7.7 Polysaccharide stabilizers and thickeners

7.7.1 Gums

Obtained from plant exudates, seaweed extracts and bacterial polymeric products, gums are high-molecular-weight polysaccharides that dissolve in water to form viscous solutions and, in some cases, gels. They are widely used in food products, usually at low levels (0.1%–1%), for many different functional reasons. Some gums have a high water-binding capacity and are used to control water migration in the finished product. These gums also act to inhibit growth of ice crystals in frozen products. In other instances the viscosity imparted by the gum is necessary for the intermediate material (e.g. a cake batter) to perform properly when baked or cooked. Gum viscosity is often used as a means of classifying gums. Table 7.4 lists the most common food-grade gums and viscosity of a 1% solution.

Table 7.4 Properties of some gums

Gum	Viscosity[a]
Low viscosity:	
arabic	2–5
ghatti	4–10
larch	2–10
Medium viscosity:	
sodium alginate	25–800
propylene glycol alginate	100–500
tragacanth	200–500
xanthan	800–1400
High viscosity:	
guar	2000–3500
karaya	2500–3500
locust bean	3000–3500
Cellulose gums:	
sodium carboxymethyl-	50–5000
hydroxypropyl methyl-	20–50 000
methyl-	10–2000
Gel formers:	
agar	gel
calcium alginate	gel
carrageenan	gel
furcelleran	gel
gellan	gel
pectin	gel

[a]Viscosity of a 1% aqueous solution of the gum, in centipoise (viscosity of water = 1 centipoise.)

Gums consist of a long polysaccharide chain with numerous side branches of sugars or oligosaccharides. Frequently, the sugar units include carboxylic acids, for example, D-glucuronic, D-mannuronic or D-galacturonic acid. In a few instances, sulfate esters provide an anionic character. Many different saccharides are found in gums, including the hexoses D-glucose, D-mannose and D-galactose, as well as the pentoses L-arabinose, D-xylose and L-rhamnose. The highly branched structure contributes to water solubility, and the anionic gums often form gels in the presence of cations such as Ca^{++}.

The low-viscosity gums (arabic, ghatti, larch) are used mainly as water-binding agents, for example, to prevent 'weeping' (water exudation) in baked meringues. They readily dissolve in water, and a concentrated solution (10–50%) of the gum is often used to emulsify hydrophobic flavor oils—citrus oil, cinnamon oil and the like. The gum forms a layer at the O/W interface, and spray drying the emulsion results in an encapsulated flavor that finds many uses in food product development.

The medium-viscosity gums (sodium alginate, propylene glycol alginate, tragacanth, xanthan), when used in foods at the usual levels, impart 'body' to the product. They also have emulsifying properties and are often found in pourable dressings (i.e. oil and water types, emulsified just before use by shaking the bottle). They can be used in making flavor oil emulsions, particularly liquid types that are often used in bakeries.

High-viscosity gums (guar, karaya, locust bean) are used as thickening or stabilizing materials. They greatly increase the viscosity of the aqueous phase of the food product and aid air incorporation in whipped toppings, for example. High-viscosity gums are often used in low-fat or fat-free pourable dressings. The enhanced viscosity (or 'body') results in a mouth feel that is somewhat similar to the fat-containing dressing.

Gel-forming gums (agar, calcium alginate, carrageenan, furcelleran, gellan, pectin) are used mainly for food gels requiring a semisolid nature—jellies, fruit fillings and the like. These gums are also said to impart freeze–thaw stability to many products. The ingredient statement for many frozen whipped toppings and ice creams include one or more of the gums listed here, particularly carrageenan or alginate.

Most dressings and many sauces contain vinegar and hence are acidic. As polysaccharides, gums are somewhat susceptible to acid hydrolysis, but again this varies—tragacanth is reasonably stable at pH 2, whereas arabic begins to depolymerize at pH 4. Most gums are acidic in nature (carboxylic acid groups or sulfate esters) and a certain sensitivity to pH would be expected. As an example, the viscosity of a sodium alginate solution is constant from pH 4 to 10, but at lower pH the carboxylate groups are converted to (unionized) carboxylic acids, and the polymer chains begin to interact, resulting in an increase of viscosity. Propylene glycol alginate ester, however, is rather insensitive to pH, and the viscosity increase at lower pH is much less than for the sodium salt.

7.7.2 Modified starch

Starch is made up of two types of molecules: linear amylose and branched amylopectin. In native starch these molecules exist partly as amorphous chains and partly in crystallites (regions of crystallized starch). During gelatinization, starch first hydrates (swells) and then the crystallites melt into the water solvent. Upon cooling, gelatinized starch retrogrades (recrystallizes) and forms a gel, the properties of which depend on the ratio of starch to water and on the ratio of amylose to amylopectin. Modification changes starch properties in ways that are useful in the food industry.

Cross-linked starch is starch treated with a reagent to form bonds between glucose residues in adjacent starch chains. Sodium trimetaphosphate or phosphorus trichloride form a phosphate diester:

$$2\,\text{starch-OH} + PO_3Cl\,(NaOH) \rightarrow \text{starch-O-P}(=O)(-O^-)-O-\text{starch Na}^+$$

Epichlorohydrin forms a diether bridge, where the central moiety is a hydroxypropyl group:

$$\text{starch}-O-CH_2-CH(-OH)CH_2-O-\text{starch}$$

The differences in reactivity of phosphate esters and ethers lead to differences in the chemical stability of the two kinds of starch; they both are generally useful and each has advantages in certain kinds of situations, such as extremes of pH, temperature and so on.

Cross-linking raises the gelatinization temperature of starch but, more important, the gel that is formed is stabilized with respect to temperature, low pH and shear. The degree of stabilization depends, of course, on the degree of cross-linking (i.e. the amount of reagent used during the modification reaction). Cross-linked starch is more resistant to loss of viscosity at low pH values than other kinds of starch. This resistance makes it useful in salad dressing, where the pH is in the range of 3–4. However, it degrades somewhat when it is cooked at a low pH, so the starch is cooked (gelatinized) at neutral pH before being combined with the oil, vinegar and spices.

Stabilized starch is derivatized without cross-linking. Reaction with propylene oxide, for example, gives hydroxypropylated starch. Sodium tripolyphosphate, under the proper reaction conditions, makes phosphate monoester derivatives; acetic anhydride yields acetylated starch and so forth. Stabilized starches have lower gelatinization temperatures and higher viscosities than does the native starch, but they are often less resistant to shear thinning. A major advantage of stabilized starches is that, because of the subtitution along the starch chains, they are less inclined to retrograde on cooling. Retrogradation (recrystallization of the gelatinized starch) renders the starch opaque (rather than translucent) and much firmer, possibly generating a gritty or rubbery mouth feel.

7.7.3 Cellulose derivatives

Cellulose per se is insoluble in water, but substitution of certain groups along the glucose backbone render it soluble. The three forms most commonly used in foods are substituted with carboxymethyl, methyl, and hydroxypropyl plus methyl groups. The degree of substitution varies depending upon the ratio of cellulose to reactant, and products with a wide range of viscosities are possible. Methylcellulose has a modest degree of surface activity and will act as an emulsifier for O/W systems. Hydroxypropyl methyl cellulose forms a film at oil–water interfaces and hence stabilizes oil emulsions. Certain grades of hydroxypropyl methyl cellulose gel even at room temperature and find use in some types of no-fat pourable dressings, contributing 'body' to the product.

Microcrystalline cellulose is a highly purified fraction of regular cellulose. It absorbs several times its own weight of water and is added to some low-fat dressings for 'body'. A suspension of microcrystalline cellulose is thixotropic, that is, it flows very slowly at low shear rates (high apparent viscosity). For this reason it is often added to condiment sauces (ketchup, barbecue sauce) to enhance 'cling'; the sauce remains where it is applied, rather than running down the sides of the food item.

7.8 Applications

7.8.1 Margarine and dairy products

7.8.1.1 Margarine
The largest food industry user of monoglycerides is the margarine industry. Monoglyceride (or, less often, lecithin) is dissolved in the margarine oil and aids in dispersing the aqueous phase to make a W/O emulsion. The monoglyceride (or lecithin) also stabilizes the emulsion in the holding tank before it is metered to the chilling machine where it is converted into solid margarine.

When the consumer uses the margarine for fying it often spatters, ejecting droplets of hot fat, sometimes with serious consequences. When the margarine is heated the vapor pressure inside the water droplets rapidly increases; the water can become superheated and then suddenly 'explode'. An emulsifier reduces the interfacial tension, allowing the water vapor to escape more readily through the water–oil interface. Thus, although the water still turns to steam it does so in a more controlled, less violent fashion. Although monoglyceride exhibits antispattering behavior, lecithin is generally considered to be more effective in this regard.

7.8.1.2 Butter
The fat droplets in milk are encased in, and stabilized by, a protein membrane. The churning of cream to make butter is a phase inversion process—the

transformation of an O/W emulsion into a W/O emulsion. During churning the fat globules are forced together, some degree of membrane removal occurs (probably as a result of mechanical forces rather than interfacial energetics) and the fat coalesces. This process continues until the fat forms a discrete clump with some entrained aqueous phase. The temperature of churning is an important processing parameter. Fat globule coalescence occurs via the liquid phase, but the solid fat stabilizes the clump and keeps the entrained aqueous droplets separate. The solid fat content (SFC) profile of milk fat varies with the season (lower in the winter, higher in the summer) and the churning operation must take this into account. If the milk fat is too hard the protein membrane does not desorb sufficiently to allow coalescence; if it is too soft the aqueous phase and air (about 3–5% of the total mass by volume) is not incorporated. (The proper water and air content is important for flavor and spreadability of the butter.)

The water droplets are stabilized by the fat protein membrane and crystals of high-melting triglycerides from the fat. When the emulsion inverts the membrane simply reverses and encases the aqueous phase rather than the fat phase. Protein desorption from an interface is normally a slow process but in churning it occurs quickly. Thus it seems likely that when two fat globules collide the membrane is physically ruptured (made easier because the fat phase is semiliquid) and the globules can make contact and coalesce. The involvement of fat crystals is somewhat less easy to explain but is not an unknown phenomenon. Solid particles tend to stabilize an interface, probably simply by interfering with contact and coalescence of the dispersed phase. Finely powdered calcium sulfate (talcum powder) will provide a modest degree of stability to an O/W emulsion, for example. The small fat crystals might well tend to form in the membrane–oil interfacial region and their solidity would then further stabilize the water drop.

7.8.1.3 Whipped cream

Foam formation is an interfacial phenomenon but the generation of a good whipped product (including ice cream, which is frozen whipped cream) is somewhat unusual in that whipping involves a controlled de-emulsification correlated with crystallization of the fat, to stabilize the air cells.

To whip properly, cream must contain sufficient fat (30% minimum), it must be held for at least 24 h in the refrigerator and the quantity must not be too large (do not overfill the mixer). Whipped cream is a foam in which the air bubbles are stabilized by agglomerated fat globules. At the early stages of mixing air is incorporated into the cream and divided into large bubbles. Fat globules concentrate at the air–water interface and stabilize the bubble. (This is due to partial removal of the membrane, exposing some fat which is forced into the air phase.) At the same time fat globules in the aqueous phase agglomerate, partly through membrane–membrane interactions and partly through binding of membrane protein to milk proteins. The fat globules must agglomerate but not coalesce (as happens during churning). Thorough cooling of the cream solidifies the fat to prevent coalescence; if the cream is too warm, the result may well be butter rather than whipped cream.

The majority of air incorporation occurs during the first stage of mixing (before significant viscosity has developed). Further mixing shears and subdivides the air bubbles, with the new air–water interface area being stabilized by more of the fat globule agglomerates from the aqueous phase. (Keep in mind that partial removal of the fat membrane exposes hydrophobic fat surfaces that collect at the air–water interface.) Increased viscosity is attributable to two factors: an increase in the number of (subdivided) bubbles, and continued agglomeration of fat globules to form a network throughout the system. As viscosity slowly builds the shear stress on the air bubbles increases, contributing to further subdivision, and the whipped cream rapidly becomes stiff (the end point of the whipping process). If the mixer is overfilled it takes longer to effect fat agglomeration and hence longer to reach the internal shear stress that contributes to completion of whipping.

Emulsifiers may be added to cream to increase overrun (the amount of air incorporated) and final stiffness. In one study using combinations of the polysorbates it was found that a 50:50 mixture of polysorbate 60 (the monostearate) and polysorbate 65 (the tristearate), used at 0.5% concentration, gave the stiffest cream with an overrun of about 200% (i.e. 2 volumes of air per volume of cream). Polysorbate 80 (the monooleate) in place of polysorbate 60 gave much lower overrun and a softer whipped cream (Min and Thomas, 1977). Studies using monoglycerides and two derivatives (the citric and the lactic acid esters) at 0.2% levels have also been reported (Sogo and Kako, 1989). Monoglycerides alone gave poor results, a blend of citric acid ester plus monoglyceride gave good results, and lactylated monoglyceride produced a very stiff foam with a high overrun. This may be due to the interfacial film-forming tendencies of LacMG. The overrun with simple whipping cream tends to be about 100%, and emulsifiers apparently increase air incorporation during the early mixing stages.

7.8.1.4 Nondairy whipped topping

Whipped toppings can be made with use of vegetable (rather than dairy) fat. Since a natural emulsifier is not present food technologists have supplied the necessary emulsification. However, the situation is more complicated than simply combining an emulsified fat with an aqueous protein solution and whipping the mixture. The fat must have the proper SFC profile so that the 'correct' amount of solid fat is present to stabilize the air bubbles. Also, the emulsifier used must interact with the protein coating the fat in such a way as to produce the necessary degree of controlled desorption of the protein layer. Finally, the emulsion is frequently spray-dried (for storage) then reconstituted with cold water before whipping. The emulsifier, protein and any other ingredients (sugar, flavors, maltodextrins, etc.) must rehydrate in the proper fashion to give an emulsion that is whippable.

A basic whipped topping powder formula is: 25% partially hydrogenated fat and 5% emulsifier (these first two are melted together); 15% maltodextrin and 5% sodium caseinate (these two are dissolved in the water) and 50% water. The two phases are mixed and homogenized, then spray-dried. The emulsion is stabilized by the combined effects of the emulsifier and protein. The powder is held at 5°C for a hour to solidify most of the fat, then stored at temperatures below 20°C. For whipping, the powder is dispersed into an equal weight of cold water. Agitation intitiates many of the same events that occur during whipping of dairy cream—fat globule agglomeration, partial desorption of protein from the fat–water interface, air incorporation (stabilized by fat globules) and air subdivision leading to formation of a stable foam.

The nature of the fat seems to be crucial. Best results are obtained with partially hydrogenated lauric fats (coconut, palm kernel) plus partially hydro-genated vegetable oils (soybean, sunflower, etc.). The hydrophobic segments of the protein, penetrating the fat globule, appear to inhibit crystallization of the fat, so the fat in the powder itself is a supercooled phase. Upon desorption (aided by the presence of emulsifier) the fat rapidly crystallizes. Crystallization concur-rent with protein desorption appears to be necessary for proper agglomeration and concentration of the fat at the air–water interface, leading to air bubble stabilization. This phenomenon appears connected to shorter-chain C_{12} fatty acids; it is observed with lauric fats but not if the fat phase contains solely the C_{18} vegetable oils. A partially hydrogenated C_{18} fat, exhibiting no supercooling phenomenon, gives a poor whipped topping, and a partially hydrogenated C_{12} fat with a low melting point (little or no fat crystal formation at the temper-ature of the mixture being whipped) also gives poor results (Barfod et al., 1989).

α-Tending emulsifiers such as propylene glycol monostearate or lactylated monostearin have been used with good success. Polysorbate 60 (the monos-tearate derivative) is frequently used in commercial nondairy whipped toppings. A common characteristic is that both types form a rather thick layer at the fat–water interface, a multilayer of emulsifier by the α-tending emulsifiers, or a layer of adsorbed water held by the polyoxyethylene chain of Polysorbate 60. These properties promote partial protein desorption yet maintain a 'sticky' surface on each fat globule, enhancing agglomeration.

7.8.1.5 Ice cream

The basic interfacial phenomena occurring during the manufacture of ice cream are similar to those occurring in whipping (Berger, 1997; Krog et al., 1989). Again, the aim is the subdivision of air bubbles, with stabilization by adsorbed fat globules. However, in this case the temperature regime is quite different, so one would expect differences in details of the types of emulsifiers that produce the best product.

After the initial ice cream emulsion is prepared it is aged for 4–24 h at approximately 5°C. This is an important step. During the aging, partial desorption of the fat globule membrane takes place and the globules agglomerate (Barfod *et al.*, 1991). Desorption is aided by the presence of emulsifiers. As in the case of the whipped toppings, it appears that the fat is supercooled, and solidification is connected with the desorption. During agitation and freezing the fat globules collect at the air–water interface and stabilize the air bubbles. Studies have shown that increasing the degree of desorption (by the addition of a monoglyceride) gives a 'drier' ice cream, with smaller average air-bubble diameter.

The emulsifiers commonly used in ice cream manufacture are monoglyceride (plastic monoglyceride and diglyceride seem to work equally well) and polysorbate 80 (the monooleate derivative). The effectiveness of the unsaturated emulsifier as compared with the saturated emulsifier used in whipped toppings is undoubtedly connected to the much lower processing temperature. The emulsifiers are used at a 0.2–0.4% level (compared with the 3–5% level in whipped toppings). Again, these details relate to the temperature differences. Gums (guar, carrageenan, tragacanth) are also frequently used in ice cream mixes. They improve the body and mouth feel of the product, but their main purpose is as crystal modifiers; they inhibit growth of ice crystals and lactose crystals in the frozen ice cream, preventing the gritty mouth feel sometimes experienced with inexpensive ice cream, particularly if it has been stored for a long period of time.

7.8.1.6 Coffee whiteners

Cream in coffee is enjoyed by many people. Today, however, the cream pitcher on the table may be replaced by a powder which, when stirred into a cup of coffee, mimics the effect of cream. The restaurant trade uses small individual packets of liquid whitener, but the powder (a spray-dried emulsion, stable at room temperature) is more popular for home use.

A wide range of formulations appears in the literature, but a typical spray-dried whitener might contain the following: vegetable fat, 37%; 42 dextrose equivalent (DE) corn syrup solids, 56%; sodium caseinate, 5%; dipotassium phosphate, 1.6%; monoglyceride (GMS), 0.3%, polysorbate 65, 0.1%; and flavor and anticaking agent as desired. The emulsifiers are added to the fat, the other materials are dissolved in water and an emulsion of the two phases is homogenized and then spray-dried. The powder is tempered at a cool temperature to solidify the fat and avoid clumping. The fat for coffee whitener usually has a melting point around 42°C–45°C but has a rather steep SFC profile.

The emulsifier used serves two purposes: it facilitates formation of fine fat globules (in the order of 1 µm average diameter) during homogenization and, in conjunction with the protein, prevents 'feathering' [which occurs if the emulsion breaks and the fat separates (oils out) when the whitener is dissolved in the coffee]. The protein helps emulsify the fat, and the proper emulsifier can stabilize the protein layer when the whitener is added to the coffee. The protein layer can be destabilized by calcium ions and organic acids in the coffee, and dipotassium phosphate counteracts this tendency. In addition to GMS and polysorbate 65, SSL has also been recommended for use in coffee whiteners.

7.8.2 Baking

The largest user of food emulsifiers of all types is the baking industry, accounting for an estimated 50% of food surfactant consumption (Kamel and Ponte, 1993; Knightly, 1996; Krog, 1981). The largest single usage category is that of mono-glycerides for antistaling purposes (Joensson and Toernaes, 1987; Knightly, 1999). The next largest category is dough strengtheners (Stauffer, 1990), fol-lowed by cake emulsifiers (Wootton *et al.*, 1967) and icings.

7.8.2.1 Antistaling
Monoglycerides retard bread staling (increased crumb firmness) by complexing with gelatinized starch and slowing the rate at which it retrogrades or recrys-tallizes (Knightly, 1996; Stauffer, 1990). The lipid–starch complex forms with any straight-chain aliphatic compound such as stearic acid (Krog, 1971), and antistaling properties are observed with a variety of such surfactants. Besides the structural requirement (i.e. a linear aliphatic chain) it also appears that the surfactant must exist in the lamellar mesophase in the dough to function effectively as a starch-complexing agent (Krog, 1975; Krog and Jensen, 1975). The hydrated GMS product is considered by many to be a more effective antistaling agent because it is already in the lamellar mesophase form. However, comparative studies using all three commercial types of monoglycerides for bak-eries (plastic monoglycerides and diglycerides, hydrated hard monoglycerides, and soft distilled monoglycerides) have shown that on an equal α-monoglyceride basis all three forms are equally effective at retarding the development of crumb firmness.

7.8.2.2 Dough strengthening
Dough strengtheners are surfactants that: (1) enhance the ability of a proofed loaf to withstand mechanical shock; and (2) increase the extent of ovenspring during baking and hence increase final loaf volume. The most widely used surfactants for this purpose are SSL and DATEM, although EMG and polysorbate 60 also show positive results in published test reports. The postulated mechanism is that the surfactant interacts with gluten molecules in the dough, somehow

making the gluten matrix more suitable ('stronger') for accomplishing the desired purpose. Unfortunately, at this time there is no clear definition of the molecular events and structures occurring in a 'strong' dough, although attempts at such a description are frequently made (Stauffer, 1998; Uthayakumaran *et al.*, 1999). It must suffice to say that SSL or DATEM, added to a dough, will increase loaf volume in any given production regime. When a variety bread—one with a lot of nonwheat flour 'weight' in the dough—is being formulated, the addition of SSL or DATEM improves the chances of having an acceptable finished product. EMG and polysorbate 60 have proven particularly useful in high-fiber breads. It is speculated that the hydrophilic polyoxyethylene chains interact with the insoluble cellulosic fibers, modifying fiber properties and making it less disruptive of the gluten matrix.

7.8.2.3 Cake emulsifiers

Successful manufacture of a good-quality layer cake requires dispersal of air into the batter and retention of these bubbles until the starch has swollen and the cake structure is set. Traditionally, air has been entrained in the shortening phase during the first stage of making a cake. This method is not an efficient way to incorporate air, as it depends primarily upon the ability of the plastic shortening to trap air bubbles during creaming. The advent in about 1930 of 'superglycerinated' shortenings containing monoglycerides enabled the baker to subdivide the air bubbles, creating smaller bubbles that are more efficiently retained by the shortening phase and that give more uniform nucleation for leavening gases throughout the batter during baking. The final crumb grain, therefore, is closer and the overall volume is larger.

This traditional approach involves at least three stages in batter preparation: in the first stage shortening and sugar are creamed vigorously to incorporate air; in the second stage eggs are blended into this whipped material; and in the third stage the flour, milk (or water), flavors and other ingredients are added and blended in to make the final batter. The development of dry cake mixes for home and industrial use dictated a different approach to air incorporation. Using a liquid vegetable oil in place of plastic shortening gives a cake with a moister eating quality and a longer shelf-life. Obviously, a liquid oil is not well suited to entrapping air bubbles. It was found that the addition of certain emulsifiers to the oil before it was incorporated into the cake mix allowed the production of high-ratio cakes with good volume, fine grain and excellent keeping qualities (Wootton *et al.*, 1967). These cakes can be mixed in one stage; all the ingredients are placed in a bowl, the liquid is added and the batter is mixed at low speed (to blend ingredients) and then at high speed to incorporate air.

The most generally useful emulsifiers for this purpose are α-tending emulsifiers (so-called because they tend to solidify in a stable α-crystalline form). The main examples in commercial use today are acetyl monoglycerides (AcMG), lactyl monoglycerides (LacMG) and propylene glycol monoesters (PGME). A number of other emulsifiers have also been used, including polysorbate 60, stearoyl lactylic acid and sucrose esters, but most commercial cake emulsifiers now offered by suppliers are based upon PGME and/or AcMG, perhaps in combination with other emulsifiers that enhance their functionality.

When dissolved in the oil phase α-tending emulsifiers lower interfacial tension, but their effectiveness in cake batters is not a result of this property. Rather, at concentrations above a certain level these emulsifiers form a solid film at the oil–water interface. This behavior is seen when one suspends a drop of water from a syringe tip in the oil, waits a few minutes for film formation, then withdraws some of the water (figure 7.5). The film appears to be a result of crystallization of the emulsifier at the interface in the α form. There is a defined relationship between temperature and the minimum bulk concentration of the emulsifier that produces a film; the minimum concentration increases as temperature increases (figure 7.6). The addition of a second surfactant enhances film formation; a mixture of PGMS/stearic acid (80/20) is a stronger film-former than pure PGMS at the same weight concentration.

In a one-stage cake batter air incorporation is primarily a function of foam stabilization by protein contributed by flour, milk and egg whites. The presence of oil inhibits this foam formation, and the solid film at the oil–water interface effectively encapsulates the oil during air incorporation, thus preventing destabilization. It should be pointed out that simply lowering cake batter density by incorporating more air is not enough to ensure a good final volume. In a comparison of two sucrose esters in a sponge cake formula batter densities

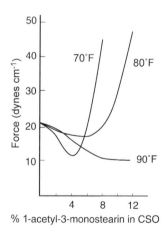

Figure 7.6 Effect of temperature on film formation by acyl monoglyceride (AcMG) at the oil–water interface. The sharp rise in tensiometer reading indicates the presence of the solid film. Note: CSO, cottonseed oil.

were 0.65 and 0.55 for F110 and F160, respectively, but F110 gave a larger finished volume than F160. Similarly, stearoyl lactylic acid decreases batter density markedly as it is increased from 0% to 4% (flour basis) but the best finished cake volume is achieved at 1.5% emulsifier. These results merely confirm the earlier remark that not only is it necessary to incorporate air in a cake batter during mixing, but also that air must remain in the batter during baking, serving as nuclei for the gases released from the chemical leavening system.

7.8.2.4   Icings

A basic icing is the butter cream icing in which softened butter and finely powdered sugar are mixed then whipped to incorporate air. (Usually, vanilla extract is added for flavor, and a little milk might be added for moistness.) Commercial bakeries are more likely to use shortening rather than butter; 2% monoglyceride in the shortening provides emulsification to enhance air incorporation. Surprisingly, a higher level of monoglyceride makes a less stable icing (more prone to loss of air and collapse). A simple sugar/fat icing is not necessarily the most appealing, and the addition of liquid (water or milk) is often desirable. The use of a high HLB emulsifier (e.g. polysorbate 60 or decaglycerol monostearate) allows formation of a smooth, stable icing. A low-viscosity, hydrated gum such as gum arabic also enhances the stability of such an icing.

7.8.3   Coatings

Confectionery coatings include those fat-based formulations that are melted, applied to some sort of solid food and then solidify to form the coating. The process, generally called enrobing, is applied to a wide variety of foods: nuts, nougats, flavored gels, cakes, donuts, cookies, sugar wafers, ice cream bars, fruit pieces and almost anything that the manufacturer thinks will be enhanced by the addition of the flavor and texture of the coating.

Coatings comprise a solid phase dispersed in a continuous fat phase. The most common solids are cocoa, sugar and milk solids. The coating fat should be rather hard at room temperature so that it does not soften or melt during storage and handling, but it must have a melting point close to body temperature so that it does not leave a waxy mouth feel. In these respects cocoa butter is considered the model, although its melting point (32–35°C) is a little low for some purposes. Other fats, tailored to mimic cocoa butter properties, are used to make 'compound coatings'. The requirements for substitute fats as well as proper handling techniques for coatings are discussed in some detail in an earlier publication (Stauffer, 1996).

Enrobing is done by passing the piece through a curtain, or 'waterfall', of melted coating. The thickness of the covering is governed by the viscosity and

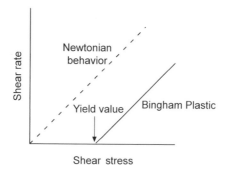

Figure 7.7 Rheological stress–strain curve of a Bingham plastic.

yield value of the liquid coating (figure 7.7). Several factors influence these rheological properties:

- fat content—increasing the amount of fat lowers the yield value
- fineness—more conching, reducing particle size, increases the viscosity and yield value
- moisture—interacting with sugar and cocoa solids increases yield value
- emulsifiers—the addition of lecithin and PGPR modifies both properties

The shear stress on the coating, as it is pumped to the 'waterfall', is above the yield value, and when coating first falls on the piece being enrobed its flow is governed by viscosity. As the layer of coating thins (because of drainage) the shear stress falls below the yield value and further drainage stops. Decreasing either factor leads to a thinner layer of coating on the enrobed piece.

Lecithin decreases coating viscosity, probably via the 'wetting' mechanism discussed in section 7.3.4. It is added to the coating at a level of 0.5–1% of the total weight. PGPR, used at 0.1–0.4% of total weight, markedly decreases yield value but has no effect on viscosity. These two emulsifiers have complementary effects and can be used together to achieve the desired thickness of coating on the enrobed piece. Of course, in the plant, minor changes in viscosity (and hence in the amount of coverage) can be made by small temperature changes of the melted coating and the piece being enrobed.

7.8.4 Dressings and sauces

7.8.4.1 Pourable salad dressings
A wide range of products is included in this category, from the simplest 'vinaigrette' (oil, vinegar and a few spices) to rather complex products marketed under a variety of names. Salad dressing oil is a vegetable oil treated to prevent development of cloudiness (arising from crystallization of high melting triglycerides) during refrigeration. A crystallization inhibitor is often added to

increase the length of time before such cloudiness can develop. Crystal inhibitors deposit on the face of the growing fat microcrystals and interfere with the further deposition of fat molecules. Two compounds specifically approved for this use are oxystearin and polyglycerol esters. Several emulsifiers (in addition to polyglycerol esters) also inhibit crystal formation. Among those reported to do this are sucrose esters, glucose esters and sorbitan tristearate.

Before a dressing is added to the salad greens the bottle is shaken to make an emulsion, with the intent of having an even distribution of the oil and water phases. Many nonstandard dressings are rather complex mixtures containing dairy products (buttermilk powder, blue cheese), spices (ground mustard, turmeric powder), dried vegetable pieces (onion, green and red peppers) and other items according to the creativity of the product developer. An emulsion stabilizer (polysorbate 60 at 0.3% level is frequently used) helps maintain uniformity of this mixture and increases the appeal of the product to the consumer.

The viscosity of a simple emulsion of 35% oil in water is almost that of water. For many dressings a higher viscosity is desired so that the dressing does not drain quickly from the salad greens and collect in the bottom of the salad bowl. Soluble gums are used at a concentration of 0.05–0.3% to give the desired viscosity. Some of the names commonly seen on dressing labels include: propylene glycol alginate, xanthan gum, modified cellulose gum and carrageenan. The main property of the gum (besides contributing the desired viscosity) is that it must be stable in an acidic environment—hydrolysis at low pH decreases viscosity.

Microcrystalline cellulose may be used at 1–2% concentration, increasing viscosity by decreasing the amount of continuous (water) phase in the formulation. It is most often seen in dressings that contain tomato products, where it gives a smooth texture and imparts a certain degree of thixotropy to the mixture so that the dressing clings to the surface of the salad greens.

7.8.4.2 Spoonable salad dressings

Spoonable dressing is a 'Bingham plastic' in rheological terms (figure 7.7). When shear stress (stirring) is applied at low levels the dressing acts like a solid; that is, it does not flow, although it may fracture (break). At some level of shear force, called the yield value, the dressing begins to act like a liquid. Put another way, if a block of dressing is simply cut the vertical edge does not flow, because the shear stress of gravity is less than the yield value. However, it will spread smoothly on a slice of bread, because the shear force applied by the knife blade is greater than the yield value. The yield value of spoonable dressing and of mayonnaise is an important factor in consumer perception of product quality.

Starch is the key component in a starch-based salad dressing, providing the desired structure as well as a creamy texture in the finished product. The modified starch most often used is a highly cross-linked, stabilized waxy maize starch. Waxy maize is essentially 100% amylopectin cross-linked with sodium trimetaphosphate. The starch requires rather high temperatures for gelatinization, then sets into a soft gel on cooling. The cross-linked starch is stable against hydrolysis in the low-pH environment of the finished dressing; without this stability the starch gel softens during storage. The preferred starch is also stabilized. The recommended derivative is a hydroxypropyl moiety, obtained by treating the starch with propylene epoxide. This stabilization interferes with recrystallization of the side chains and maintains a creamy texture in the finished dressing during storage, particularly if the dressing is held at refrigerator temperatures or even frozen.

Spoonable salad dressing is made in two stages. First, vegetable oil is emulsified with egg yolk and some of the water and vinegar. The egg yolk lipoproteins are the surface active materials that stabilize this emulsion, which is rather coarse (relatively large oil-drop diameter). Meanwhile the starch is gelatinized by heating with water; after it is cooled the vinegar, sugar, salt, spices and so on are mixed into this paste. The cooled paste is then fed into the premixer and blended into the oil–egg emulsion. This soft mass is then pumped to a colloid mill where the final high viscosity and smooth texture is generated. The gelatinization operation is the crucial step in attaining good final product quality, and the details of time and temperature depend to a large degree on the specific starch being used. Information about the proper operation of the starch cooker is best obtained from the starch supplier.

Mayonnaise contains vegetable oil (minimum 65% by weight), acidifiers (vinegar, lemon or lime juice), egg yolks (liquid, frozen or as whole egg) and other ingredients (salt, sugars, spices, monosodium glutamate, sequestrants, citric and/or malic acid, and crystallization inhibitors). Mayonnaise is an O/W emulsion, stabilized by the lipoprotein components of egg yolk. The legal minimum for oil content is 65%, but mayonnaise at this oil level is rather thin (low viscosity) and the usual commercial product today contains 77–82% oil. Liquid egg yolk (45% solids) is the emulsifier, at 5.3–5.8% of total formula weight. Sometimes whole eggs (25% solids) are substituted for egg yolks on a total solids basis (i.e. whole eggs at 9.5–10.4% of the total formula weight); this gives a somewhat 'stiffer' product than that with egg yolks because of denaturation of egg albumin at the interface. The denatured albumin forms a matrix in the aqueous phase, increasing the Bingham yield value of the product.

Oil is the internal phase in the emulsion. If all the droplets are spherical, incompressible and have the same diameter, the maximum volume percentage of oil is 74.05%. (For example, if a 1000 ml container were filled with small, uniform ball bearings, then 260 ml of water would be needed to fill the space between the bearings.) If the droplets differ in diameter, then small droplets can fill in the spaces between larger drops. This is the case with mayonnaise,

8 Role of milk fat in hard and semihard cheeses

Timothy P. Guinee and Barry A. Law

8.1 Introduction

Cheese is a dairy product of major economic importance, accounting for around 30% of total milk usage. World production of cheese is around 15×10^6 tonnes per annum and has been increasing at a rate of around 2.3% per annum since 1985. Approximately 7% of total production is traded on the global market, the major suppliers being the EU (*c.* 50%), New Zealand (*c.* 16%) and Australia (*c.* 11%) (Sørensen, 1997). The increase in global cheese consumption may be attributed to the versatility it offers in terms of taste and texture. Cheese may be consumed directly as a table food and as a functional ingredient in an extensive array of cooking applications, including pizza, omelettes, quiches, sauces, chicken, cordon bleu and pasta dishes. Cheese is also used extensively in the industrial food sector for the preparation of ready-to-use grated or shredded cheeses and cheese blends and for the mass production of cheese-based ingredients such as pasteurized processed-cheese products (PCPs), cheese powders and enzyme-modified cheeses (EMCs). These ingredients are in turn used by the food service industry (such as burger outlets, pizzerias and restaurants) and by the manufacturers of formulated foods such as soups, sauces and ready-prepared meals.

When used as an ingredient in foods, cheese is required to perform one or more functions. In the unheated state, the cheese may be required to exhibit a number of rheological properties to facilitate its size reduction and use in the preparation of various dishes—for example, the ability to crumble easily, to slice or to shred cleanly or to bend when in slice form. The rheological properties also determine the textural properties of the cheese during mastication. Cheese is generally required to contribute to the sensory characteristics (i.e. taste, aroma and texture) of the food in which it is an ingredient. On grilling or baking, the cheese may be required to melt, flow, brown, oil-off and/or stretch to varying degrees.

The fat content of cheese differs markedly with variety—for example from *c.* <1 wt% in uncreamed cottage cheese and fromage frais, to *c.* 54 wt% in mascarpone (table 8.1). However, in most varieties fat is a major component and has a marked influence on the yield of cheese (IDF, 1991) and on its rheological and functional properties (Fenelon and Guinee, 2000; Guinee *et al.*, 2000a,

Table 8.1 Fat content of different cheese varieties

Variety	Dry matter (wt%)	Fat (wt%)	Fat in dry matter (wt%)
Uncreamed low-fat cottage cheese	18.0	0.5	2.8
Speise quarg	14.0	1.0	7.1
Half-fat Cheddar	n.a.	16.0	27.3
Low moisture part-skim mozzarella	51.4	17.1	33.3
Feta	43.5	20.2	46.4
Reduced-fat Cheddar	n.a.	23.5	40.5
Camembert	49.3	23.7	48.1
Danish blue	56.2	25.4	45.2
Edam	56.2	25.4	45.2
Tilsiter	57.1	26.0	45.5
Brie	51.4	26.9	52.3
Gouda	58.5	27.4	46.9
Swiss	62.8	27.5	43.7
Romano	70.2	28.7	40.8
Roquefort	60.6	30.6	50.5
Full-fat Cheddar	63.3	33.1	52.3
Double cream cheese	46.3	34.9	75.4
Swiss Gruyère	64.4	35.5	55.1
Old Amsterdam	68.7	37.0	53.9
Cambazola	61.3	44.5	72.7
Mascarpone cheese	61.1	54.1	88.6

n.a. Not available.
Data from Fenelon *et al.*, 2000a; Fox and Guinee, 1987; Holland *et al.*, 1991; Kosikowski and Mistry, 1997; USDA, 1976; Guinee, unpublished data; other published data.

2000b; Olson and Bogenrief, 1995; Tunick and Shieh, 1995; Tunick *et al.*, 1993, 1995). Fat also contributes directly to flavour and indirectly via lipolysis, the hydrolysis of triglycerides to free fatty acids (FFAs). Moreover, in some varieties FFAs serve as precursors for other flavour compounds—for example, in blue-type cheeses where FFAs are oxidised to methyl ketones which in turn are reduced to secondary alcohols (Fox *et al.*, 1996; Kilcawley *et al.*, 1998).

Dietary fat has been shown to be associated with an increased risk of obesity, arteriosclerosis, coronary heart disease, elevated blood pressure and tissue injury diseases associated with oxidation of unsaturated fats (Bonorden and Pariza, 1994; Hodis *et al.*, 1991; McNamara, 1995; Simon *et al.*, 1996). Although it is difficult to establish unequivocally precise relationships between human health and the type and levels of dietary fat, the consumer has become increasingly concerned with the amount of fat in the diet. This heightened consumer awareness has led to an increased demand for low-fat foods (Dexheimer, 1992). However, the consumption of low-fat and reduced-fat cheeses remains relatively low (e.g. 8% total cheese in the UK). The low consumption has been attributed to poor consumer perceptions of the products based on taste and texture (Anonymous, 1996; Bullens, 1994; Olson and Johnson, 1990). The textural defects include increased firmness, rubberiness, elasticity, hardness, dryness and graininess. The negative flavour attributes associated with reduced-fat Cheddar include bitterness (Ardö and Manssön, 1990) and low intensities of typical Cheddar cheese aroma and flavour (Banks *et al.*, 1989; Jameson, 1990). Approaches employed to improve the quality of reduced-fat cheese include:

- alterations to the cheese-making procedure to lower the calcium-to-casein ratio, to increase the moisture-to-protein ratio and to decrease the extent of *para*-casein aggregation; for example via pasteurization temperature, pH at setting and whey drainage, and firmness at cutting
- the use of specialized starter cultures and the use of starter culture adjuncts in conjunction with normal starter cultures
- the addition of fat mimetics to the milk

These approaches have been extensively reviewed (Ardö, 1997; Fenelon, 2000; Fenelon and Guinee, 1997; Jameson, 1990).

The focus of this chapter is on the generic effects of fat on the composition, structure, yield, flavour, rheology and functionality of hard and semihard cheeses and pasteurized processed-cheese products.

8.2 Effect of fat on cheese composition

8.2.1 *Fat level*

Numerous studies have investigated the effects of fat content on the composition of several cheese types, including Cheddar and mozzarella (Bryant *et al.*, 1995; Drake *et al.*, 1996b; Fenelon and Guinee, 1999; Gilles and Lawrence, 1985; Nauth and Ruffie, 1995; Tunick *et al.*, 1991; Rudan *et al.*, 1999). In studies where cheese-making conditions are held constant, fat reduction is paralleled by increases in the levels of moisture and protein and reductions in the levels of fat in dry matter (FDM) and moisture in nonfat substance (MNFS) (tables 8.2 and 8.3). For cheddar cheese the unit changes in these compositional parameters on reducing fat content in the range 33 wt% to 6 wt% were found to be: $+0.36$ g moisture per g fat, $+0.55$ g protein per g fat, $+0.05$ g ash per g fat, -0.2 g MNFS per g fat, and -1.5 g FDM per g fat (figure 8.1).

Small increases in MNFS (2–4 wt%) lead to a relatively large increase in free availability of water, which in turn leads to increases in the activity of microorganisms and enzymes and the degree of proteolysis in cheese (Creamer, 1971; Lawrence and Gilles, 1980; Lawrence *et al.*, 1987; Pearce and Gilles, 1979). Hence, normalisation of the content of the MNFS is considered especially important to improve the quality of reduced-fat cheeses. Consequently, in commercial cheese manufacture and in many studies relating to the improvement of the quality of reduced-fat cheese, cheese-making procedures are frequently altered so as to give reduced-fat cheeses with levels of MNFS similar to that of

Table 8.2 Composition, proteolysis and pH in retail cheddar cheeses of different fat content

	Fat category[a]		
	13–18 wt%	19–24 wt%	30–36 wt%
Composition			
Fat	15.7	21.7	33.6
Salt in moisture (wt%)	4.3	4.4	5.6
Moisture in nonfat cheese substance (wt%)	53.6	54.3	53.7
Proteolysis			
pH-4.6-soluble nitrogen (%)[b]	19.1	18.2	22
Phosphotungstic-acid-soluble nitrogen (%)[b]	5.3	4.3	4.2
Free amino acids (mg kg^{-1})	1418	1217	1004

[a]The retail cheese samples were arbitrarily classed into three categories based on fat content.
[b]As a percentage of total nitrogen.
Data from Fenelon et al., 2000a.

Table 8.3 Composition of Cheddar and mozzarella cheeses with different fat contents

Cheese type	Fat (wt%)	Moisture (wt%)	Protein (wt%)	MNFS MNFS (wt%)	Calcium (mg per 100 g)	Phosphorus (mg per 100 g)	pH (at 180 days)
Cheddar[a]							
low-fat	7.2	46.1	38.5	49.6	1097	839	5.52
half-fat	17.2	43.0	33.3	51.9	937	680	5.45
reduced-fat	21.9	40.9	31.0	52.4	872	639	5.37
full-fat	30.4	37.8	26.4	57.0	742	533	5.25
Mozzarella[b]							
reduced-fat	12.3	48.5	32.8	55.3	n.a.	n.a.	n.a.
reference	21.2	47.0	25.5	59.6	n.a.	n.a.	n.a.
Mozzarella[c]							
low-fat	2.2	51.2	24.6	64.5	n.a.	n.a.	n.a.
low-fat	5.0	62.5	30.4	64.5	n.a.	n.a.	n.a.
part-skim	19.3	63.6	30.1	65.0	n.a.	n.a.	n.a.
Mozzarella[d]							
low-fat	9.9	54.0	n.a.	60.0	n.a.	n.a.	n.a.
high-fat	24.4	48.5	n.a.	64.5	n.a.	n.a.	n.a.

n.a. Not available.
[a]Data from Guinee et al., 2000a.
[b]Data from Poduval and Mistry, 1999.
[c]Data from Fife et al., 1996.
[d]Data from Tunick et al., 1995.
Note: MNFS, moisture in nonfat cheese substance.

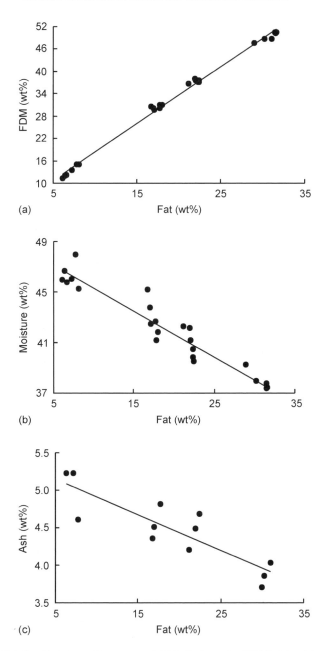

Figure 8.1 Relationship between fat content and: (a) fat in dry matter (FDM), (b) moisture, (c) ash, (d) moisture in nonfat substance (MNFS), (e) protein and (f) calcium, for Cheddar cheese. Adapted from Fenelon and Guinee, 1999.

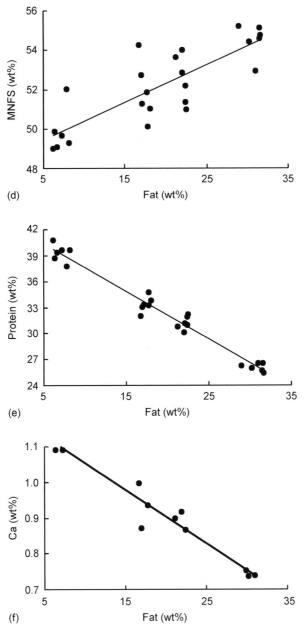

Figure 8.1 (continued)

the full-fat equivalent (Ardö, 1993; Banks *et al.*, 1989). Hence, the fat content of retail cheddar cheeses is inversely correlated with the levels of moisture and protein but it does not significantly affect the level of MNFS (Banks *et al.*, 1992; Fenelon *et al.*, 2000a).

8.2.2 Effect of degree of fat emulsification

The degree of fat emulsification in cheese milk, as affected by the use of homogenization or microfluidization, has a marked influence on the composition of cheeses (e.g. for Cheddar and mozzarella see table 8.4).

Table 8.4 Effect of homogenization and microfluidization on the composition of cheese

Milk treatment	Composition (wt%)				
	fat (wt%)	FDM (wt%)	protein (wt%)	moisture (wt%)	MNFS (wt%)
Cheddar:					
homogenization[a]					
0 MPa	29.9	49.2	25.3	39.2	55.9
30 MPa	30.6	51.2	23.9	40.2	57.9
microfluidization[b]					
0 MPa	34.4	52.9	na	35.0	53.4
7 MPa	34.0	54.5	na	37.6	57.0
Half-fat Cheddar:					
homogenization[c]					
0 MPa	18.2	31.8	33.5	42.7	52.2
15 MPa	18.0	32.1	32.6	44.0	53.6
homogenization[d]					
0 MPa	18.1	32.5	31.3	44.2	54.0
20.7 MPa	17.8	32.7	30	45.7	55.6
Mozzarella:					
homogenization[e]					
0 MPa	25.9	48.2	na	46.3	62.4
10.3 MPa	26.9	51.7	na	47.9	65.5
17.2 MPa	27.8	52.2	na	46.8	64.8
homogenization[f]					
0 MPa	25.8	47.8	21.5	46.0	62.0
0.5 MPa	26.5	54.6	19.0	51.4	69.9

na, Not available.
[a] Data from Guinee *et al.*, 2000b.
[b] Data from Lemay *et al.*, 1994.
[c] Data from Fenelon, 2000.
[d] Data from Metzger and Mistry, 1995.
[e] Data from Tunick *et al.*, 1993.
[f] Data from Jana and Upadhyay, 1991.
Note: FDM, fat in dry matter; MNFS, moisture in nonfat substance.

Homogenization of milk is practised in the manufacture of some cheese varieties where lipolysis is important for flavour development (e.g. in blue cheese), the objective being to increase the accessibility of the fat to mould lipases and thus to increase the formation of fatty acids and their derivatives (e.g. methyl ketones). Moreover, homogenization is a central part of the manufacturing process for cheeses from recombined milks. Homogenization of milk reduces the fat globule size and increases the surface area of the fat by a factor of 5–6 (McPherson et al., 1989). The newly formed fat globules are coated with a layer, denoted the recombined fat-globule membrane (RFGM), which consists of casein micelles, submicelles, whey protein and some of the original fat-globule membrane (Keenan et al., 1988; Walstra and Jenness, 1984). Owing to the protein membrane, the newly formed emulsified fat globules (EFG) interact and complex with the main gel-forming protein, casein, during rennet or acid gelation of milk. The emulsified fat globules thereby become an integral part of the casein matrix in the cheese (Green et al., 1983; Lelievre et al., 1990; Tunick et al., 1997; van Vliet and Dentener-Kikkert, 1982). Microfluidization is also a particle size reduction process that is applied mainly in the manufacture of products such as antibiotic dispersions, parenteral emulsions and diagnostics. It is generally accepted that for the application of equivalent pressures to the milk microfluidization gives a lower mean fat-globule diameter (e.g. 0.03–0.3 μm compared with 0.5–1.0 μm) and a narrower size distribution than homogenization. Moreover, the fat-globule membrane in microfluidized milk has a higher proportion of fragmented casein micelles and little or no whey protein compared with that in homogenized milk.

The effects of milk homogenization on cheese composition and quality have been reviewed (Jana and Upadhyay, 1992). It is generally accepted that homogenization impedes the aggregation of casein particles, curd fusion and curd syneresis and thereby gives cheese with higher levels of moisture. Cheeses for which increases in moisture or MNFS have been reported include Cheddar cheeses of different fat contents (Emmons et al., 1980; Mayes et al., 1994; Metzger and Mistry, 1994), Edam (Amer et al., 1977), Gouda (Versteeg et al., 1998) and mozzarella (Jana and Upadhyay, 1992; Tunick et al., 1993; Rudan et al., 1998a). The extent of the moisture increase varies, depending on homogenization pressure and cheese-making practices (Jana and Upadhyay, 1992; table 8.4). Microfluidization of milk (at 7 MPa) or cream (at 14 or 69 MPa) has also been found to give a higher moisture content in Cheddar cheese (Lemay et al., 1994).

The impaired syneresis in homogenized milk curd or cheese may be associated with the increased interaction between the casein and the fat, which reduces the surface area of the casein micelles available for mutual interaction (Green et al., 1983). Curd syneresis may be defined as the physical expulsion of whey, which accompanies contraction of the protein matrix. Matrix contraction may be viewed at the microstructural level as an increased aggregation and joining of adjacent casein strands into larger aggregates both within curd particles and at curd particle junctions (Kimber *et al.*, 1974). The reduction in casein hydration and increase in matrix contraction that parallels casein aggregation reduces the ability of the matrix to retain whey. Hence, a lower degree of casein–casein interaction in homogenized milk curds would lead to higher moisture content. The increase in the fineness of the rennet-induced milk gels that accompanies milk homogenization (Green *et al.*, 1983) may also be a contributory factor to the higher moisture of homogenized milk cheeses; finer gels have a lower gel porosity than coarse-structured gels, a factor that would be expected to impede moisture expulsion in cheese curds from homogenized milk.

8.3 Proteolysis

Proteolysis in cheese has been extensively studied and reviewed (Fox, 1989; Fox and McSweeney, 1996; Fox and Wallace, 1997). It plays a major role in the development of texture and flavour in most rennet curd cheese varieties during ripening. It contributes directly to flavour via formation of peptides and free amino acids (FAA), and indirectly via the breakdown of free amino acids to various compounds, including amines, acids and thiols. Proteolysis affects the level of intact casein, which is a major determinant of the fracture and functional properties of cheese (Guinee *et al.*, 2000a; Prentice *et al.*, 1993). Most of the published studies on proteolysis have focused on full-fat or on half-fat or reduced-fat cheeses; comparatively little has been reported on the effects of incremental fat reduction, especially in cheeses other than Cheddar.

Rank (1985) investigated the effects of fat level, in the range 13.5–30.6 wt%, on proteolysis in Colby cheeses where alterations were made to the manufacturing protocol of the low-fat cheese so as to give a content of MNFS similar to that in the full-fat cheese. Although reducing fat content had little effect on the overall proteolysis, as detected by polyacrylamide gel electrophoresis (PAGE), the concentration of α_{s1}-casein was slightly lower in the low-fat cheeses after 6 or 8 months storage at 4°C. Similar results were noted by Banks *et al.* (1989), who found that a 50% reduction in the fat content of Cheddar cheese only slightly reduced the water-soluble nitrogen level at 2 months (9.9 g per 100 g N compared with 11.8 g per 100 g N) or 4 months (15.3 g per 100 g N compared with 17.8 g per 100 g N). Likewise, Ardö (1993) found that proteolysis, as measured by nitrogen solubility at pH 4.4 and in 5% phosphotungstic acid, in Herrgårds and Drabant cheeses at 3 or 14 weeks was scarcely affected by a 40% reduction in fat level. The absence of large differences in proteolysis in these studies is surprising considering the effect of fat per se on composition (table 8.2) and the major effect of composition on proteolysis (Creamer, 1971; Thomas and Pearce, 1981). The relatively minor effect of fat on proteolysis in these studies may be attributed in part to the alterations to the cheese-making procedures, which minimized the differences in the levels of MNFS between the full-fat and reduced-fat cheeses. Hence, studies on retail Cheddar cheeses have found no significant relationship between the levels of fat and MNFS or between fat content and levels of primary or secondary proteolysis (Banks *et al.*, 1992; Fenelon *et al.*, 2000a). This trend suggests that cheese-making recipes are designed to minimize the differences in MNFS between cheeses of different fat content during the commercial manufacture of Cheddar.

8.3.1   Primary proteolysis (where moisture in nonfat solids differs)

In contrast to the above, a recent study (Fenelon *et al.*, 2000b) found that fat level, in the range 6–33 wt%, had a marked influence on the level of proteolysis in Cheddar cheeses manufactured using identical conditions and with different levels of MNFS (tables 8.2 and 8.3). The mean level of primary proteolysis throughout the 225 day ripening period, as measured by the percentage of total nitrogen soluble at pH 4.6 per 100 g nitrogen, decreased significantly as the fat level was reduced [figure 8.2(a)]. The decrease in proteolysis was expected owing to the parallel decreases in the levels of MNFS and residual rennet activity (figure 8.1; table 8.3; Fenelon and Guinee, 2000). However, the level of nitrogen soluble at pH 4.6 per 100 g cheese was not significantly influenced by fat content [figure 8.2(b)]. This trend suggests that the reduction in nitrogen soluble at pH 4.6 per 100 g nitrogen as the fat content decreased was compensated for by the concomitant increase in protein content.

Variation in fat level also leads to differences in the levels of primary proteolysis as monitored by PAGE. Reducing the fat content resulted in higher levels of intact α_{s1}-casein and β-casein in both Cheddar (Fenelon and Guinee, 2000) and mozzarella cheese (Tunick *et al.*, 1993, 1995). This trend was expected because of the inverse relationship between the fat and protein levels in the cheese. However, for a given protein content, reduction in fat level led to a more extensive degradation of β-casein and accumulation of γ-caseins (Fenelon and Guinee, 2000). The increase in the concentration of γ-caseins was attributed to the higher pH in the reduced-fat cheeses, which would enhance the activity of the native milk alkaline proteinase plasmin (Grufferty and Fox, 1988). The increase in pH as the fat content increased may also affect the degree of hydration and aggregation of the β-casein (Creamer, 1985), which could affect its susceptibility to hydrolysis by chymosin and plasmin. In contrast to the trend noted for β-casein, the degradation of α_{s1}-casein decreased as the fat content was reduced.

This trend may be the result of a number of associated effects:

- the decrease in level of residual rennet activity per unit weight of protein (Fenelon and Guinee, 2000)
- the decrease in the level of MNFS (tables 8.2 and 8.3)
- the higher pH of the lower-fat cheeses, which would be less favourable to the proteolytic activity of residual rennet (Tam and Whitaker, 1972; Vanderpoorten and Weckx, 1972)

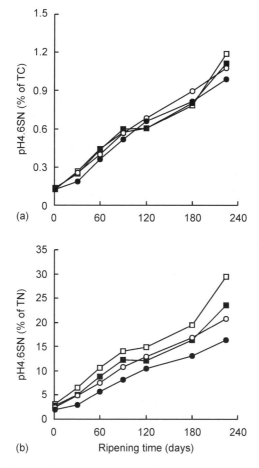

Figure 8.2 Cheddar cheese: effect of fat content on: (a) level of nitrogen soluble at pH4.6SN as a percentage of total cheese [pH4.6SN (% of TC)]; (b) the level of nitrogen soluble at pH4.6SN as a percentage of total nitrogen [pH4.6SN (% of TN)]; (c) amino acid nitrogen (AAN) as a percentage of total cheese [AAN (% of TC)]; (d) AAN as a percentage of total nitrogen [AAN (% of TN)]. —□— full-fat cheese (30.4 wt% fat); —■— reduced-fat cheese (21.9 wt% fat); —○— half-fat cheese (17.2 wt% fat); —●— low-fat cheese (7.2 wt% fat). Adapted from Fenelon *et al.*, 2000b.

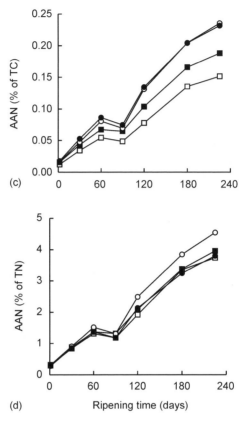

Figure 8.2 (continued)

8.3.2 Secondary proteolysis (where moisture in nonfat substance differs)

The effect of reducing fat level on the degree of secondary proteolysis in Cheddar cheese is the opposite to that noted for primary proteolysis (Fenelon et al., 2000a,b). Reducing the fat level led to a significant increase in the level of amino acid nitrogen (AAN) per 100 g cheese but did not significantly affect the concentration of AAN per 100 g nitrogen [figures 8.2(c) and 8.2(d)]. Hence, the increase in AAN per 100 g cheese as the fat content decreased is at least partly due to the concomitant increase in the total concentration of protein. Consistent with the higher level of AAN per 100 g cheese in the reduced-fat cheeses, the level of low molecular mass peptides (less than 1 kDa) in the pH 4.6 cheese extracts increased as the fat content of the cheese decreased (Altemueller and Rosenberg, 1996; Fenelon *et al.*, 2000a,b).

8.4 Effect of fat on cheese microstructure

8.4.1 Microstructure of rennet-curd cheese

The microstructure of milk gels and cheeses has been studied extensively, with most emphasis on cheeses of standard fat content (Bryant *et al.*, 1995; de Jong, 1978b; Desai and Nolting, 1995; Green *et al.*, 1981, 1983; Guinee *et al.*, 1998, 1999, 2000a; Hall and Creamer, 1972; Kaláb, 1977; Kiely *et al.*, 1992, 1993; Kimber *et al.*, 1974; Mistry and Anderson, 1993). The protein matrices of both acid-curd and rennet-curd cheese are particulate, being composed of entangled chains of partially fused casein or *para*-casein aggregates (in turn formed from fused casein or *para*-casein micelles). The integrity of the matrix is maintained by various intra-aggregate and inter-aggregate hydrophobic and electrostatic attractions (Walstra and van Vliet, 1986). The network is essentially continuous, extending in all directions, although some discontinuities exist at the microstructural and macrostructural levels.

Discontinuities at the macrostructural level exist in the form of curd granule junctions or curd chip junctions (in Cheddar and related dry-salted varieties) (Kaláb, 1979; Lowrie *et al.*, 1982; Paquet and Kaláb, 1988). Curd granule junctions in low-moisture mozzarella are well-defined, *c.* 3–5 μm wide and appear as veins running along the perimeters of neighbouring curd particles (Kaláb, 1977). Unlike the interior of the curd particles, the junctions comprise mainly casein, being almost devoid of fat. Factors that contribute to the formation of these junctions include casein dehydration and matrix contraction at the curd particle surfaces with the consequent loss of fat. Chip junctions in Cheddar and related dry-salted varieties are clearly discernible on examination of the cheese by light microscopy and, like curd granule junctions, have a higher casein-to-fat ratio than the interior of the chip (Brooker, 1979). The difference in cheese composition between the interior and surface of curd particles (or chips) probably leads to differences in the molecular attractions between contiguous *para*-casein layers in the interior and exterior of curd particles and thus to differences in structure–function relationships.

The matrix occludes, within its pores, fat globules (clumped or coalesced to varying degrees), moisture and its dissolved solutes (minerals, lactic acid, peptides and amino acids) and enzymes (e.g. residual rennet, proteinases and peptidases from starter and nonstarter microorganisms) (Guinee *et al.*, 2000a; Kimber *et al.*, 1974; Laloy *et al.*, 1996). The starter and nonstarter bacteria appear to attach themselves, via filaments from their cell walls, to the casein matrix (Kimber *et al.*, 1974) and are concentrated near the fat–casein interface

(Haque *et al.*, 1997; Laloy *et al.*, 1996). The concentration of bacteria near the fat–casein interface may have several potential consequences (Laloy *et al.*, 1996):

- concentration of intracellular bacterial peptidases in the vicinity of the protein–fat interface following bacterial autolysis
- a more heterogeneous distribution of starter bacteria in cheese as the fat level is reduced owing to the concomitant reduction in fat–casein surface area
- a more uneven distribution of starter-cell proteinases or peptidase activity in reduced-fat cheese and possible restricted access of substrates (poly-peptides or peptides released by the action of coagulant, etc.) to enzymes

Hence, the location of bacteria in cheese at the fat–casein interface may be of importance in relation to the growth dynamics of starter and nonstarter bacteria in cheese and their effects on cheese maturation (see section 8.5).

8.4.1.1 *Microstructure of the fat phase*

The enmeshed fat globules occupy the spaces between the protein strands and physically impede the aggregation of the *para*-casein matrix, to a degree dependent on their volume fraction and size distribution. Some clumping and/or coalescence of fat globules generally occurs in most cheese varieties.

Evidence for the clumping of fat globules in Cheddar cheese has been clearly demonstrated by both scanning electron microscopy (SEM) and transmission electron microscopy (TEM) (Bryant *et al.*, 1995; Hall and Creamer, 1972; Kaláb, 1979; Metzger and Mistry, 1995; Mistry and Anderson, 1993) and confocal laser scanning microscopy (CLSM) (Auty *et al.*, 1999; Guinee *et al.*, 1999). Fissures, or irregular-shaped openings, in the *para*-casein matrix, which remain after removal of fat during sample preparation, are evident in SEM micrographs (figure 8.3). TEM micrographs taken over the course of Cheddar manufacture show clearly the aggregation of fat globules, which is first notable at maximum scald and increases with progression of cheese-making as the protein network shrinks and forces the fat globules into closer proximity (Kaláb, 1995; Kimber *et al.*, 1974; Laloy *et al.*, 1996). Clumping of fat globules is also pronounced in mozzarella and string cheeses, where the clumped fat globules coexist with moisture in the form of long channels between the *para*-casein fibres (Guinee *et al.*, 1999; Kaláb, 1995; Kiely *et al.*, 1992; McMahon *et al.*, 1999; Taneya *et al.*, 1992). In contrast to Cheddar and mozzarella, relatively little clumping and coalescence of fat globules is evident in other cheese types such as Cheshire, Gouda (Hall and Creamer, 1972) and Meshanger cheese (de Jong, 1978b).

The clumping and coalescence of fat in mozzarella and Cheddar appears to have a major effect on the melt properties of the cooked or baked cheese, as discussed in section 8.9. Fat coalescence may also affect cheese flavour, as leakage of nonglobular fat into the cheese matrix may increase its exposure

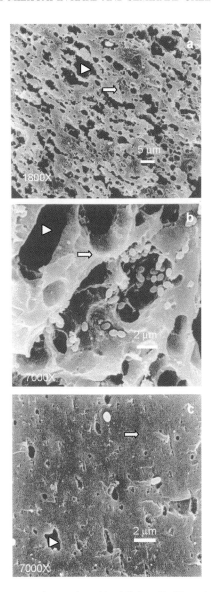

Figure 8.3 Scanning electron micrographs: (a) full-fat Cheddar cheese (33.0 wt% fat) at 1800×magnification; (b) full-fat Cheddar cheese at 7000×magnification; (c) low-fat Cheddar cheese (3.9 wt% fat) at 7000 × magnification. The full arrows indicate the *para*-casein matrix and the arrowheads indicate the areas occupied by fat and free serum prior to their removal during sample preparation; bacteria (most likely starter lactococci) are visible in part (b), being concentrated mainly at the fat–*para*-casein interface. Part (a) adapted from Guinee *et al.*, 1998 (reproduced by permission of the Society of Dairy Technology); parts (b) and (c) adapted from Fenelon *et al.*, 1999.

to lipases and thus to potential flavour-generating reactions (Wijesundera and Drury, 1999; Wijesundera *et al.*, 1998). Moreover, fat coalescence is indicative of damage to the native milk fat globule membrane (NMFGM), which, in milk, and cheese, protects the enclosed triglycerides against attack from lipase (Walstra and Jenness, 1984). Damage to the NMFGM and its replacement by caseins and whey proteins impairs this protection and the fat becomes more susceptible to attack from lipase activities.

Factors contributing to the clumping and coalescence of fat globules in cheese include:

- the shearing of the NMFGM during cheese manufacture
- dehydration and contraction of the *para*-casein matrix during manufacture, which forces the occluded globules into closer contact (Kimber *et al.*, 1974)
- a possible increase in the permeability of the NMFGM during maturation arising from storage-related hydrolysis of membrane components by lipoprotein lipase activity (Deeth, 1997; Sugimoto *et al.*, 1983).

At the temperatures used in the manufacture of cheese (*c.* 30–50°C) much or all of the milk fat is liquid (Norris *et al.*, 1973) and therefore flows and aggregates on the application of stress. The more extensive degree of fat coalescence in mozzarella and string cheeses compared with that in Cheddar reflects the shearing of the NMFGM during plasticization and elongation of the curd. Plasticization, which involves kneading and heating of curd to *c.* 57°C in hot water, is conducive to deformation and disruption of the NMFGM and aggregation of globular and nonglobular fat. Similarly, elongation of plasticized curd in the manufacture of string cheese results in the deformation (and probably coalescence) of the fat globules lying between contiguous layers of the *para*-casein matrix, to a degree that increases with elongation (Taneya *et al.*, 1992).

Some coalescence of fat globules probably also occurs during the ripening of Cheddar, mozzarella and other varieties, as reflected by:

- the increases in level of fat that can be expressed from the cheese when subjected to hydraulic pressure at room temperature [figure 8.4(a); Guinee *et al.*, 1997, 2000a; Thierry *et al.*, 1998] or by extraction of the melted cheese (at *c.* 74°C) with a water:methanol mixture followed by centrifugation [figure 8.4(b); Yun *et al.*, 1993b]
- the increase in the degree of aggregation of fat as revealed by SEM (Tunick and Shieh, 1995)

The proportion of liquid fat at the ripening temperatures employed for Cheddar or mozzarella (*c.* 4–7°C) is *c.* 20–30% total (Norris *et al.*, 1973). It has been suggested that the age-related increase in fat coalescence may be attributed to proteolytic breakdown of the casein matrix, which holds the fat globules in place (Tunick and Shieh, 1995).

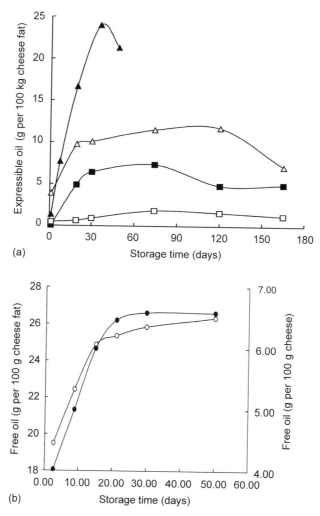

Figure 8.4 (a) Changes in the levels of oil expressed on subjecting; Cheddar and mozzarella cheese to hydraulic pressure (2.3 MPa) for 3 h at 21°C. Cheddar: —△—, full-fat (30 wt% fat); —■—, reduced-fat (21.9 wt% fat); —□—, half-fat (17.2 wt% fat). Mozzarella: —▲—, low-moisture. (b) Changes in the levels of oil extracted from melted low-moisture mozzarella cheese (at *c.* 57.5°C) with a water:methanol mixture followed by centrifugation. —○—, per g 100 g cheese fat; —●—, per g 100 g cheese. Part (a): data from Guinee *et al.*, 2000b (reprinted with permission from Elsevier Science, copyright 2000). Part (b): data from Yun *et al.*, 1995.

8.4.2 Microstructure of pasteurized processed cheese products and analogue cheese products

Microstructural studies on PCPs or analogue cheese products (ACPs) indicate that the structure represents a concentrated emulsion of discrete, rounded fat droplets of varying size in a protein matrix (Guinee *et al.*, 1999; Heertje *et al.*, 1981; Kaláb, 1995; Kaláb *et al.*, 1987; Kimura *et al.*, 1979; Lee *et al.*, 1981; Rayan *et al.*, 1980; Tamime *et al.*, 1990; Taneya *et al.*, 1980; Savello *et al.*, 1989). The fat and *para*-casein are more homogeneously distributed than they are in natural cheese. Compared with natural cheese, there is less clumping or coalescence of fat globules; consequently, the mean fat globule size tends to be generally smaller, though it varies depending on emulsifying salt type and processing conditions (figure 8.5; Kaláb *et al.*, 1987; Rayan *et al.*, 1980; Tamime *et al.*, 1990). The *para*-caseinate membranes of the emulsified fat globules appear to attach to the matrix strands and thereby contribute to an increase in the continuity of the protein matrix. High resolution TEM (60, 000 ×) with negative staining of the protein phase reveals that the matrix consists of strands that are finer than those of natural cheese and appear to be composed of *para*-caseinate particles (20–30 nm diameter) joined end to end. It has been suggested that these particles may correspond to casein submicelles released from the *para*-casein micelles in the matrix of the natural cheese as a result of calcium chelation by the emulsifying salts (Heertje *et al.*, 1981; Kimura *et al.*, 1979; Taneya *et al.*, 1980).

8.4.3 Effect of fat level

A number of studies have evaluated the effects of fat content on the microstructure of Cheddar (Bryant *et al.*, 1995; Bullens *et al.*, 1994; Desai and

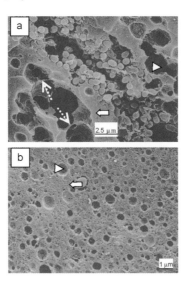

Figure 8.5 Scanning electron micrograph of (a) natural Cheddar cheese; (b) processed Cheddar cheese. The micrographs show protein (long arrows), fat (short arrows) and residues of fat globule membrane (broken arrows). Adapted from Tamime *et al.*, 1990 with permission from Scanning Microscopy International.

fat liquefaction and coalescence, and its seepage from the cheese mass. The heat-induced coalescence of fat in cheese suggests a tendency towards phase separation on heating and is consistent with the increase in fat leakage or oiling-off that occurs on baking or grilling of cheeses (figure 8.4).

Changes in the fat distribution on baking half-fat Cheddar (figure 8.7) and reduced-fat (15 wt%) mozzarella are similar to those noted for their full-fat counterparts. However, the degree of aggregation of fat globules and the sizes of the coalesced fat particles in the reduced-fat cheeses are generally smaller than in the full-fat cheeses. This trend undoubtedly reflects the lower volume fraction of the fat phase in the reduced-fat cheeses (van Boekel and Walstra, 1995).

In contrast to natural cheeses prepared from homogenized milk, heating generally has little influence on the microstructures (distributions of fat and protein) in pasteurized PCPs (Paquet and Kaláb, 1988) or on full-fat Cheddar cheese prepared from homogenized milk (figures 8.6 and 8.7) (Guinee et al., 2000b). The relatively high thermostability against fat coalescence in these products probably resides in the higher degree of emulsification prior to heating and a higher heat stability of the RFGM compared with the NMFGM.

8.5 Effect of fat on cheese microbiology

The population of starter bacteria generally decreases during maturation whereas that of nonstarter lactic acid bacteria (NSLAB) generally increases; these changes are well documented for the full-fat versions of many rennet-curd cheese varieties, for example Cheddar (Cromie et al., 1987; Fox et al., 2000; Haque et al., 1997; Jordan and Cogan, 1993; Lane et al., 1997; McSweeney et al., 1993). Comparatively little information is available on the effect of fat content on the dynamics of starter and NSLAB populations in cheese. Laloy et al. (1996) found that the population of starter cells in full-fat curd prior to pressing were four fold to ten fold higher than the corresponding populations in fat-free Cheddar curd, depending on the starter strain type. They suggested that the higher starter

Table 8.5 Effect of fat content on recovery and yield characteristics of Cheddar cheese

Fat in milk (wt%)	0.54	1.50	2.00	3.33
Fat in cheese (wt%)	6.88	17.41	22.06	30.77
Fat recovered in cheese (percentage total milk fat)	80.84	87.16	89.48	87.48
Solids-nonfat-nonprotein	0.10	0.11	0.13	0.16
Actual yield (kg per 100 kg)	6.37	7.49	8.09	9.50
Dry matter yield (kg per 100 kg)	3.43	4.29	4.79	5.92

Data from Fenelon and Guinee, 1999.

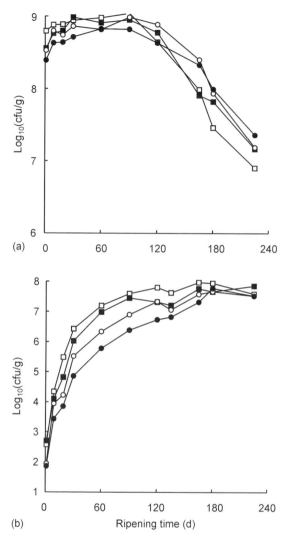

Figure 8.8 Effect of fat content on populations of (a) starter and (b) nonstarter bacteria in Cheddar cheese. —□—, full-fat (30.4 wt%) cheese; —■—, reduced-fat (21.9 wt%) cheese; —○—, half-fat (17.2 wt% cheese); —●—, low-fat (7.2 wt%) cheese. The values presented are the means of three replicate trials. Note: cfu, colony-forming units. Reproduced with permission from Fenelon *et al.*, 2000b.

cell numbers in the full-fat cheese might be attributable to:

- the association between starter lactococci and fat globules and the higher retention of fat globules in the full-fat curd [it is noteworthy that the recovery of milk fat during Cheddar manufacture decreases as the fat content of the cheese is reduced from 22 wt% to 7 wt% (table 8.5)]

- the greater physical impedance to syneresis by the fat globules, which in effect act as 'stoppers' in the pores of the *para*-casein matrix and thereby reduce the loss of starter cells in the whey exuding from the curd; the retained starter cells aggregate around the fat globules

In agreement with the trend noted by Laloy *et al.* (1996), Fenelon *et al.* (2000b) reported that the starter cell count in full-fat Cheddar (FFC; 32.5 wt%) at 1 day was significantly higher than that in low-fat Cheddar (LFC; 6.3 wt%). However, the starter population in the FFC declined more rapidly and was significantly lower than that in the LFC at 180 days [figure 8.8(a)]. In contrast, Haque *et al.* (1997) reported similar populations (*c.* $10^{8.5}$ cfu g^{-1}) of starter lactococci in both LFC and FFC at 1 day, but counts in the LFC decreased more rapidly during maturation; the populations in the LFC and FFC at 180 days were *c.* $10^{3.9}$ and $10^{4.4}$ cfu g^{-1}, respectively. (Note: cfu=colony-forming units.)

The growth of NSLAB in Cheddar cheese is also influenced by fat content [figure 8.8(b)]. The populations of NSLAB in Cheddar decrease with fat content, with those in LFC (5 wt%) being significantly lower than those in FFC (33 wt%) during ripening (Fenelon *et al.*, 2000b; Haque *et al.*, 1997). The decrease in NSLAB numbers as the fat content of cheese is reduced may be the result of a number of factors, including:

- the reduction in the concentration of milk fat globule membrane associated glycoproteins; mesophilic lactobacilli possess glycoside hydrolase activity and may be capable of releasing sugars from glycoproteins associated with the fat globule membrane (Williams and Banks 1997); the released sugars may be then used as a carbon source to provide energy for propagation (Fox *et al.*, 1998)
- the lower level of MNFS (Lane *et al.*, 1997), which is expected to give lower water activity and microbial growth (Lawrence and Gilles, 1980)

8.6   Effect of fat on cheese flavour

Fat clearly has a major influence on the perception of flavours in all fat-containing foods, but its role may be direct or indirect. This is because fat acts both as a substrate for flavour-forming reactions and as a solvent or dispersant for important aroma compounds. In cheese, fat certainly has a direct effect as a source of flavour and aroma compounds, but its role as a flavour carrier is more speculative. Thus we know that low-fat versions of hard and semihard cheeses have poorer flavour than their normal fat equivalents (Johnson and Law, 1999) but it is also true that the low-fat cheeses lack the typical texture of the full-fat versions. They therefore present a very different mouth feel and mastication pattern to the consumer, and it is logical to assume that flavour compounds within the cheese matrix will reach olfactory sensors at different rates and in different concentrations from normal-fat and lower-fat cheeses (Piggot, 2000). However, there is as yet no research base to provide data that cheese technologists can use to improve flavour delivery through fat manipulation, so we will confine this section to the known direct contributions of cheese fat to the properties and quality of the product.

Fat breakdown appears to make an essential contribution to the flavour profile of ripened cheese varieties, though this role is more obvious in some than in others. For example, the characteristic flavour of the blue-vein cheese varieties ripened with the blue mould, *Penicillium roqueforti*, is almost entirely fat-derived. The basic acidity and open structure is created by lactic cultures, but the major flavour impact in the ripened cheese is from methyl ketones (particularly heptan-2-one and nonan-2-one) formed by the oxidation of the corresponding fatty acid released by penicillium lipase from milk fat.

This is a well-documented mechanism of cheese flavour production, summarized by Law (1982), together with other similar pathways to known cheese flavour compounds. Indeed the combination of lipase-mediated fat breakdown and fat oxidation by mould spores is used in cheese flavour ingredients technology to produce concentrated blue-cheese flavour from milk fat for snack-food applications (Balcao and Malcata, 1998).

The white mould *Penicillium camemberti* is also a producer of fat-derived cheese flavour and aroma on the surface of French soft cheeses such as Brie and Camembert. Gripon (1997) has comprehensively reviewed this topic, and only a summary is presented here. Thus, although the extent of lipolysis by the surface mould microflora in these cheeses is much lower than it is in blue cheese, the subsequent metabolism of the released fatty acids is vital to the development of characteristic flavour both by the penicillium itself, and by the accompanying and succeeding microflora. The basic mild mushroom-like aroma of Brie and Camembert is attributable to the formation of oct-1-en-3-ol by the oxidative degradation of linoleic acid (*P. camemberti*) produced in large quantities by a particular lipase of *Geotrichum candidum*, the other major mould culture used to make these surface-ripened cheeses.

The subtle but obvious levels of mould-induced lipolysis are also important for the production of other flavour notes in surface-mould cheese varieties. Recent multidisciplinary research in the European Union has established the central role of S-methyl thioesters in the aroma of surface smear cheeses and Camembert flavour and aroma (Berger *et al.*, 1999; Khan *et al.*, 1999) and these potent flavour compounds are produced by synergy between lipase-producing moulds and aerobic bacteria (Brevibacteria, Arthrobacter, 'micrococci') that can form the ester from activated fatty acids and methane thiol (a product of amino acid catabolism in cheese) (Berger *et al.*, 1999; Lamberet *et al.*, 1997).

Alkyl esters of cheese fatty acids are also important flavour compounds, though the fruity ethyl and hexyl esters of short-chain acids can give flavour defects in Cheddar. Some hard-cheese varieties, especially those of Italian origin (e.g. Parmesan) depend directly on lipolysis of cheese fat to short-chain fatty acids for their characteristic piquant, peppery flavour profiles. The lipase is present in the 'rennet paste' used to coagulate the milk and, like most animal esterases, produces predominantly short-chain fatty acids with low flavour thresholds, which combine to give the pleasant characteristic notes of this type of cheese.

Because the fat in cheese is a source of flavour compounds, manipulation of lipolysis by enzymes presents a possible route to technology for accelerating the maturation of cheese. This area of cheese ripening technology is rather complex and involves many different approaches, including other types of enzymes, new types of microbial cultures and novel processing technology (Law, 1999), but lipases have not been particularly successful for a number of reasons. This is not because of lack of available commercial food-grade lipases. These are available from many sources (Godfrey and West, 1996), including the 'traditional' calf and lamb lipases and pregastric esterases and the food-approved fungal lipases from Aspergillus and Rhizomucor. However, although putting these enzymes into semihard and hard cheese varieties accelerates lipolysis very nicely, flavour formation is thrown out of balance by the sweaty, rancid and soapy flavour defects produced (Law, 1999). Even novel lipases, developed specifically for Cheddar ripening (Arbige et al., 1986), have failed to solve the flavour defect problems associated with the other lipases on the market.

In contrast to the lack of success of lipase applications in the 'table cheese' sector, they are vital for the production of 'enzyme-modified cheese' (EMC) products used in the processed-cheese and snack-food industry as concentrated cheese flavour ingredients. Here, the lipases are used with proteases to ripen semiliquid cheese slurries at relatively high temperatures, by rapid and extensive breakdown of both milk fat and milk protein. The resulting EMC does not itself taste of cheese but it imparts a generic cheese flavour to other products (Kilcawley et al., 1998).

8.7 Effect of fat on cheese yield

The fat content of milk, and therefore the fat level in cheese, has a direct influence on fat recovery and cheese yield (table 8.5). The direct contribution of fat to cheese yield is reflected by prediction equations that relate cheese yield to component concentrations and recoveries. An example of such an equation is the van Slyke formula (Fenelon and Guinee, 1999):

$$Y_p = \left(\frac{F^{cm} R^{fat}}{100} + C^{cm} - a + \frac{W^{unpas} DW}{100} \right) \left(\frac{1 + S^{NFP}}{1 - \frac{M^r}{100}} \right) \qquad (8.1)$$

where Y_p is the predicted yield; F^{cm} and C^{cm} are the percentages of fat and casein in the cheesemilk (with added starter culture), R^{fat} is the percentage recovery of milk fat into cheese; a is a coefficient for casein loss (typically 4% of total casein); W^{unpas} is the percentage of whey protein in the unpasteurised milk; DW is the percentage of total whey protein denatured on pasteurisation; $(1+S^{NFP})$ is a coefficient to account for nonfat, nonprotein cheese solids (e.g. lactates, ash); and M^r is the reference moisture content.

It has been found that the fat recovery in Cheddar cheese increased significantly on raising the fat content in the cheese milk from 0.5 wt% to 2.7 wt% and thereafter decreased as the fat was raised to 3.3 wt% (table 8.5; Fenelon and Guinee, 1999). A similar trend was noted by Banks and Tamime (1987) who noted that fat recovery during Cheddar cheese manufacture increased to a maximum as the casein-to-fat ratio (CFR) was raised from 0.65 to 0.72, and thereafter decreased as the CFR was raised further to 0.75. The increase in fat recovery with milk fat content to 2.7 wt% may be due to the associated increase in the extent of clumping and coalescence of fat globules in the cheese milk during gelation and in the curd throughout cheese-making (figures 8.3 and 8.6; section 8.4.1.1). The probable consequence of this partial clumping is an increase in the effective size of the fat globules (clumps), which in effect impedes their flow and escape through the pores of the *para*-casein matrix, to the whey. A tentative explanation for the reduction in fat recovery at the higher fat levels (greater than 2.7 wt% fat) is excessive clumping, which leads to coalescence and the formation of free fat that easily permeates the *para*-casein matrix and is lost to the cheese whey.

The level of fat recovery during cheese manufacture is also influenced by cheese variety, which determines the type and level of processes to which the curd is subjected, which in turn influence the level of damage to the NMFGM. Hence the level of fat recovery reported for mozzarella is markedly lower than that of Cheddar (e.g. 80 compared with 88 in pilot-scale studies; Fenelon and Guinee, 1999; Guinee *et al.*, 2000c). The lower fat recovery for low-moisture mozzarella cheese compared with Cheddar may be attributed to the high loss of fat during the kneading and stretching of the curd in hot water (plasticization). The high fat losses during the plasticization process are consistent with the increase in the degree of fat coalescence that accompanies the shearing of the curd, as observed by confocal laser scanning microscopy (see Fox *et al.*, 2000).

Fat also contributes indirectly to cheese yield. Although fat on its own has little water-holding capacity, its presence in the *para*-casein matrix affects the degree of matrix contraction and hence moisture content and cheese yield. The occluded fat globules physically limit contraction of the surrounding *para*-casein network and therefore reduce the extent of syneresis. Thus, it becomes more difficult to expel moisture as the fat content of the curd is increased. Consequently, the moisture-to-casein ratio generally increases unless the cheesemaking process is modified to enhance casein aggregation, for example by increasing the scald temperature (Fenelon and Guinee, 1999; Gilles and Lawrence, 1985). Owing to its negative effect on syneresis, fat indirectly contributes more than its own weight to actual cheese yield (i.e. Cheddar cheese yield increases by *c.* 1.16 kg kg^{-1} milk fat). The greater than pro rata increase is a result of the concomitant increase in the level of moisture associated with the cheese protein as reflected by the positive relationship between milk fat level and MNFS (figure 8.1). Hence, although the percentage moisture in Cheddar cheese is inversely related to its fat content, the weight of cheese moisture from a given weight of cheesemilk increases (at a rate of *c.* 0.24 kg moisture from 100 kg milk per kg fat in milk) as the fat content of the cheese increases (figure 8.9). Moreover, the increase in the moisture-to-protein ratio with fat content contributes indirectly to cheese yield because of the presence of dissolved solids, which include native whey proteins, κ-casein glycomacropeptide, lactate and soluble salts. However, if the content of MNFS is maintained constant (e.g. by process modifications), fat contributes less than its own weight to cheese yield (i.e. *c.* 0.9 kg kg^{-1} for Cheddar), because of the fact that *c.* 8–10% of the milk fat is normally lost in the whey during Cheddar manufacture.

The dry matter cheese yield (Y_{dm}) increases with milk fat level but at a lower rate than actual yield (Y_a); that is, *c.* 0.93 compared with 1.16 kg kg^{-1} milk fat for Cheddar (table 8.5). The difference between the rates of increase in Y_a and Y_{dm} per unit weight of fat in milk (i.e. 0.23 kg kg^{-1} milk fat for Cheddar) is a result of the fact that Y_{dm} eliminates the contribution of milk fat to cheese moisture (i.e. 0.24 kg kg^{-1} fat for Cheddar) whereas Y_a incorporates it. However, the increase in Y_{dm} per kilogram of milk fat is greater than that expected based

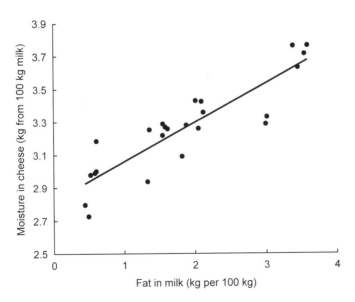

Figure 8.9 Effect of fat level in cheese milk (as measured by kg fat in 100 kg of milk) on the weight (kg) of Cheddar cheese moisture obtained from 100 kg cheese milk. Reproduced, with permission, from Fenelon and Guinee, 1999.

on the corresponding increase in the weight of cheese fat per kilogram of milk fat (i.e. $0.90\,kg\,kg^{-1}$; table 8.5). The difference (i.e. $0.03\,kg\,kg^{-1}$ milk fat) in the rates of increase between Y_{dm} and the weight of fat in cheese per unit weight of fat in the milk may be attributed to the increased weight of the soluble portion of the nonfat nonprotein solids (which form a major part of the dissolved solids) in the cheese as the fat content increases (table 8.5). The latter trend in turn is a result of the increase in cheese moisture per kilogram cheese milk as the milk fat level increases (figure 8.9). However, the direct contribution of fat to Y_{dm} is less than its own weight in milk because of the loss of fat in cheese whey (c. 11% total).

8.8   Effect of fat on cheese texture and rheology

8.8.1   Contribution of fat to cheese elasticity and fluidity

Cheese is a viscoelastic food material, exhibiting the characteristics of both an elastic solid and a Newtonian fluid. The relationship between stress and strain for these materials is linear at very low strains. The stress or strain at which linearity is lost is referred to as the critical stress or strain, which for most foods, including cheese, is relatively small (e.g. a strain of c. 0.02–0.05). At a strain less than the critical strain, the rheological quantities of viscoelastic materials are time-dependent. The typical change in strain with time on the application of a constant stress is referred to as a creep curve (Ma et al., 1997) and shows three distinct regions (figure 8.10):

- elastic deformation (region A–B), where the strain, referred to as elastic compliance, is instantaneous and fully reversible (J_0)
- viscoelastic deformation (region B–C), where the strain, referred to as retarded elastic compliance (J_R), is partly elastic and partly viscous, increases slowly and is only partially recoverable (elastic component)
- viscous deformation (region C–D), where the deformation increases linearly with time and is permanent and nonrecoverable; the creep compliance is referred to as Newtonian (JN)

On removal of the strain at point D, the strain recovery follows a similar sequence to strain creep, with three regions evident in the recovery curve: an instantaneous elastic recovery (D–E), a delayed elastic recovery (E–F) and an eventual flattening (figure 8.10). In the elastic region, the strands of the cheese matrix absorb and store the applied stress energy, which is instantly released on removal of the stress, enabling the cheese to regain its original dimensions (van Vliet, 1991; Walstra and van Vliet, 1982). The extent and duration of the elastic region depends on the magnitude of the applied stress and the structural and compositional characteristics of the cheese. On the application of a stress to a cheese, the volume fraction of the para-casein matrix is the major determinant of the extent of elastic deformation. As the concentration of casein increases,

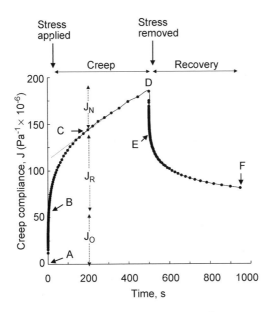

Figure 8.10 Creep compliance and recovery of a three month-old Cheddar cheese. Region A-B, region of elastic deformation; region B-C, region of viscoelastic deformation region C-D, region of viscous deformation; region D-E, region of instantaneous elastic recovery; region E-F, region of delayed elastic recovery. J_0, J_R and J_N refer to the creep compliance in the regions of elastic deformation, viscoelastic deformation and viscous deformation, respectively. Adapted with permission, from Fox *et al.*, 2000, © 2000, Aspen Publishers, Inc.

the number of stress-bearing strands per unit volume of the gel increases, as do the intrastrand and interstrand linkages (Walstra and van Vliet, 1986).

Fat, depending on the temperature of the cheese, may contribute to elastic, viscous or viscoelastic deformation. The effect of temperature is clearly demonstrated by the decrease in the magnitude of the storage modulus, G', of Cheddar cheese (figure 8.11), as the temperature is raised from 4°C, where milk fat is predominantly solid, to 40°C, where milk fat is predominantly liquid. At low temperatures (e.g. 4°C), the fat globules encased within the *para*-casein network are solid and augment the elastic contribution of the casein matrix. The solid fat globules limit the deformation of the casein matrix, as deformation of this matrix would also require deformation of the fat globules enmeshed within its pores. However, the contribution of fat to cheese elasticity decreases rapidly as the ratio of solid-to-liquid fat decreases with increasing temperature and is very low at 40°C. At higher temperatures, the fat behaves more as a fluid and confers viscosity rather than elasticity or rigidity to the cheese. On the application of a stress at 40°C the fat globules flow to an extent dependent on, among other factors, the solid-to-liquid fat ratio and the extent of deformation of the casein matrix. Hence, the magnitude of G' for half-fat Cheddar is lower

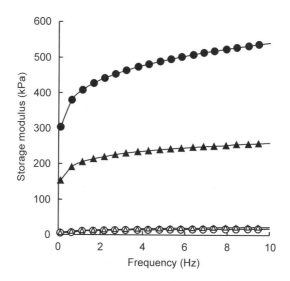

Figure 8.11 Elastic shear modulus as a function of frequency for 120-day-old full-fat and half-fat Cheddar cheese at 4°C and 40°C. Samples were tempered at the assay temperature for 20 min and subjected to a low-amplitude oscillating strain of 0.06. —●—, full-fat cheese at 4°C; —○—, full-fat cheese at 40°C; —▲—, half-fat cheese at 4°C; —△—, half-fat cheese at 40°C.

than that of full-fat Cheddar at 4°C when the fat is solid. However, the values G' for the half-fat and full-fat cheeses is similar at 40°C when fat is liquid, even though the dry matter content of full-fat cheese is higher than that of half-fat cheese (figure 8.11). In agreement with the latter trend, other studies on Cheddar cheese have also shown that G' at 40°C decreased with fat content, in the range 34 wt% to 1.3 wt% (Guinee *et al.*, 2000b; Ustanol *et al.*, 1995). In contrast to the foregoing, Ma *et al.* (1997) reported that G' for full-fat Cheddar at 20°C was higher than that of half-fat Cheddar.

From a sensory viewpoint, it may be assumed that cheese rapidly approaches body temperature (i.e. *c.* 37°C) on ingestion. At this temperature the cheese fat is almost fully liquid (Norris *et al.*, 1973) and therefore confers fluidity and lubrication to the cheese, and mouth coating, on mastication.

8.8.2 *Effect of fat level on fracture-related properties, as measured by using large strain deformation*

In most applications, cheese is subjected to size-reduction operations. Cheese may be portioned (e.g. for consumer packs), sliced, crumbled (e.g. feta, Stilton; in salads), shredded (e.g. for sandwiches, pizza), diced (for salads) or grated (e.g. dried Parmesan), comminuted (e.g., in preparation of sauces, processed cheese products) or compressed and sheared during mastication and consumption. During these applications the high stresses (e.g. greater than 600 kPa) and strains (e.g. much greater than 0.02) applied generally result in fracture in a part (e.g. along a knife cut) or over the whole (e.g. during comminution) of the cheese mass.

It is difficult to quantify the direct effect of fat on the rheological character-istics of cheese. In natural cheeses, the concentrations of different components generally vary simultaneously, especially where large changes in the concentra-tion of a particular component occur and in the absence of process intervention. Hence, a reduction in fat content is paralleled by increases in moisture, protein, intact casein and calcium (see figure 8.1 and tables 8.2 and 8.3). Therefore, rather than establishing the exact effect fat level per se on cheese texture and rheology, it is possible only to determine the overall effect of fat, which consists of the interactive effects of changes in fat, moisture and protein.

Altering the fat content has marked effects on the fracture-related prop-erties of different cheese varieties, including Cheddar (Bryant *et al.*, 1995; Emmons *et al.*, 1980; Fenelon and Guinee, 2000; Mackey and Desai, 1995), mozzarella (Tunick and Shieh, 1995; Tunick *et al.*, 1991, 1993, 1995) and cottage cheese (Rosenberg *et al.*, 1995). Such effects are expected because of the differences in the viscoelastic contributions of fat and casein (section 8.8.1), and the effect of fat on cheese composition (section 8.2). Increasing the fat level of Cheddar cheese results in decreases in elasticity, fracture stress, fracture strain, firmness, cohesiveness, springiness, chewiness and guminess and an increase in adhesiveness (figure 8.12). The latter trends are expected because of the concomitant reduction in the concentration of intact casein and its contribution to cheese elasticity (Walstra and van Vliet, 1986). Moreover, liquid fat acts as a lubricant on fracture surfaces of the casein matrix and thereby reduces the stress required to fracture the matrix (Marshall, 1990; Prentice *et al.*, 1993). Thus, although Chen *et al.* (1979) found no significant relationship between hardness and fat content of different varieties, the hardness generally tended to increase as the fat level was reduced.

Similar trends have been noted for the effect of fat content on the rhe-ological properties of Mozzarella cheese (Tunick and Shieh, 1995; Tunick *et al.*, 1993). Reduction in the fat content (e.g. 21–25 wt% to *c.* 9–11 wt%) of low-moisture (47.7–51.8 wt%) and high-moisture (52.2–57.4 wt%) mozzarella cheeses resulted in significant increases in hardness and springiness, with the magnitude of the effect being most pronounced for hardness. There was a significant effect of the interaction between scald temperature and fat content on hardness, with the effect of fat reduction on hardness being more pronounced as the scald temperature was raised from 32 to 45.9°C.

8.8.3   *Effect of degree of fat emulsification on fracture-related properties, as measured by using large strain deformation*

It is generally accepted that homogenisation of milk results in higher moisture levels and an associated decrease in the magnitude of texture-related parameters

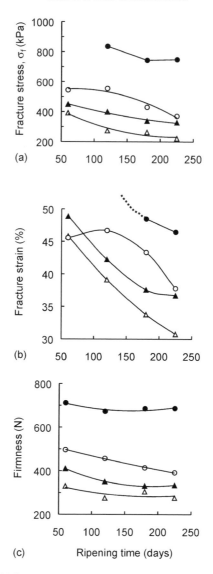

Figure 8.12 Changes in (a) fracture stress, (b) fracture strain and (c) firmness in full-fat (30.45 wt%, —△—), reduced-fat (21.9 wt%, —▲—), half-fat (17.2 wt%, —○—), and low-fat (7.2 wt%, —●—) Cheddar cheese. The broken line indicates that the sample did not fracture on compression at the early ripening times. Reprinted from Fenelon and Guinee, 2000, with permission from Elsevier Science, copyright 2000.

such as firmness, hardness and fracture stress (Jana and Upadhyay, 1992). Metzger and Mistry (1994) found that homogenisation of milk (at first-stage and second-stage pressures of 17.6 and 3.5 MPa, respectively) resulted in significant decreases in the hardness and fracture stress of half-fat Cheddar at 13 weeks but did not significantly influence springiness or cohesiveness. Similar trends were

Table 8.6 Effect of homogenization on the textural and rheological characteristics of mozzarella cheese

| Homogenization conditions | | Cheese code | Composition (wt%) | | | | Texture and rheology[a] | | | |
pressure (MPa)	temperature (°C)		moisture	protein	fat	MNFS	hardness	springiness	cohesiveness	chewiness
Milk[b]										
none	50	CL	46.5	21.5	25.5	62.0	3.1	5.3	0.48	7.7
0.5		HM	51.4	19.0	26.5	69.9	1.7	4.6	0.40	3.2
Milk[c]										
none	70	CL	53.6	31.5	9.0	58.8	73.9	5.3	0.65	n.a.
$P_1 = 13.8, P_2 = 3.45$		HM	54.1	31.4	8.1	58.9	68.2	5.6	0.77	n.a.
Cream (20 wt% fat)[c]										
$P_1 = 13.8, P_2 = 3.45$	70	HC	54.3	30.2	9.1	59.8	68.9	5.5	0.78	n.a.
Milk[d]										
none	55	CL	48.51	32.57	12.3	55.32	27.6	n.a.	n.a.	n.a.
Cream[d]										
$P_1 = 17.25, P_2 = 3.4$		HC	48.85	32.07	12.7	55.96	28.0	n.a.	n.a.	n.a.
Milk[e]										
none	63	CL	54.0	n.a.	9.9	60.0	105	8.59	n.a.	n.a.
$P_1 = 6.87, P_2 = 3.43$		HM	55.9	n.a.	9.5	61.8	96	8.85	n.a.	n.a.
		CL	48.5	n.a.	24.4	64.5	49	7.94	n.a.	n.a.
		HM	47.4	n.a.	26.5	64.5	61	8.29	n.a.	n.a.
Milk[f]										
none		CL	46.3	n.a.	25.9	62.4	55	8.04	n.a.	n.a.
10.3	63	HM	47.9	n.a.	26.9	65.5	76	8.31	n.a.	n.a.
17.2	63	HM	46.8	n.a.	27.8	64.8	82	7.55	n.a.	n.a.

n.a. Data not available.

[a] Empirical units as defined in the reference cited; [b] Jana and Upadhyay, 1991; [c] Rudan et al., 1998a; [d] Poduval and Mistry, 1999; [e] Tunick et al., 1993; [f] Tunick et al., 1995.

Note: CL, control cheese from nonhomogenized milk; HM, cheese from homogenized milk; HC, cheese from homogenized cream; P_1, first-stage homogenization pressure; P_2, second-stage homogenization pressure; FDM, fat in dry matter; MNFS, moisture in nonfat cheese substance.

previously reported by others for Cheddar (Emmons *et al.*, 1980; Green *et al.*, 1983).

The effect of homogenization of milk or cream on the composition and texture of mozzarella cheese has been studied in detail and is summarized in table 8.6. Jana and Upadhyay (1991) reported that homogenization of milk for full-fat mozzarella (*c.* 22 wt% fat) cheese, at total pressures of 250 kPa or 500 kPa, resulted in significant decreases in hardness and springiness and in an increase in cohesiveness; simultaneously there were nonsignificant decreases in gumminess and chewiness. The magnitude of the effect, which increased with homogenization pressure, coincided with a decrease in protein content and with increases in the content of moisture and MNFS. However, Rudan *et al.* (1998a) reported that homogenization of cheese milk or cream (first-stage and second-stage pressure 13.8 MPa and 3.45 MPa, respectively) did not significantly affect hardness or springiness of reduced-fat (*c.* 8 wt%) mozzarella cheese at 30 days. In contrast, homogenization of cream resulted in a significant increase in cohesiveness at 30 day. An opposite trend to the above was noted by Tunick *et al.* (1995) who reported that two-stage homogenization of milk, at 10.3 MPa or 17.2 MPa, resulted in a general increase in hardness (at 23°C) of low-fat or high-fat mozzarella after storage for 1–6 weeks (table 8.6). There was a significant effect of the interaction between homogenization pressure and scald temperature used in cheese manufacture, with the increase in hardness being more pronounced in the higher-scald-temperature cheeses [figures 8.13(a) and 8.13(b)]. The magnitude of the effect also tended to be more pronounced at the lower fat content. Likewise, there was significant interaction between fat content and homogenization pressure on springiness [figures 8.13(c) and 8.13(d); Tunick *et al.*, 1995]; the increase in springiness as the fat content was reduced tended to diminish with increasing homogenization pressure.

Discrepancies between the above studies *vis-à-vis* the effects of homogenization on the textural and rheological characteristics of cheese may be associated with the differences in homogenization conditions, measuring conditions, age of cheese and fat content (see Fox *et al.*, 2000).

8.9 Effect of fat on the functional properties of heated cheese

Cheese is used extensively in cooking applications owing to its heat-induced functionality, which is a composite of different attributes, including softening (melting), stretchability, flowability, apparent viscosity (AV) and tendency to brown. These attributes have a major impact on the quality of products in which cheese is used (e.g. grilled cheese sandwiches, pizza, cheeseburgers, pasta dishes and sauces). Depending on the application, one or more functional attributes may be required. The various functional properties of heated cheese and their interpretation have been discussed by Fox *et al.* (2000) and are summarised in table 8.7). Some of the main functional properties (e.g. flow, stretch, melt resistance, AV) involve displacement (strain) of contiguous planes of the *para*-casein matrix on the application of stress. Stress may take the form of shear as applied by a viscometer, or as an extension force, as applied manually or by instrument (e.g. during stretching). Displacement probably also occurs when the fat globules or pools, which when solid (e.g. at 4°C) contribute to rigidity and physical support of the matrix, collapse, flow and coalesce on heating.

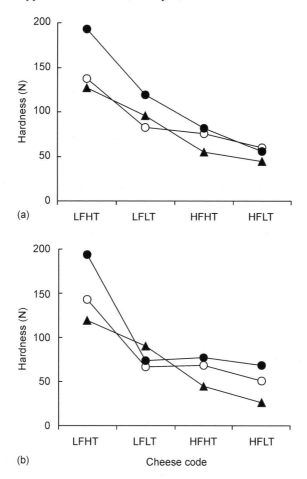

Figure 8.13 The effect of homogenisation pressure on the rheological properties of mozzarella cheese with different levels of fat-in-dry matter (FDM) and manufactured using low or high curd scalding temperatures: (a) hardness at I week; (b) hardness at 6 weeks; (c) springiness at 1 week; (d) springiness at 6 weeks. Note: LFHT, low FDM (*c.* 22 wt%) and high scalding temperature (495.9°C); LFLT, low FDM and low scalding temperature (32°C); HFHT, high FDM (*c.* 49 wt%) and high scalding temperature; HFLT, high FDM and low scalding temperature. The milk was homogenized at total pressures of 0 MPa (control, —o—), 10.3 MPa (—▲—) and 17.2 MPa (—●—). Data from Tunick *et al.*, 1993.

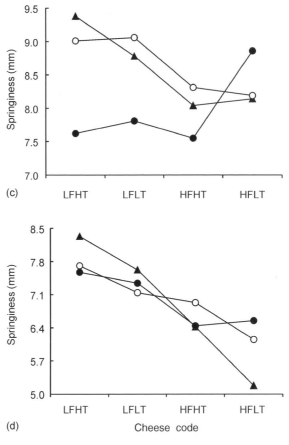

Figure 8.13 (continued)

Lower fat content and acceptable cooking performance are important quality attributes of cheese that affect usage appeal (MTI, 1998). Consequently, extensive research has been undertaken on the effect of various treatments on the quality of reduced-fat cheese. However, most studies have considered the effects of these treatments on quality aspects of the unheated cheese only (sections 8.4–8.8). Fewer studies have investigated the effects of fat reduction per se on the heat-induced functionality of cheese, especially for cheeses other than mozzarella, for example Cheddar (Fife *et al.*, 1996; Guinee *et al.*, 2000b; Olson and Bogenrief, 1995; Poduval and Mistry, 1999; Rudan *et al.*, 1998a,b, 1999; Tunick *et al.*, 1995; Ustanol *et al.*, 1994). The high level of interest in Cheddar and mozzarella is expected as the production of these varieties (*c.* 1.6 million tonnes per year and 1.2 million tonnes per year, respectively in 1997) account for *c.* 20% of total cheese production. Moreover, these cheeses are consumed mainly in the USA and the UK, where the use of cheese as an ingredient is particularly high.

Table 8.7 Functional properties of grilled and baked cheese that influence its functionality as an ingredient

Property	Definition	Cheeses that generally display this property	Property related to physicochemical state
Meltability	The ability of cheese to soften to a molten cohesive mass on heating	Most cheeses, after a given storage period, depending on the variety, (e.g. PCPs, APCs, cream cheese)	Fat liquifaction, fat coalescence
Flowability	The ability of the melted cheese to flow	Most cheeses, after a given storage period, depending on the variety, (e.g. PCPs, OACs, cream cheese)	Fat liquifaction, casein hydration, high degree of fat coalescence, limited oiling-off
Stretchability	The ability of the melted cheese to form cohesive fibres, strings or sheets when extended	Low-moisture mozzarella, Kashkaval, young Cheddar (i.e. 15 days old)	Moderate degree of casein hydration and casein aggregation, level and type of molecular attractions between *para*-casein molecules
Flow resistance (often referred to as melt-resistance)	The resistance to flow of melted cheese	Paneer, PCPs, OACs, natural cheeses from high heat-treated milk	Absence of fat coalescence, heat-induced gelation of a particular (or several) component(s) (e.g. whey proteins), thermo-irreversibility of gel system in the uncooked product on heating
Chewiness (rubberiness, toughness, elasticity)	High resistance to breakdown on mastication	Low-moisture mozzarella, Kashkaval, young Cheddar (i.e. 15 days old)	As for stretchability
Viscosity (soupiness)	Low resistance of melted cheese to breakdown on mastication	Mature Cheddar, aged mozzarella, cream cheese, PCPs, OACs	Relatively high level of casein hydration

Table 8.7 (continued)

Property	Definition	Cheeses that generally display this property	Property related to physicochemical state
Limited cooling-off	Ability of cheese to express a little free-oil on heating, so as to reduce cheese dehydration; maintains succulence of and imparts surface sheen to melted cheese	Most natural cheeses (if not very mature or very young), PCPs, APCs	Limited degree of fat coalescence
Desirable surface appearance	Desired degree of surface sheen with few if any dry, scorched black or brown particles	Mature Cheddar, aged mozzarella, cream cheese, PCPs, OACs (depending on formulation)	Adequate degree of casein hydration and of oiling off during baking; low level of residual reducing sugars (e.g. lactose, galactose) and Maillard browning

Note: PCP, processed cheese product; APC, analogue pizza cheese; OAC, other analogue cheese.
Data from Fox et al., 2000.

8.9.1   Effect of fat level

A recent study showed that reducing the fat level of Cheddar in the range 33 wt%
to 6 wt% resulted in significant increases in the mean AV and reductions in
flowability and stretchability over the 180-days maturation period (figure 8.14;
Guinee *et al.*, 2000a,b). The reduction in flowability with fat level was essentially

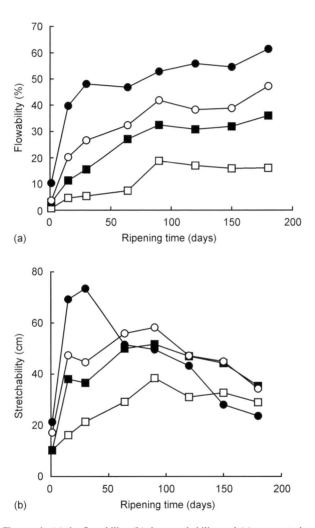

Figure 8.14 Changes in (a) the flowability, (b) the stretchability and (c) apparent viscosity of baked
Cheddar cheese of different fat content. —□—, low-fat (7.2 wt%); —■—, half-fat (17.2 wt%); —○—, reduced-
fat (21.9 wt%); —●—, full-fat (30.5 wt%). The cheeses were baked at 280°C for 4 min. The broken
line indicates that the apparent viscosity exceeded 1000 Pa s. Adapted from Guinee *et al.*, 2000a with
permission from Elsevier Science, copyright 2000.

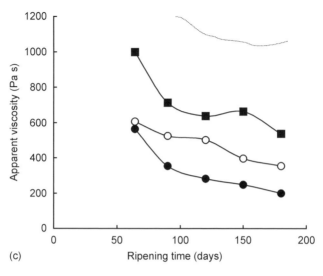

Figure 8.14 (continued)

linear at all ripening times. However, Olson and Bogenrief, (1995) noted that the difference in flowability on melting for 5 min between full-fat Cheddar and reduced-fat Cheddar cheese (75% or 50% normal fat level) decreased with ripening time and had disappeared after 200 days storage at 7°C. Indeed, when the cheese was melted for 12 min, the flowability of the reduced-fat cheese attained a value similar to that of the control at 100 days storage and surpassed that of the control at 100–280 days storage. The differences in the foregoing studies may arise from differences in the measurement technique, which affected the propensity of the cheese to dehydration during cooking [e.g. a cylinder of cheese in a Pyrex glass tube in a water bath at 94°C was used by Olson and Bogenrief (1995), and a disc of cheese placed on a stainless steel surface in a convection oven at 280°C for 4 min was used by Guinee et al., 2000a, 2000b].

A reduction in the fat content (c. 25–9%) of mozzarella cheese from nonhomogenized milk resulted in significant decreases (c. 56%) in the flowability at 1 week and 6 weeks (figure 8.15; Tunick and Shieh, 1995; Tunick et al., 1993, 1995). In contrast, lowering the fat content of cheese from homogenized milk had only a minor impact on the melt properties. Similarly, McMahon et al. (1999) reported that the flowability of low-fat (8 wt%) mozzarella was significantly lower than that of the control (19.2 wt% fat) at 21 days. The adverse effects of fat reduction on flow may be attributed to a number of factors, as discussed below.

8.9.1.1 Increased content of intact casein
Large reductions in fat content are paralleled by increases in the volume fraction of the casein matrix and the level of intact casein (table 8.3; figures 8.1, 8.3

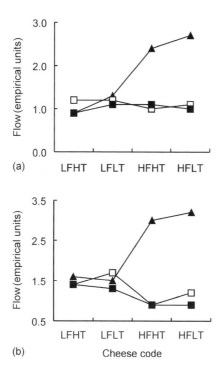

Figure 8.15 The effect of homogenization pressure on the flowability of mozzarella cheese with different levels of fat in dry matter (FDM) and manufactured using low or high curd-scalding temperatures: (a) 1 week; (b) 6 weeks. Note: LFHT, low FDM (*c.* 22 wt%) and high scalding temperature (45.9°C); LFLT, low FDM and low scalding temperature (32°C); HFHT, high FDM (*c.* 49 wt%) and high scalding temperature; HFLT, high FDM and low scalding temperature. The milk was homogenized at total pressures of 0 MPa (control, —▲—), 10.3 MPa (—□—) and 17.2 MPa (—■—). Data from Tunick *et al.*, 1993.

and 8.6), which is positively correlated with AV and negatively correlated with flowability (Guinee *et al.*, 2000a). The concomitant reduction in the number of fat globules is conducive to a higher degree of aggregation and fusion of casein strands during gel formation (McMahon *et al.*, 1993). Fat in the gel exists as globules occluded within the casein network, which physically impede casein aggregation. The higher degree of casein aggregation is unfavourable to heat-induced flow because of:

- a probable decrease in the degree of heat-induced slippage of contiguous casein layers
- the reductions in the content of MNFS (table 8.2)
- lower degree of casein hydration in the resultant cheese

The low degree of casein hydration is less favourable to moisture retention during baking and results in dehydration, crusting and poor flow (Kindstedt,

1995; Kindstedt and Guo, 1997). Moisture acts as a lubricant between the protein layers and between protein and fat layers during melting and thereby facilitates heat-induced slippage of different parts of the matrix (McMahon *et al.*, 1993; Prentice *et al.*, 1993).

8.9.1.2 Lower degree of proteolysis

The level of primary proteolysis in cheese decreases as the fat level is lowered (section 8.3). However, several studies have shown a positive relationship between the level of proteolysis and flowability in cheeses with a fixed fat content; for example, mozzarella (Madsen and Qvist, 1998; Yun *et al.*, 1993a, 1993b), Cheddar (Arnott *et al.*, 1957; Bogenrief and Olson, 1995) and model acid-type cheeses (Lazaridis *et al.*, 1981). The positive effect of proteolysis may be associated with a number of concomitant changes including the increased water-binding capacity (Kindstedt, 1995), and an increase in the number of discontinuities or 'breaks' in the *para*-casein matrix at the microstructural level (de Jong, 1978a). The latter changes are expected to promote a decrease in casein aggregation, an occurrence that should enhance heat-induced displacement of adjoining layers of the casein matrix.

8.9.1.3 Lower moisture-to-protein ratio

A reduction in fat level results in a decrease in the moisture-to-protein ratio, as reflected by the lower content of MNFS (figure 8.1; table 8.3). However, flowability of rennet-curd cheeses is positively correlated with MNFS content (McMahon *et al.*, 1993; Rüegg *et al.*, 1991), an effect which may in part be a result of the concomitant increase in casein hydration and the lubrication effect of moisture.

8.9.1.4 Lower degree of heat-induced fat coalescence

The degree of fat globule clumping and coalescence in both unheated and heated (to 90°C) Cheddar cheese decrease as the fat level is reduced (section 8.4). Fat coalescence results in an increase in the continuity of the fat phase and in the free oil (FO) on heating the cheese. FO forms a layer on the surface of melting cheese, which limits moisture evaporation during cooking and the occurrence of associated defects such as skin formation, scorching and impaired flow and stretch (Rudan and Barbano, 1998). Moreover, FO lubricates the displacement of adjoining layers of the casein matrix and thereby contributes positively to flow and stretch. An increase in FO may also alter the polarity of the solvent system (water and oil) in contact with the *para*-casein matrix and thereby increase the degree of solvation of the *para*-casein per se. A study on model analogue cheese systems (Hokes *et al.*, 1982) showed that the addition of water to a dispersion of calcium caseinate in heated vegetable oil (or other polar solvents such as dioxane) resulted in gelation of the calcium caseinate. This effect was attributed to an increase in the polarity of the solvent system. Taking note of this observation, it is conceivable that FO formed on heating of cheese may be conducive to a structural rearrangement (e.g. exposure of hydrophobic groups) and increase in level of solvation of the *para*-casein molecules. An increase in casein solvation is in turn expected to favour an increase in heat-induced fluidity and flowability of the cheese. Hence, it is noteworthy that:

- the increases in phase angle, and thus fluidity, on heating low-fat Cheddar cheese or full-fat Cheddar cheese from homogenized milk are much lower than that on heating full-fat Cheddar cheese from nonhomogenized milk (figure 8.16)
- the flowability of melted Cheddar and mozzarella increases with fat content (figures 8.14 and 8.15)

8.9.2   Effect of milk homogenization and degree of fat emulsification

Increasing the degree of emulsification (DE) of fat by high-pressure homogenization of the cheese milk has a marked effect on the functionality of cheese. A recent study found that the flowability, fluidity and stretchability of full-fat Cheddar cheese were significantly impaired by homogenization of milk

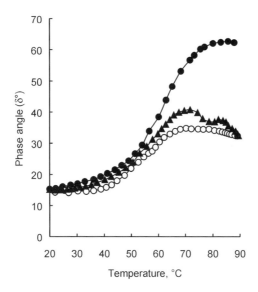

Figure 8.16 Phase angle, δ, as a function of temperature for five-day-old Cheddar-type cheese of different fat content, from nonhomogenized and homogenized milk. Cheese milk: —○—, nonhomogenized, low-fat cheese (1.35 wt%); —▲—, nonhomogenized, full-fat cheese (30.0 wt%); —●—, homogenized, full-fat cheese (30.6 wt%). The phase angle, which represents the phase lag between the stress and the applied sinusoidal shear strain, was measured using low amplitude oscillation rheometry. Adapted from Guinee *et al.*, 2000b.

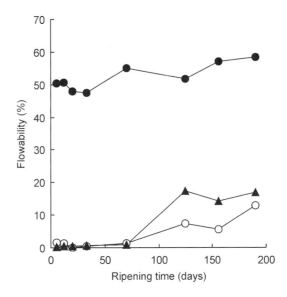

Figure 8.17 Flowability of Cheddar type cheeses of different fat contents, as a function of ageing time, from non-homogenized and homogenized milk. Cheese milk: —o—, nonhomogenized, low-fat cheese (1.35 wt%); —▲—, nonhomogenized, full-fat cheese (30.0 wt%); —●—, homogenized, full-fat cheese (30.6 wt%). Adapted from Guinee *et al.* (2000b).

at first-stage and second-stage pressures of 25 MPa and 5 MPa, respectively (figures 8.16 and 8.17). The effect of homogenization was similar to reducing the fat content of Cheddar from 30 wt% to 1.3 wt% (Guinee *et al.*, 2000b).

Similarly, increasing the homogenization pressure of milk, in the range 0.4 MPa to 6.7 MPa, leads to a progressive deterioration in the flowability and stretchability of halloumi cheese prepared from fresh milk or from reconstituted low-heat skim milk powder (RSMP) and anhydrous milk fat (AMF) (Lelievre *et al.*, 1990). However, the adverse effects of homogenization at high pressure (6.7 MPa) on the melt characteristics of halloumi were reversed by the addition of lecithin to the RSMP/AMF blend during the preparation of the recombined milk.

Homogenization has also been found to impair flowability and to reduce the level of FO released by full-fat mozzarella (*c.* 25 wt%) on baking, with the effect becoming more pronounced as the homogenization pressure was increased from 250 kPa to 500 kPa (Jana and Upadhyay, 1991). In contrast, Rudan *et al.* (1998a,b) found that homogenization of milk or cream, at respective first-stage and second-stage pressures of 13.8 MPa and 3.45 MPa, did not significantly affect the mean flowability or AV of the low-fat (*c.* 8.5 wt%) mozzarella over a 45 days ripening period. However, similar to the findings of Jana and Upadhyay

(1991), it was noted that there was a significant reduction in the content of FO released on baking. The absence of differences in flowability and AV in low-fat cheeses from homogenized milk despite the differences in FO was probably because of the very low level of FO in all cheeses. The FO as a percentage of total fat in cheese was *c*. 0.5, 05 and 3.9 for the homogenized milk cheese, homogenized cream cheese and the control, respectively, at 40 days. Using the same methodology of measurement, the FO values for commercial low-moisture part-skim mozzarella ranged from *c*.10% to 40% of total fat in cheese depending on the FDM content and age (Kindstedt, 1993; Kindstedt and Rippe, 1990). Thus, it is noteworthy that Tunick *et al.* (1993) reported that the interaction between fat content and homogenization pressure had a significant effect on flowability of mozzarella, ripened for 1 week or 6 weeks (figure 8.15). Homogenization of milk had little effect on the melt properties of low-fat mozzarella (*c.* 10 wt% fat) but markedly impaired those of mozzarella with higher fat levels (*c.* 25 wt% fat).

Increasing the DE of fat in PCPs, by selective use of emulsifying salts and extending the duration of processing, results in a marked reduction in the flowability (Rayan *et al.*, 1980). Similarly, the fluidity of melted analogue pizza cheese, as reflected by magnitude of the loss tangent (tan δ) at 80°C, was negatively correlated with DE (Neville, 1998). The adverse effects of increasing the level of fat emulsification on the melt properties of cheese may be attributed to a number of factors, as discussed below.

8.9.2.1 *Lower degree of heat-induced fat coalescence*
Homogenization of cheese milk results in a marked reduction in the fat globule size and degree of fat globule clumping or coalescence in both unheated and heated cheeses (figures 8.6 and 8.7) and thereby renders the cheese more susceptible to dehydration during baking (see section 8.9.1).

8.9.2.2 *Interaction of new fat globule membrane with* para-*casein matrix*
On homogenization of milk, the NMFGM is largely displaced and replaced by an adsorbed layer comprising casein micelles, submicelles and whey proteins (Keenan *et al.*, 1988; Walstra and Jenness, 1984). The newly formed membrane interacts with the 'free' micelles and thereby enables the EFG to become an integral part of the matrix formed during cheese-making, rather than being otherwise occluded within the matrix as inert particles (van Vliet and Dentener-Kikkert, 1982; Walstra and van Vliet, 1986). The incorporation of the EFG into the casein matrix increases the effective protein concentration and the overall level of protein–protein interactions. Consequently, it is expected that functional properties relying on displacement of contiguous layers of the casein matrix (e.g. flow, stretch) would be impaired by homogenization of cheese milk. However,

the adverse effects of homogenisation on flowability and stretchability may be reduced (Lelievere *et al.*, 1990) by:

- lowering the homogenization pressure, which has the effect of reducing the surface area of the fat phase and the number of EFGs in the milk and thereby the number of cross-links between the matrix and EFGs in the cheese
- preventing the casein micelles adsorbing at the fat–water interface by using a surface film of lecithin

8.10   Conclusions

Fat is a major component in most cheeses, apart from some acid-curd varieties (e.g. Quarg) and low-fat variants of different varieties. In natural rennet-curd cheese varieties, fat exists as globules occluded within the pores of the *para*-casein matrix (gel). Dehydration and contraction of the matrix during cheese-making results in an increase in the volume fraction of the enclosed fat globules. Concomitant with the increase in volume fraction and physical damage to the NMFGM, clumping and coalescence of the fat globules occur during manufacture, at least in the case of Cheddar and mozzarella. Coalescence in these cheeses appears to continue during storage, as indicated by the age-related increases in the level of free oil, expressed from the unheated cheese when subjected to centrifugation or hydraulic pressure, or extracted from the melted cheese (at *c.* 74°C) with a water:methanol mixture followed by centrifugation. Age-related changes in fat coalescence occur possibly as a consequence of enzymatic degradation of the NMFGM via microbial lipoprotein lipase activity, or the physical flow of nonglobular fat that is liquid at storage temperatures (e.g. *c.* 40% total at 10°C) through the matrix. Heating of cheese to temperatures applied during grilling and baking (i.e. 90–100°C) accentuates coalescence and results in oiling-off to a degree dependent on fat level and nature of fat globule membrane. The degree of coalescence of fat globules in the raw and heated cheese decrease on lowering the fat level and if the cheesemilk is homogenized.

Homogenization of cheese milk is optional for some varieties such as cream cheese and blue cheese and results in the replacement of the phosopholipid–protein NMFGM with an RFGM that consists of casein micelles, submicelles, whey protein and some of the original fat globule membrane. The RFGM participates in gel formation by complexing with the casein micelles, and the homogenized fat globules thereby become an integral part of the gel matrix. In contrast to the NMFGM, the RFGM is much more stable as reflected by the markedly lower tendency of the fat globules in the homogenized milk cheeses to clump and coalesce, as evidenced from microscopical and FO measurements. Moreover, the marked reduction in the level of heat-induced coalescence of fat in cheeses when the cheese milk is homogenized suggests that the RFGM is more heat-stable than the NMFGM in the cheese environment.

Fat affects several aspects of cheese, including microstructure, composition, yield, microbiology, proteolysis, flavour, rheology and melting properties. However, it is difficult to quantify the direct effects of fat per se on cheese because a change in the level of fat results in simultaneous changes in the concentrations of moisture, protein and calcium. Moreover, for a given fat content, the effects of fat on some characteristics (e.g. rheology and melting properties) depend on the ratio of solid-to-liquid fat, as affected by temperature, and the degree of fat coalescence.

Fat coalescence would appear to have several consequences for cheese quality. In most heated cheese dishes (e.g. pizza, lasagna), limited heat-induced destabilization of the fat phase is generally a prerequisite for good functionality (cooking quality). The consequent release of free oil contributes positively to surface sheen, succulence, flowability and fluidity of the melted cheese and restricts dehydration-associated defects. Hence, homogenization of cheese milk markedly impairs the functionality of heated cheese, with defects such as poor flowability and crusting or burning being prevalent. In milk, the NMFGM protects the enclosed triglycerides against attack by lipases (Walstra and Jenness, 1984). Consequently, the displacement of the NMFGM by homogenization of the cheese milk leads to an elevated degree of lipolysis and higher levels of free fatty acids in many cheese varieties (Jana and Upadhyay, 1992). Likewise, the displacement of the NMFGM by coalescence of fat globules is expected to contribute to the lipolysis, which occurs in the many cheese varieties prepared from nonhomogenized milk, for example Cheddar (Farkye and Fox, 1990; Harboe, 1994).

References

Altemueller, A.G. and Rosenberg, M. (1996) Monitoring proteolysis during ripening of full-fat and low-fat cheddar cheese by reverse-phase HPLC. *J. Food Sci.*, **61**, 295-298.

Amer, S.N., EL-Koussy, L. and S.M. Ewais (1977) Studies on making baby Edam cheese with low-fat content. *Egyptian J. Dairy Sci.*, **5**, 215-221.

Anonymous (1996) UK lite hard cheese market. *Low and Lite Digest*, **1**(7), 8-11.

Arbige, M.V., Freund, P.R., Silver, S.C. and Zelco, J.T. (1986) Novel lipase for Cheddar cheese flavour development. *Food Technol.* (April), 91-98.

Ardö, Y. (1997) Flavour and texture in low-fat cheese, in *Microbiology and Biochemistry of Cheese and Fermented Milk*, 2nd edn (ed. B.A. Law), Blackie Academic & Professional, London, pp. 207-218.

Ardö, Y. (1993) Characterizing ripening in low-fat, semi-hard-eyed cheese made with undefined mesophilic DL-starter. *Int. Dairy J.*, **3**, 343-357.

Ardö, Y. and Manssön, H.L. (1990) Heat treated lactobacilli develop desirable aroma in low-fat cheese. *Scandinavian Dairy Info.*, **1**, 38-40.

Arnott, D.R., Morris, H.A. and Combs, W.B. (1957) Effect of certain chemical factors on the melting quality of process cheese. *J. Dairy Sci.*, **40**, 957-963.

Auty, M.A.E., Fenelon, M.A., Guinee, T.P., Mullins, C. and Mulvihill, D.M. (1999) Dynamic confocal scanning laser microscopy methods for studying milk protein gelation and cheese melting. *Scanning*, **21**, 299-304.

Balcao, V.M. and Malcata, F.X. (1998) Lipase catalysed modification of milk fat. *Biotechnol. Advances*, **16**, 309-341.

Banks, J.M. and Tamime, A.Y. (1987) Seasonal trends in the efficiency of recovery of milk fat and casein in cheese manufacture. *J. Soc. Dairy Technol.*, **43**, 64-66.

Banks, J.M., Brechany, E.Y. and Christie, W.W. (1989) The production of low-fat Cheddar cheese. *J. Soc. Dairy Technol.*, **42**, 6-9.

Banks, J.M., Muir, D.D., Brechany, E.Y. and Law, A.J.R. (1992) The production of low-fat hard ripened cheese, in *3rd Cheese Symposium, Moorepark* (ed. T.M. Cogan), National Dairy Products Research Centre, Moorepark, Fermoy, Co. Cork, Ireland, pp. 67-80.

Berger, C., Martin, N., Collin, S. *et al.* (1999) Combinatorial approach to flavour synthesis: 2. Olfactory analysis of a library of *S*-methyl thioesters and sensory evaluation of selected compounds. *J. Agric. Food. Chem.*, **47**, 3374-3379.

Brooker, B. (1979) Milk and its products, in *Food Microscopy* (ed. J.G. Vaughan), Academic Press, London, pp. 273-311.

Bogenrief, D.D. and Olson, N.F. (1995) Hydrolysis of β-casein increases Cheddar cheese meltability. *Milchwissenschaft*, **50**, 678-682.

Bryant, A., Ustanol Z and Steffe, J. (1995) Texture of Cheddar cheese as influenced by fat reduction. *J. Food Sci.*, **60**, 1216-1219.

Bullens, C., Krawczyk, G. and Geithman, L. (1994) Reduced-fat cheese products using carageenan and microcrystalline cellulose. *Food Technol.*, **48**(1), 79-81.

Carić, M., Gantar, M. and Kaláb, M. (1985) Effects of emulsifying salts on the microstructure and other characteristics of process cheese—a review. *Food Microstructure*, **4**, 297-312.

Chen, A.H., Larkin, J.W., Clark, C.J. and Irwin, W.E. (1979) Texture analysis of cheese. *J. Dairy Sci.*, **62**, 901-907.

Creamer, L.K. (1971) Beta-casein hydrolysis in cheddar cheese ripening. *NZ J. Dairy Sci. Technol.*, **6**, 91.

Creamer, L.K. (1985) Water absorption by renneted casein micelles. *Milchwissenschaft*, **40**, 589-591.

Cromie, S.J., Giles, J.E. and Dulley, J.R. (1987) Effect of elevated ripening temperatures on the microflora of Cheddar cheese. *J. Dairy Res.*, **54**, 69-76.

Deeth, H.C. (1997) The role of phospholipids in the stability of the milk fat globules. *Aus. J. Dairy Technol.*, **52**, 44-46.

de Jong, L. (1978a) The influence of moisture content on the consistency and protein breakdown of cheese. *Neth. Milk Dairy J.*, **32**, 1-14.

de Jong, L. (1978b) Protein breakdown in soft cheese and its relation to consistency: 3. The micellar structure of Meshanger cheese. *Neth. Milk Dairy J.*, **32**, 15-25.

Desai, N. and Nolting, J. (1995) Microstructure studies of reduced-fat cheeses containing fat substitute, in *Chemistry of Structure–Function Relationships in Cheese* (eds. E.L. Malin and M.H. Tunick), Plenum Press, New York, pp. 295-302.

Dexheimer, E. (1992) On the fat track. *Dairy Foods*, **93**(5), 38-50.

Drake, M.A., Boylston, T.D. and Swanson, B.G. (1996a) Fat mimetics in low-fat Cheddar cheese. *J. Food Sci.*, **61**, 1267-1270, 1288.

Drake, M.A., Herrett, W., Boylston, T.D. and Swanson, B.G. (1996b) Lecithin improves texture of reduced-fat cheeses. *J. Food Sci.*, **61**(3), 639-642.

Emmons, D.B., Kaláb, M., Larmond, E. and Lowrie, R.J. (1980) Milk gel structure: X. Texture and microstructure in Cheddar cheese made from whole milk and from homogenised low-fat milk. *J. Text. Stud.*, **11**, 15-34.

Farkye, N.Y. and Fox, P.F. (1990) Objective indices of cheese ripening. *Trends Food Sci. Technol.*, **1**, 37-40.

Fenelon, M.A. (2000) *Studies on the Role of Fat in Cheese and the Improvement of Half-fat Cheddar Cheese Quality* PhD thesis, National University of Ireland, Cork.

Fenelon, M.A. and Guinee, T.P. (1997) The compositional, textural and maturation characteristics of reduced-fat Cheddar made from milk containing added Dairy-Lo™. *Milchwissenschaft*, **52**, 385-389.

Fenelon, M.A. and Guinee, T.P. (1999) The effect of milk fat on Cheddar cheese yield and its prediction, using modifications of the van Slyke cheese yield formula. *J. Dairy Sci.*, **82**, 1-13.

Fenelon, M.A. and Guinee, T.P. (2000) Primary proteolysis and textural changes during ripening in Cheddar cheeses manufactured to different fat contents. *Internat. Dairy J.*, **10**, 151-158.

Fenelon, M.A., Guinee, T.P. and Reville, W.J. (1999) Characteristics of reduced-fat Cheddar prepared from blending of full-fat and skim cheese curds at whey drainage. *Milchwissenschaft*, **54**, 506-510.

Fenelon, M.A., Guinee, T.P., Delahunty, C., Murray, J. and Crowe, F. (2000a) Composition and sensory attributes of retail Cheddar cheeses with different fat contents. *J. Food Composition and Analysis*, **13**, 13-26.

Fenelon, M.A., O'Connor, P. and Guinee, T.P. (2000b) The effect of fat content on the microbiology and proteolysis in Cheddar cheese during ripening. *J. Dairy Sci.*, **83**, 2173-2183.

Fife, L.F., McMahon, D.J. and Oberg, C.J. (1996) Functionality of low-fat mozzarella cheese. *J. Dairy Sci.*, **79**, 1903-1910.

Fox, P.F. (1989) Proteolysis during cheese manufacture and ripening. *J. Dairy Sci.*, **72**, 1379-1400.

Fox, P.F. and Guinee, T.P. (1987) Italian cheeses, in *Cheese: Chemistry, Physics and Microbiology*. 2. *Major Cheese Varieties* (ed. P.F. Fox), Elsevier Science Publishers Ltd, London, pp. 221-255.

Fox, P.F. and McSweeney, P.L.H. (1996) Proteolysis in cheese during ripening. *Good Reviews International*, **12**, 457-509.

Fox, P.F. and Wallace, J.M. (1997) Formation of flavour compounds in cheese. *Adv. Appl. Microbiol.*, **45**, 17-85.

Fox, P.F., Guinee, T.P., Cogan, T.M. and McSweeney, P.L.H. (2000) *Fundamentals of Cheese Science*, Aspen Publishers, Gaithersburg, MD.

Fox, P.F., O'Connor, T.P., McSweeney, P.L.H., Guinee, T.P. and O'Brien N.M. (1996) Cheese: physical, biochemical, and nutritional aspects. *Adv. Food Nutr. Res.*, **39**, 163-328.

Fox, P.F., McSweeney, P.L.H. and Lynch, C.M. (1998) Significance of non-starter lactic acid bacteria in Cheddar cheese. *Aust. J. Dairy Technol.*, **53**, 83-89.

Gilles, J. and Lawrence, R.C. (1985) The yield of cheese. *NZ J. Dairy Sci. Technol.*, **20**, 205-214.

Ginzinger, W. (1995) Innovationen bei Käse. *Milchwirschaftliche Berichte aus den Bundesanstalten, Wolfpassing und Rotholz*, **124**, 144-146.

Godfrey, T. and West, S.I. (1996) *Industrial Enzymology*, 2nd edn, Stokton Press, NY, and Macmillan, London.

Green, M.L., Glover, F.A., Scurlock, E.M.W., Marshall, R.J. and Hatfield, D.S. (1981) Development of structure and texture in Cheddar cheese. *J. Dairy Res.*, **48**, 333-341.

Green, M.L., Marshall, R.J. and Glover, F.A. (1983) Influence of homogenization of concentrated milks on the structure and properties of rennet curds. *J. Dairy Res.*, **50**, 341-348.

Gripon, J.-C. (1997) Flavour and texture in soft cheese, in *Microbiology and Biochemistry of Cheese and Fermented Milk*, 2nd edn (ed. B.A. Law), Blackie Academic & Professional, London, pp. 193-206.

Grufferty, M.B. and Fox, P.F. (1988) Milk alkaline proteinase. *J. Dairy Res.*, **55**, 609-630.

Guinee, T.P., Auty, M.A.E. and Fenelon, M.A. (2000a) The effect of fat on the rheology, microstructure and heat-induced functional characteristics of Cheddar cheese. *Int. Dairy J.*, **10**, 277-288.

Guinee, T.P., Auty, M.A.E. and Mullins, C. (1999) Observations on the microstructure and heat-induced changes in the viscoelasticity of commercial cheeses. *Aust. J. Dairy Technol.*, **54**, 84-89.

Guinee, T.P., Auty, M.A.E., Mullins, C., Corcoran, M.O. and Mulholland, E.O. (2000b) Preliminary observations on effects of fat content and degree of fat emulsification on the structure–functional relationship of Cheddar-type cheese. *J. Text. Studies.*, **31**, 645-663.

Guinee, T.P., Fenelon, M.A., Mulholland, E.O., O'Kennedy, B.T., O'Brien, N. and Reville, W.J. (1998) The influence of milk pasteurization temperature and pH at curd milling on the composition, texture and maturation of reduced-fat Cheddar cheese. *Int. J. Dairy Technol.*, **51**, 1-10.

Guinee, T.P., Mulholland, E.O., Mullins, C. and Corcoran, M.O. (1997) Functionality of low moisture mozzarella cheese during ripening. *Proceedings of the 5th Cheese Symposium* (eds. T.M. Cogan, P.F. Fox and R.P. Ross), Teagasc, Dublin, pp. 15-23.

Guinee, T.P., Mulholland, E.O., Mullins, C. and Corcoran, M.O. (2000c) Effect of salting method on the composition, yield and functionality of low moisture mozzarella cheese. *Milchwissenschaft*, **55**, 135-138.

Hall, D.M. and Creamer, L.K. (1972) A study of the sub-microscopic structure of Cheddar, Cheshire and Gouda cheese by electron microscopy. *NZ J. Dairy Sci. Technol.*, **7**, 95-102.

Haque, Z.U., Kucukoner, E. and Aryana, K.J. (1997) Aging-induced changes in populations of lactococci, lactobacilli, and aerobic micro-organisms in low-fat and full-fat Cheddar cheese. *J. Food Prot.*, **60**, 1095-1098.

Harboe, M.K. (1994) Use of lipases in cheesemaking. *International Dairy Federation Bulletin*, **294**, 11-16.

Heertje, I., Boskamp, M.J., van Kleef, F. and Gortemaker, F.H. (1981) The microstructure of processed cheese. *Neth. Milk Dairy J.*, **35**, 177-179.

Hodis, H.N., Crawford, D.W. and Sevanian, A. (1991) Cholesterol feeding increases plasma and aortic tissue cholesterol oxide levels in parallel: further evidence for the role of cholesterol oxidation in artherosclerosis. *Artherosclerosis*, **89**, 117-126.

Hokes, J.C., Mangino, M.E. and Hansen, P.M.T. (1982) A model system for curd formation and melting properties of calcium caseinates. *J. Food Sci.*, **47**, 1235-1240, 1249.

Holland, B., Welch, A.A., Unwin, I.D., Buss, D.H., Paul, A.A. and Southgate, D.A.T. (1991) *McCance and Widdowson: The Composition of Foods*, 5th edition, Royal Society of Chemistry, Ministry of Agriculture, Fisheries and Food, Cambridge, UK, pp. 86-92.

Horne, D.S., Banks, J.M., Leaver, J. and Law, A.J.R. (1994) Dynamic mechanical spectroscopy of Cheddar cheese, in *Cheese Yield and Factors Affecting its Control*, Special Issue 9402, International Dairy Federation, Brussels, pp. 507-512.

IDF (1991) *Factors Affecting the Yield of Cheese*, Special Issue 9301, International Fairy Federation, Brussels.

IDF (1994) *Cheese Yield and Factors affecting its Control: Proceedings of the IDF Seminar, Cork, Ireland, 1993*, International Dairy Federation, Brussels.

Jameson, G.W. (1990) Cheese with less fat. *Aust. J. Dairy Technol.*, **45**(2), 93-98.

Jana, A.H. and Upadhyay, K.G. (1991) The effects of homogenization conditions on the textural and baking characteristics of buffalo milk mozzarella cheese. *Aust. J. Dairy Technol.*, **46**, 27-30.

Jana, A.H. and Upadhyay, K.G. (1992) Homogenisation of milk for cheesemaking—a review. *Aust. J. Dairy Technol.*, **47**, 72-79.

Johnson, M. and Law, B.A. (1999) The origins, development and basic operations of cheesemaking technology, in *Technology of Cheesemaking* (ed. B.A. Law), Sheffield Academic Press, Sheffield, pp. 1-32.

Jordan, K.N. and Cogan, T.M. (1993) Identification and growth of non-starter lactic bacteria in Irish Cheddar cheese. *Ir. J. Agric. and Food Res.*, **32**, 47-55.

Kaláb, M. (1977) Milk gel structure: VI. Cheese texture and microstructure. *Milchwissenschaft*, **32**, 449-457.

Kaláb, M. (1979) Microstructure of dairy foods: 1. Milk products based on protein. *J. Dairy Sci.*, **62**, 1352-1364.

Kaláb, M. (1995) Practical aspects of electron microscopy in cheese research, in *Chemistry of Structure–Function Relationships in Cheese* (eds. E.L. Malin and M.H. Tunick), Plenum Press, New York, pp. 247-276.

Kaláb, M. and Modler, H.W. (1985) Milk gel structure: XV. Electron microscopy of whey protein-based cream cheese spread. *Milchwissenschaft*, **40**, 193-196.

Kaláb, M., Yun, J. and Hing Yiu, S. (1987) Textural properties and microstructure of process cheese food rework. *Food Microstruct.*, **6**, 181-192.

Keenan, T.W., Maher, I.H. and Dylewski, D.P. (1988) Physical equilibria: lipid phase, in *Fundamentals of Dairy Chemistry* (ed. N.P. Wong), Van Nostrand Reinbold, New York, pp. 511-582.

Keogh, M.K. and Morrissey, A. (1990) Anhydrous milk fat: 6. Baked goods. *Ir. J. Food Sci. and Technol.*, **14**, 69-83.

Khan, J.A., Gijs, L., Berger, C. *et al.* (1999) Combinatorial approach to flavour analysis: I. Preparation and characterisation of an *S*-methyl thioester library. *J. Agric. Food Chem.*, **47**, 3269-3273.

Kiely, L.J., Kindstedt, P.S., Hendricks, G.M., Levis, J.E., Yun, J.J. and Barbano, D.M. (1992) Effect of pH on the development of curd structure during the manufacture of mozzarella cheese. *Food Struct.*, **11**, 217-224.

Kiely, L.J., Kindstedt, P.S., Hendricks, G.M., Levis, J.E., Yun, J.J. and Barbano, D.M. (1993) Age related changes in the microstructure of mozzarella cheese. *Food Struct.*, **12**, 13-20.

Kilcawley, K.N., Wilkinson, M.G. and Fox, P.F. (1998) Enzyme-modified cheese. *Int. Dairy J.*, **8**, 1-10.

Kimber, A.M., Brooker, B.E., Hobbs, D.G. and Prentice, J.H. (1974) Electron microscope studies of the development of structure in Cheddar cheese. *J. Dairy Res.*, **41**, 389-396.

Kimura, T., Taneya, S. and Furuichi, E. (1979) Electron microscopic observation of casein particles in processed cheese, in *20th International Dairy Congress (France)*, Congrilait, Paris, pp. 239-240.

Kindstedt, P.S. (1993) Mozzarella and pizza cheese: factors affecting the functional characteristics of unmelted and melted mozzarella cheese, in *Cheese: Chemistry, Physics and Microbiology, Volume 2: Major Cheese Groups*, 2nd edn (ed. P.F. Fox), Chapman & Hall, London, pp. 337-362.

Kindstedt, P.S. (1995) Factors affecting the functional characteristics of unmelted and melted mozzarella cheese, in *Chemistry of Structure–Function Relationships in Cheese* (eds. E.L. Malin and M.H. Tunick), Plenum Press, New York, pp. 27-41.

Kindstedt, P.S. and Fox, P.F. (1991) Modified Gerber test for free oil in melted mozzarella cheese. *J. Food Sci.*, **56**, 1115-1116.

Kindstedt, P.S. and Guo, M.R. (1997) Recent developments in the science and technology of pizza cheese. *Aust. J. Dairy Technol.*, **52**, 41-43.

Kindstedt, P.S. and Rippe, J.K. (1990) Rapid quantitative test for free oil (oiling off) in melted mozzarella cheese. *J. Dairy Sci.*, **73**, 867-873.

Kosikowski, F.V. and Mistry, V.V. (1997) *Cheese and Fermented Milk Foods, Volume 1: Origins and Principles*, F.V. Kosikowski LLC, Wesport, CT.

Laloy, E., Vuillemard, J.C., El Soda, M. and Simard, R.E. (1996) Influence of the fat content of Cheddar cheese on retention and localization of starters. *Int. Dairy J.*, **6**, 729-740.

Lamberet, G., Auberger, B. and Bergere, J.L. (1997) Aptitude of cheese bacteria for volatile *s*-methyl thioester synthesis: 11. Comparison of coryneform bacteria, micrococcaceae and lactic acid bacteria starters. *Appl. Microbiol. Biotechnol.*, **48**, 393-397.

Lane, C.N., Fox, P.F., Walsh, E.M., Folhersma, B. and McSweeney P.L.H. (1997) Effect of compositional and environmental factors on the growth of indigenous non-starter lactic acid bacteria in Cheddar cheese. *Lait*, **77**, 561-573.

Law, B.A. (1982) Cheeses, in *Fermented Foods: Economic Microbiology Volume 7* (ed. A.H. Rose), Academic Press, London, pp. 147-198.

Law, B.A. (1999) Cheese ripening and cheese flavour technology, in *Technology of Cheesemaking* (ed. B.A. Law), Sheffield Academic Press, Sheffield, pp. 163-192.

Lawrence, R.C. and Gilles, J. (1980) The assessment of potential quality of young Cheddar cheese. *NZ J. Dairy Sci. Technol.*, **15**, 1-12.

Lawrence, R.C., Creamer, L.K. and Gilles, J. (1987) Texture development during cheese ripening. *J. Dairy Sci.*, **70**, 1748-1760.

Lazaridis, H.N., Rosenau, J.R. and Mahoney, R.R. (1981) Enzymatic control of meltability in a direct acidified cheese product. *J. Food Sci.*, **46**, 332-335, 339.

Lee, B.O., Kilbertus, G. and Alais, C. (1981) Ultrastructural study of processed cheese: effect of different parameters. *Milchwissenschaft*, **36**, 343-348.

Lelievre, J., Shaker, R.R. and Taylor, M.W. (1990) The role of homogenization in the manufacture of halloumi and mozzarella cheese from recombined milk. *J. Soc. Dairy Technol.*, **43**, 21-24.

Lemay, A., Paquin, P. and Lacroix, C. (1994) Influence of milk microfluidization on Cheddar cheese composition, quality and yields, in *Cheese Yield and Factors affecting its Control: Proceedings of the IDF Seminar, Cork, Ireland, 1993*, International Dairy Federation, Brussels, pp. 288-292.

Lowrie, R.J., Kaláb, M. and Nichols, D. (1982) Curd granule and milled curd junction patterns in Cheddar cheese made by traditional and mechanized processes. *J. Dairy Sci.*, **65**, 1122-1129.

Ma, L., Drake, M.A., Barbosa-Cánovas, G.V. and Swanson, B.G. (1997) Rheology of full-fat and low-fat Cheddar cheeses as related to type of fat mimetic. *J. Food Sci.*, **62**, 748-752.

McEwan, J.A., Moore, J.D. and Colwill, J.S. (1989) The sensory characteristics of Cheddar cheese and their relationship with acceptability. *J. Soc. Dairy Technol.*, **4**, 112-117.

Mackey, K.L. and Desai, N. (1995) Rheology of reduced-fat cheese containing fat substitute, in *Chemistry of Structure–Function Relationships in Cheese* (eds. E.L. Malin and M.H. Tunick), Plenum Press, New York, pp. 21-26.

McMahon, D.J., Alleyne, M.C., Fife, R.L. and Oberg, C.J. (1996) Use of fat replacers in low-fat mozzarella cheese. *J. Dairy Sci.*, **79**, 1911-1921.

McMahon, D.J., Fife, R.L. and Oberg, C.J. (1999) Water partitioning in mozzarella cheese and its relationship to cheese meltability. *J. Dairy Sci.*, **82**, 1361-1369.

McMahon, D.J., Oberg, C.J. and McManus, W. (1993) Functionality of mozzarella cheese. *Aust. J. Dairy Technol.*, **49**, 99-104.

McNamara, D.J. (1995) Dietary cholesterol and the optimal diet for reducing risk of artherosclerosis. *Can. J. Cardiology*, **11** (Supplement G), 123G-126G.

McPherson, A.V., Dash, A.V. and Kitchen, B.J. (1989) Isolation and composition of milk fat globule membrane material: II. From homogenized and ultra heat treated milks. *J. Dairy Res.*, **51**, 289-297.

McSweeney, P.L.H., Fox, P.F., Lucey, J.A., Jordon, K.N. and Cogan T.M. (1993) Contribution of the indigenous mircroflora to the maturation of cheddar cheese. *Int. Dairy J.*, **3**, 613-634.

Madsen, J.S. and Qvist, K.B. (1998) The effect of added proteolytic enzymes on meltability of mozzarella cheese manufactured by ultrafiltration. *Le Lait*, **78**, 258-272.

Marshall, R.J. (1990) Composition, structure, rheological properties and sensory texture of processed cheese analogues. *J. Sci. Food and Agric.*, **50**, 237-252.

Mayes, J.J., Urbach, G. and Sutherland, B.J. (1994) Does addition of buttermilk affect the organoleptic properties of low-fat Cheddar cheese? *Aust. J. Dairy Technol.*, **49**, 39-41.

Metzger, L.E. and Mistry, V.V. (1994) A new approach using homogenization of cream in the manufacture of reduced-fat Cheddar cheese: 1. Manufacture, composition and yield. *J. Dairy Sci.*, **77**, 3506-3515.

Metzger, L.E. and Mistry, V.V. (1995) A new approach using homogenization of cream in the manufacture of reduced-fat Cheddar cheese: 2. Microstructure, fat globule distribution, and free oil. *J. Dairy Sci.*, **78**, 1883-1895-3515.

Mistry, V.V. and Anderson, D.L. (1993) Composition and microstructure of commercial full-fat and low-fat cheeses. *Food Struct.*, **12**, 259-266.

MTI (1998) New product and packaging development, in *The International Cheese Market 1999–2003*, Market Tracking International Ltd, London, pp. 27-61.

Nauth, K.R. and Ruffie, D. (1995) Microbiology and biochemistry of reduced-fat cheese, in *Chemistry of Structure–Function Relationships in Cheese* (eds. E.L. Malin and M.H. Tunick), Plenum Press, New York, pp. 345-357.

Neville, D.P. (1998) Studies on the melting properties of cheese analogues, MSc thesis, National University of Ireland, Cork.

Norris, G.E., Gray, I.K. and Dolby, R.M. (1973) Seasonal variations in the composition and thermal properties of New Zealand milk fat. *J. Dairy Res.*, **40**, 311-321.

Oberg, C.J., McManus, W.R. and McMahon, D.J. (1993) Microstructure of mozzarella cheese during manufacture. *Food Struct.*, **12**, 251-258.

Olson, N.F. and Bogenrief, D.D. (1995) Functionality of mozzarella and Cheddar cheese, in *4th Cheese Symposium* (eds. T.M. Cogan, P.F. Fox and R.P. Ross), National Dairy Products Research Centre, Teagasc, Moorepark, Fermoy, Cork, Ireland, pp. 81-89.

Olson, N.F. and Johnson, M.E. (1990) Light cheese products: characteristics and economics. *Food Technol.*, **44**(10), 93-96.

Paquet, A. and Kaláb, M. (1988) Amino acid composition and structure of cheese baked as a pizza ingredient in conventional and microwave ovens. *Food Microstruct.*, **7**, 93-103.

Pearce, K.N. and Gilles, J. (1979) Composition and grade of cheddar cheese manufactured over three seasons. *NZ J. Dairy Sci. and Technol.*, **14**, 63-71.

Piggot, J.R. (2000) Dynamism in flavour science and sensory methodology. *Food Res. Int.*, **33**, 191-97.

Poduval, V.S. and Mistry, V.V. (1999) Manufacture of reduced-fat mozzarella cheese using ultrafiltered sweet buttermilk and homogenized cream. *J. Dairy Sci.*, **82**, 1-9.

Prentice, J.H., Langley, K.R. and Marshall, R.J. (1993) Cheese rheology, in *Cheese: Chemistry, Physics and Microbiology, Volume 1: General Aspects* 2nd edn (ed. P.F. Fox), Chapman & Hall, London, pp. 303-340.

Rank, T.C. (1985) Proteolysis and flavour development in low-fat and whole milk Colby and Cheddar-type cheeses. PhD Thesis, University of Wisconsin, Madison.

Rayan, A.A., Kaláb, M. and Ernstrom, C.A. (1980) Microstructure and rheology of process cheese. *Scanning Electron Microscopy*, **III**, 635-643.

Rosenberg, M., Wang, Z., Sulzer, G. and Cole, P. (1995) Liquid drainage and firmness in full-fat, low-fat, and fat-free cottage cheese. *J. Food Sci.*, **60**, 698-702.

Rowney, M., Roupas, P., Hickey, M.W. and Everett, D.W. (1999) Factors affecting the functionality of mozzarella cheese. *Aust. J. Dairy Technol.*, **54**, 94-102.

Rudan, M.A. and Barbano, D.M. (1998) A dynamic model for melting and browning of mozzarella cheese during pizza baking. *Aust. J. Dairy Technol.*, **53**, 95-97.

Rudan, M.A., Barbano, D.M., Guo, M.R. and Kindstedt, P.S. (1998a) Effect of modification of fat particle size by homogenization on composition, proteolysis, functionality, and appearance of reduced-fat mozzarella cheese. *J. Dairy Sci.*, **81**, 2065-2076.

Rudan, M.A., Barbano, D.M. and Kindstedt, P.S. (1998b) Effect of fat replacer (Salatrim) on chemical composition, proteolysis, functionality, appearance and yield of reduced-fat mozzarella cheese. *J. Dairy Sci.*, **81**, 2077-2088.

Rudan, M.A., Barbano, D.M., Yun, J.J. and Kindstedt, P.S. (1999) Effect of fat reduction on chemical composition, proteolysis, functionality, and yield of mozzarella cheese. *J. Dairy Sci.*, **82**, 661-672.

Rüegg, M., Eberhard, P., Popplewell, L.M. and Peleg, M. (1991) Melting properties of cheese, in *Rheological and Fracture Properties of Cheese*, Bulletin 268, International Dairy Federation, Brussels, pp. 36-43.

Savello, P.A., Ernstrom, C.A. and Kaláb, M. (1989) Microstructure and meltability of model process cheese made with rennet and acid casein. *J. Dairy Sci.*, **72**, 1-11.

Simon, J.A., Fong, J. and Bernert Jr, J.T. (1996) Serum fatty acids and blood pressure. *Hypertension*, **27**, 303-307.

Sørensen, H.H. (1997) *The World Market for Cheese*, Bulletin 326, International Dairy Federation, Brussels, pp. 8-17.

Sugimoto, I., Sato, Y. and Umemoto, Y. (1983) Hydrolysis of phosphatidyl ethanolamine by cell fractions of *Streptococcus lactis*. *Agric. Biolog. Chem.*, **47**, 1201-1206.

Tamime, A.Y., Kaláb, M., Davies, G. and Younis, M.F. (1990) Microstructure and firmness of processed cheese manufactured from Cheddar cheese and skim milk powder cheese base. *Food Struct.*, **9**, 23-37.

Taneya, S., Izutsu, T., Kimura, T. and Shioya, T. (1992) Structure and rheology of string cheese. *Food Struct.*, **11**, 61-71.

Taneya, S., Kimura, T., Izutsu, T. and Bucheim, W. (1980) The submicroscopic structure of processed cheese with different melting properties. *Milchwissenschaft*, **35**, 479-481.

Thierry, A., Salvat-Brunaud, D., Madec, M.-M., Michel, F. and Maubois, J.-L. (1998) Affinage de l'emmental: dynamique des populations bactériennes et évolution de la composition de la phase aquese. *Lait*, **78**, 521-542.

Thomas, T.D. and Pearce, K.N. (1981) Influence of salt on lactose fermentation and proteolysis in Cheddar cheese. *NZ J. Dairy Sci. Technol.*, **16**, 253-259.

Tunick, M.H. and Shieh, J.J. (1995) Rheology of reduced-fat mozzarella, in *Chemistry of Structure–Function Relationships in Cheese* (eds. E.L. Malin and M.H. Tunick). Plenum Press, New York. pp. 7-19.

Tunick, M.H., Cooke, P.H., Malin, E.L., Smith, P.W. and Holsinger, V.H. (1997) Reorganization of casein submicelles in mozzarella cheese during storage. *Int. Dairy J.*, **7**, 149-155.

Tunick, M.H., Mackey, K.L., Smith, P.W. and Holsinger, V.H. (1991) Effects of composition and storage on the texture of mozzarella cheese. *Neth. Milk Dairy J.*, **45**, 117-125.

Tunick, M.H., Malin, E.L., Smith, P.W., *et al.* (1993) Proteolysis and rheology of low-fat and full-fat mozzarella cheeses prepared from homogenized milk. *J. Dairy Sci.*, **76**, 3621-3628.

Tunick, M.H., Malin, E.L., Smith, P.W. and Holsinger, V.H. (1995) Effects of skim milk homogenization on proteolysis and rheology of mozzarella cheese. *Internat. Dairy J.*, **5**, 483-491.

United States Department of Agriculture, Agriculture Research Service (USDA) (1976) *Agriculture Handbook No. 8-1. Composition of Foods: Dairy and Egg Products, Raw, Processed, Prepared*, US Government Printing Office, Washington, DC.

Ustanol, Z., Kawachi, K. and Steffe, J. (1994) Arnott test correlates with dynamic rheological properties for determining Cheddar cheese meltability. *J. Food Sci.*, **59**, 970-971.

Ustanol, Z., Kawachi, K. and Steffe, J. (1995) Rheological properties of Cheddar cheese as influenced by fat reduction and ripening time. *J. Food Sci.*, **60**, 1208-1210.

van Boekel, M.A.J.S. and Walstra, P. (1995) Effect of heat treatment on chemical and physical changes to milkfat globules, in *Heat Induced Changes in Milk*, 2nd edn (ed. P.F. Fox), International Dairy Federation, Brussels, pp. 51-65.

Vanderpoorten R. and Weckx, M. (1972) Breakdown of casein by rennet and microbial milk clotting enzymes. *Neth. Milk Dairy J.*, **26**, 47-59.

van Vliet, T. (1991) Terminology to be used in cheese rheology, in *Rheological and Fracture Properties of Cheese*, Bulletin 268, International Dairy Federation, Brussels, pp. 5-15.

van Vliet, T. and Dentener-Kikkert, A. (1982) Influence of the composition of the milk fat globule membrane on the rheological properties of acid milk gels. *Neth. Milk Dairy J.*, **36**, 261-265.

Versteeg, C., Ballintyne, P.C., McAuley, C.M., Tan, S.E., Alexander, M. and Bromme, M.C. (1998) Control of reduced-fat cheese quality. *Aust. J. Dairy Technol.*, **53**, 106.

Walstra, P. and Jenness, R. (1984) *Dairy Chemistry and Physics*, John Wiley, New York.

Walstra, P. and van Vliet, T. (1982) Rheology of cheese, in Bulletin No. 153, International Dairy Federation, Brussels, pp. 22-27.

Walstra, P. and van Vliet, T. (1986) The physical chemistry of curd making. *Neth. Milk Dairy J.*, **40**, 241-259.

Wijesundera, C. and Drury, L. (1999) Role of fat in production Cheddar cheese flavour using a fat-substituted cheese model. *Australian J. Dairy Technol.*, **54**, 28-35.

Wijesundera, C., Drury, L., Muthuku-marappan, K., Gunasekaran, S. and Everett, D.W. (1998) Flavour development and distribution of fat globule size in Cheddar-type cheeses made from skim milk homogenised with AMF or its fractions. *Australian J. Dairy Technol.*, **53**, 107.

Williams, A.G. and J.M. Banks. (1997) Proteolytic and other hydrolytic enzyme activities in nonstarter lactic acid bacteria (NSLAB) isolated from Cheddar cheese manufactured in the United Kingdom. *Int. Dairy J.*, **7**, 763-774.

Yun, J.J., Barbano, D.M. and Kindstedt, P.S. (1993a) Mozzarella cheese: impact of coagulant type on chemical composition and proteolysis. *J. Dairy Sci.*, **76**, 3648-3656.

Yun, J.J., Barbano, D.M., Kindstedt, P.S. and Larose, K.L (1995) Mozzarella cheese: impact of whey pH at draining on chemical composition, proteolysis, and functional properties. *J. Dairy Sci.*, **78**, 1-7.

Yun, J.J., Kiely, L.J., Kindstedt, P.S. and Barbano D.M. (1993b) Mozzarella cheese: impact of coagulant type on functional properties. *J. Dairy Sci.*, **76**, 3657-3663.

9 Culinary fats: solid and liquid frying oils and speciality oils

John Podmore

9.1 Introduction

The use of hot and cold oils in food preparation has been known for thousands of years. The oils and fats can be of animal or vegetable origin, depending on geographical and historical factors. The oil or fat used influences the flavour and texture of the food. This is particularly relevant in fried food where, though reaction flavours from proteins and carbohydrates account for much of the flavour, the condition of the oil used to fry the food also has an influence on flavour. A crisp outer texture is a characteristic of most fried foods, caused by rapid surface dehydration, which again has an influence on flavour perception.

Deep fat frying of food has gained enormously in popularity because of its speed, operational simplicity and ability to supply a desirable flavour, golden colour and crisp texture to the food.

The physical and chemical stresses placed on a frying fat during the deep fat frying process are highly complex, leading to the oil decomposing thermally and oxidatively to form a wide variety of volatile and nonvolatile decomposition products, which ultimately alter the nutritive and functional properties of the oil or fat. The way in which these changes take place has been the subject of considerable research during the past 30 years.

The importance of ghee and vanaspati in Asia requires detailed consideration, as these are major culinary fats in a number of countries—principally India and Pakistan.

Developments in the purification of oils have made available salad oils and cooking oils that are both clear and bland. When these are used for shallow frying, the fat or oil not only acts to prevent the food sticking to the pan but also contributes to the mouth feel and texture of the food. In the case of animal fats, they also contribute a characteristic flavour.

There has been a developing interest in so-called 'speciality oils' which often are used in their natural state to retain their full colour and flavour. They are used in a range of culinary applications to impart flavour, colour and an oily sheen to the food.

This chapter will consider the attributes required of these oils and fats for use in food preparation.

9.2   Salad and cooking oils

Salad oils and cooking oils can be either fully liquid at room temperature or solid or semisolid. Examples of liquid oils are soybean, sunflower, rapeseed, groundnut, corn, cottonseed and olive oil. Examples of solid and semisolid oils are lard, tallow, palm oil, coconut oil and palm kernel oil.

Traditionally, in northern Europe cooking oils were derived from domestic animals, so that lard, tallow and butter were popular; in Mediterranean countries, Asia and Africa liquid oils were more popular. In North America, though tallow was popular in the early part of the twentieth century it has now been largely replaced by vegetable oils such as soybean, cottonseed, sunflower and corn oils.

There has now been a significant shift in the composition of salad oils and cooking oils as a result of the greater global movement of people and an increased interest in nutrition and health issues that highlight the relationship between saturated fatty acids and cardiovascular disease. Thus cooking and salad oils are almost exclusively vegetable oils.

Cooking oils can be used in the natural state or after processing. The majority of cooking oil is now used as processed oil. Oils used in the natural state tend to be those oils found in the speciality sector of the market.

Ghee and vanaspati represent a specific type of cooking oil traditionally used in Asia, though liquid vegetable oils are also used extensively in Asian cooking.

Modern salad and cooking oils are purified by way of refining and deodorisation to give a clear, bland, stable and 'low-coloured' oil. The distinction between salad oils and cooking oils is the difference in their oxidative stability, with cooking oils being more stable at frying temperatures. Also, the term 'salad oil' is applied to those oils that remain clear at refrigerator temperatures. The criteria for clarity at refrigerator temperature is defined by the cold test described in AOCS (cc 11–53) (1981).

Oils that are likely to deposit crystals or waxes after prolonged storage at 0°C are subjected to a winterisation process. This is a modified fractionation process that crystallises high melting triacylglycerols and waxes so that they can be filtered off in order to meet the salad oil criteria. Sunflower oil and corn oil need dewaxing before they meet the salad oil criteria, whereas soybean oil and low-erucic-acid rapeseed oil both naturally conform to the requirements of the cold test. Cottonseed oil, because of its high content of palmitic acid, requires winterisation in order to be certified as a salad oil. Groundnut oil, because of the noncrystallinity of the high melting fraction, cannot be made to conform to the salad oil standards. An oil that has gained prominence, particularly in Asian countries, is rice bran oil. This requires winterising to remove saturated tryglycerides and the 2–6% waxes that occur in the oil. Owing to the presence of relatively high levels of unsaponifiables containing sterols and orzyganols, rice bran oil has excellent oxidative stability.

Salad oils and cooking oils are interchangeable when the criterium of good oxidative stability is achieved and when clarity at refrigeration is maintained. This has been achieved with oils such as soybean and rapeseed oil. These are lightly hydrogenated (i.e. the iodine value is reduced by approximately 20 units), which reduces the linolenic acid content by more than half and so improves the oxidative stability. The oil must be then winterised to remove the solid triglycerides created in the hydrogenation [figures 9.1(a) and 9.1(b)]. This process provides an oil that has all the characteristics necessary for use in salads and deep fat frying. Where the use is limited to deep fat frying then the hydrogenated oil can be processed into a pourable slurry.

The oxidative stability of salad and cooking oils is heavily influenced by the degree of unsaturation, particularly by the linolenic acid content as well as the levels of natural antioxidants such as tocopherols, sterols and phospholipids. Other factors affecting quality are the initial quality of the crude oil, processing and handling, and conditions of transport and storage.

The shelf-life of salad oils can be enhanced by the inclusion of synthetic antioxidants such as butylated hydroxyanisole (BHA) and butylated hydrox-ytoluene (BHT). Although these antioxidants slow down the rate of oxidative degradation during storage they have little effect on oil stability during the frying process.

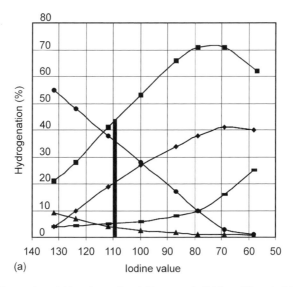

(a)

Figure 9.1 Hydrogenation of (a) soybean oil and (b) rapeseed oil. The solid vertical line indicates the point at which the reduction in linolenic acid is maximised and the increase in saturated and trans fatty acids is minimised. ━■━ C18:0; ━■━ C18:1; ━●━ C18:2; ━▲━ C18:3; ━◆━ trans.

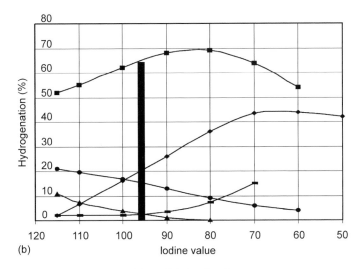

Figure 9.1 (continued)

Salad oils, as the name implies, are used as dressings for salads. They can either be added directly at the table or they can be used in the preparation of, for example, French dressing, which is basically oil, vinegar and spices. Salad dressings, either as mayonnaise or as salad cream, are emulsified and thickened by the use of gums, egg yolk and emulsifiers. Sauces of the hollandaise and bernaise variety represent another use of salad and cooking oils. These sauces contain about 30% oil and are designed for dressing hot food in order to add succulence and flavour. Though oils are used in sauces of this kind, butter is favoured in these recipes as it contributes a special richness and flavour.

Olive oil has always been considered the outstanding salad oil, though it is considered as a speciality oil when used in gourmet cooking. However, its use has grown considerably in Western societies in recent years since it was shown that the cholesterol-lowering effect of monounsaturated fatty acids compared with saturated fatty acids is similar to that of polyunsaturated fatty acids. The increased appearance of olive oil on supermarket shelves suggests it can be dealt with as a salad oil.

Table 9.1 shows the composition of olive fruit; however, the oil content can vary in the range 15–30%. Table 9.1 also shows how widely the fatty-acid composition can vary, being affected by region, climate, fruit variety and maturity at the time of harvest. For example, Spanish, Italian and Greek olive oils are generally low in linoleic acid and high in oleic acid content, whereas Tunisian olive oils are high in linoleic and palmitic acids and lower in oleic acid content.

Though nutritional concerns have helped to drive up the usage of olive oil, consumers traditionally have selected it for its unique and distinctive flavour,

Table 9.1 Composition of olives and olive oil

	Percentage
Olive fruit:	
Moisture	50.0
Protein	1.6
Oil	22.0
Carbohydrate	19.1
Cellulose	5.8
Ash	1.5
Olive oil:[a]	
Oleic acid	55–83
Linoleic acid	3.5–21
Palmitic acid	7.5–20
Stearic acid	0.5–5

[a]Fatty-acid composition; there are also traces of myristic, palmitoleic, heptadecanoic, heptadecenoic, linolenic, arachidic, behenic lignoceric and eicosenoic acid.

which is altered by regional and climatic conditions. There has been a considerable volume of work published on the flavour components of olive oil and how they are varied by the maturity of the fruit at the time of harvesting and the variety of the fruit. Major flavour components have been shown to be hexenal, *cis*-3-hexenal, *trans*-2-hexenal, hexanol, *cis*-3-hexenol and 2-hexenol and the corresponding esters.

Regional differences in flavour are well known; for example Spanish olive oil has a strong 'peppery' note, Greek oils are thought to be 'grassy' and Italian olive oil is 'fruity'. There are then variants of these basic flavour characteristics. In the modern marketplace, unless an olive oil is sold as being specific to a particular region, it will usually be a blend of oils of various varieties of olive and from various regions. This enables the supplier to ensure a more consistent flavour.

Olive oil is sold in a number of grades, each of which has a strict definition and chemical specification. Olive oil is obtained by crushing the fruit and pressing the resulting paste. The yield from the pressing is a mixture of oil and water. This mixture is then separated centrifugally.

The first pressing of the olive paste gives 'virgin' oil, which is then graded on acidity and flavour. The criteria for defining olive oil quality has now been strictly regularised, particularly in the European Union. The grades are defined in EC Regulation 2568/91 and its amendments (1991).

The grades can be defined as follows:

- extra virgin olive oil: this is the premium grade and must have an acidity less than 1% and a flavour of closely defined character
- virgin olive oil: in this grade the acidity can be up to 2% but flavour criteria are as described above

- ordinary virgin olive oil: this oil must have an 'acceptable flavour' and a maximum acidity of 3.3%
- lampante virgin olive oil: this is the lowest grade, with a poor flavour and usually with an acidity of greater than 3.3%

Other grades of olive oil are based on olive pomace oil. After pressing there is a 3–8% residue of oil in the pulp. This is solvent extracted to give crude olive pomace oil. Crude pomace oil and lampante oil are usually refined and deodorised for use in edible products.

The grades of oil usually found in the retail market are as follows:

- extra virgin olive oil and virgin olive oil, sold in the quality described above
- ordinary olive oil (this is a blend of virgin olive oil and olive oil refined and deodorised to have an acidity less than 0.5%; the proportions of each oil are varied in order to give the desired flavour intensity, colour and acidity, though the flavour of this oil will always be milder than that of virgin olive oil)
- olive pomace oil (this is a blend of virgin oil and refined solvent-extracted pomace oil), similar in appearance and flavour to olive oil

A new grade of olive oil, which first appeared on the US market, was 'extra-light' olive oil. Fully refined olive oil is blended with treated olive oil to reduce the flavour so that the consumer can have the health benefits of olive oil but not the typical flavour.

The health benefits have been exploited in spreads that contain olive oil, but generally not the flavour. There are now cooking and salad oils available that are blends of sunflower oil and olive oil, as well as of rapeseed oil and olive oil. There are mildly flavoured but have a mix of the nutritionally acceptable monounsaturated and polyunsaturated fatty acids.

Olive oil is used widely as a culinary salad oil. The oil can be 'drizzled' onto salad and vegetables and used in the cooking of classic Mediterranean and Asian dishes, particularly where vegetables are being cooked. In Indian cooking, olive oil is often flavoured with garlic when used in cooking; for example, in Goa it is used to cook beef roulade. It is now possible to purchase in retail outlets olive oil flavoured with a variety of herbs and spices.

9.3   Frying fats

There are two major methods of frying: shallow (pan) frying and deep fat frying. The two frying methods put entirely different demands on the fat used. In shallow frying the oil or fat is generally used only once so that its resistance to breakdown during frying is unimportant; in deep fat frying the reuse of the frying fat and its effect on fat quality is of the greatest importance.

9.3.1   Shallow (pan) frying

In the case of pan frying the fat or oil used is present mainly to prevent the food sticking to the pan while it is being cooked by the heat applied. This frying ...d has been popular for hundreds of years and is a quick method of food ...ation. During the frying the food surface absorbs some of the heated fat. ...ay transfer distinctive flavours to the food and improves the palatability, ... the food an oily surface.

Solid fats based on lard, tallow and palm oil have been used extensively in shallow frying but have been largely replaced by liquid vegetable cooking oils. Where distinctive odours and flavours are required then butter and margarine can be used. The presence of water causes spattering, though in the case of margarine the presence of certain emulsifiers can alleviate the effect. The presence of milk solids gives the fried food colour and flavour; however, it can also cause the food to stick to the pan. There are also now available flavoured oils and liquid margarines designed for use in frying and griddling, and there are gaining in popularity.

Consideration of shallow frying would not be complete without noting the contribution of the Chinese, who have used both shallow and deep frying in food preparation for many centuries. Stir-frying (Hom, 1990) is a highly popular method of preparing many Chinese dishes and is done in a wok. This is a conical shaped pan that is set on top of a brazier so that heat can be spread over the whole surface. This means large amounts of food can be cooked rapidly by stir frying, as the food always returns to the hottest part of the wok after stirring. Woks are also used for deep fat frying, for which they are very efficient in that they provide depth of oil with relatively small amounts of oil so that heating and temperature maintenance are easy.

In Chinese cuisine the oil most favoured is groundnut oil because of its mild pleasant flavour of roasted peanuts. A range of other oils are used, including corn oil, rapeseed oil, cottonseed oil, soybean oil and sunflower oil. Sesame oil is occasionally used for stir frying; however, its principal use is as a condiment because of its strong and distinctive aroma and flavour.

9.3.2   Deep frying

Deep fat frying has become one of the most important methods of food preparation. It is used domestically and in the food service, snack and baking industries. The frying fat plays such a unique role in the frying process that frying has become the focus of an enormous volume of research such that the reactions taking place during the frying of food are now more clearly understood so that food quality is better maintained and nutritional standards improved.

There is no ideal frying fat suitable for all frying applications. Factors such as the process used, the food being fried, storage, shelf-life of the finished product and cost impact on frying fat selection. In the process, the frying temperature and turnover have a major impact on the maintenance of quality during use, and whether the process is batch or continuous has an effect on the fat used. The product to be fried also influences the frying fat to be used, as the oil affects the surface texture of the food, its flavour and the shelf-life.

The cost of a frying fat is very important in commercial operation, not only in terms of purchase price, but also cost in use. Resistance to degradation and hence to increased rates of absorption may be required, justifying the use of a highly priced hydrogenated fat to ensure longer frying life.

The selection of frying media is now becoming heavily influenced by nutritional issues. The findings that high intakes of saturated and trans fatty acids are implicated in arterial and cardiovascular disorders, and that monounsaturated and polyunsaturated fatty acid are beneficial, have placed an increasing emphasis on the use of natural more highly unsaturated oils and away from solids and hydrogenated fats. Balancing nutritional and health requirements with those of cost and thermal stability has become a major task for suppliers.

9.3.2.1   The chemistry of frying fats

Deep fat frying is a heat-transfer process in which food is cooked and dried and in which the heat-transfer medium—the oil—is absorbed by the food particles. The reactions taking place in this process are shown in figure 9.2, and the

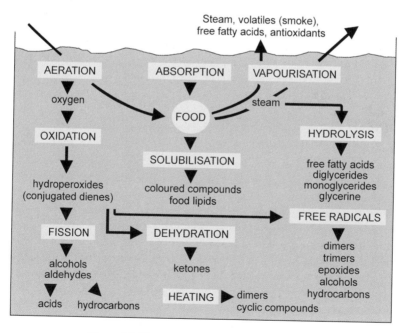

Figure 9.2 Changes occurring during deep fat frying.

physical and chemical changes taking place are shown in table 9.2. These changes are caused by hydrolysis, oxidation and polymerisation. The degradation products, both volatile and nonvolatile, affect the physical properties of the medium.

The thermal degradation of frying oil is complex, with many variables. Oil, food and process variables are shown in table 9.3. These are the factors that affect the hydrolysis, oxidation and polymerisation processes and the frying oil deterioration; thus the management of these factors controls the rate of degradation of a frying oil. For example, it is preferred that fresh oil of good quality is used; that is, oil with no prior oxidation and with low levels of

Table 9.2 Effects of physical and chemical reactions during deep fat frying (Warner, 1998)

Physical changes:
  increased viscosity, colour and foaming
  decreased smoke-point

Chemical changes:
  increased free fatty acids, carbonyl compounds and high molecular weight products
  decreased unsaturation, flavour quality and nutritive value (e.g. from essential fatty acids)

Table 9.3 Factors affecting frying oil degradation

Oil or food factors:

  Unsaturation of fatty acids
  Type of oil
  Type of food
  Metals in oil or food
  Initial oil quality
  Degradation products in oil
  Antioxidants
  Antifoam additives

Process factors:

  Oil temperature
  Frying time
  Aeration or oxygen absorption
  Frying equipment
  Continuous or intermittent heating or frying
  Frying rate
  Heat transfer
  Turnover rate;[a] addition of makeup oil
  Filtering of oil or fryer cleaning

[a]Turnover is the ratio of the volumetric capacity of the fryer to the rate at which fresh frying oil is added to replenish the fryer.

Table 9.4 Basic specification for frying fat (Brinkmann, 2000)

Criterion	Specification
Colour	Light
Taste	Bland
Flavour	Bland
Free fatty acids (wt%)	0.1[a]
Peroxid value (meq O_2 kg^{-1})[c]	1[a]
Smoke-point (°C)	220[b]
Moisture (%)	0.1[a]
Linolenic acid (%)	2[a]
Melting point	To fit the application

[a]Maximum.
[b]Minimum.
[c]meq O_2 kg^{-1}, milliequivalents of oxygen per kilogram of oil.

polyunsaturation. These features are shown in the basic specification given in table 9.4.

The rate and extent of degradation can be controlled by management of frying conditions, such as temperature, time, exposure to oxygen, filtration and turnover. Intermittent frying and heating has been found to increase the rate of oil degradation compared with continuous heating (Perkins and Van Akkeren, 1965).

Turnover has always been seen as critical to maintaining frying oil quality, where turnover is the ratio of the fryer's volumetric capacity to the rate at which fresh frying oil is added to replenish the fryer (Banks, 1996). It can be seen that the higher the turnover, and so the greater the replenishment with fresh oil, the more the frying oil is maintained in its best condition. Where the turnover is slower it is impossible to avoid discarding used frying fat in order to maintain quality. Low turnover leads to the accumulation of degradation products in the frying oil and their eventual incorporation in the fried food—an area of concern in nutritional and flavour terms.

The many thermal and oxidative reactions involving oil, protein, carbohydrates and minor food constituents that take place during frying leading to the development of volatile and nonvolatile degradation products have been the subject of much research (figure 9.3).

During the frying process the volatile degradation products are in part responsible for flavour being formed in the oil or food particle. The degradation products can be detrimental to the oil and food as well as making the flavour more attractive. The nonvolatile degradation products shown in table 9.5 are likely to promote further degradation of the oil, which in turn affects the long-term flavour stability of the fried food.

The breakdown of frying fats in use is believed to follow a distinct pattern of degradation. Blumenthal has developed a 'frying oil quality curve', describing

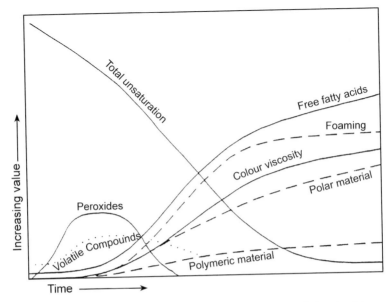

Figure 9.3 Changes in volatile and nonvolatile decomposition products during the frying process (Warner, 1998).

Table 9.5 Volatile and nonvolatile degradation products from frying oil (Warner, 1998)

Nonvolatile products	Volatile products
Monoglycerols	Hydrocarbons
Diglycerols	Ketones
Oxidised triacylglycerols	Aldehydes
Triacylglycerol dimers	Alcohols
Triacylglycerol trimers	Esters
Triacylglycerol polymers	Lactones
Free fatty acids	

the stages the frying oil passes through during use, (figure 9.4) (Blumenthal and Stiers, 1991). The various stages are quantified in table 9.6 for a mixed-use fast-food service fryer.

The measurement of the quality indicators of used frying fats has been the subject of considerable investigation and has been reviewed by Gertz (2000), who looked at physical and chemical parameters. The control of quality in the production of fried food is an area of great concern as the achievement of consistent quality with good control of cost is very important to the fryer. Quality control tests need to be simple, easy to carry out and preferably avoid the use of chemicals.

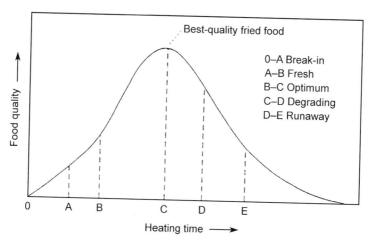

Figure 9.4 Frying oil quality curve showing the five phases that a frying oil passes through during the degradation process (Blumenthal and Stiers, 1991).

Table 9.6 Effects of oil quality on the characteristics of fried potato strips (Blumenthal and Stiers, 1991)

Oil quality	Characteristics of fried potato strips
'Break-in'	White, raw interior; no rich potato odour; surface not crisp
Fresh	Slight darkening of surface; some crust formation; interior not fully cooked
Optimum	Golden brown; oily surface; fully cooked centre (ringing gel); rich potato odour
Degraded	Oily surface; darkening and spotting of surface; surface hardening; excess oil; centre not fully cooked
'Runaway'	Excessively oily (greasy); dark and hardened surface; walls beginning to collapse; centre hollow

9.3.3   Selection of frying media

Though the basic selection requirements are the same irrespective of the specific application, as shown in table 9.4, the type of fryer and operational demands will always affect the choice. Table 9.7 lists those features that are of greatest interest when a frying medium is being selected. The major areas of deep fat frying will be considered individually.

9.3.3.1   Fast-food-service frying

In fast-food restaurants the rate of turnover in the fryer is of the order of 25%–30%, which is the result of relatively low fat absorption and periods of fryer 'idling'. The low turnover and necessity to fry a range of food types means that used frying fat has to be discarded frequently. The timing of the point when the fat is discarded is important economically and in terms of food quality.

Table 9.7 Features to be considered when selecting a frying fat

No contribution of off flavours to the food
Long frying life to make the operation economical
Ability to produce an appetising, golden brown, nongreasy surface on the food during its fry life
Resistance to excessive smoking after continued use
Ability to produce food with good taste and texture
Resistance to gumming (polymer formation)
Resistance to rancidity
Uniform in quality
Ease of use, including form and packaging

Discarding fat too early means unnecessarily high frying fat costs and too late means poor-quality food.

Many restaurants either have a fixed period for discard or fix the weight of food fried before discard, otherwise the restaurants use parameters such as colour, evidence of smoking, condition of the fried food and evidence of foaming. None of these approaches is completely satisfactory as they lack objectivity and accuracy and so can lead to variable food quality. The search for a quick and easy test for quality is still continuing, as discussed earlier (section 9.3.2.1).

Up to about 1985 the fats used in food-service applications were usually solid plasticised fats of animal and/or vegetable origin; for example, in the USA blends of tallow and cottonseed oil were very popular, and in Europe palm oil was widely used. Hydrogenated vegetable oils such as soybean or rapeseed oil were also popular, with the oil hydrogenated to a slip melting point of 30°C–34°C to ensure the virtual elimination of linolenic acid content. Where fast-food service was of a lighter duty then liquid vegetable oils were popular, such as soybean, rapeseed, corn and cottonseed oil.

Liquid vegetable oils increased in popularity when the effectiveness of dimethyl polysiloxane, an antifoaming agent, was demonstrated to reduce the oxidative breakdown of frying oils when added at as little as 1 ppm. Frying stability has been found to increase by three to ten times compared with original frying stability, as studied under laboratory-controlled conditions (O'Brien, 1998). Work by Freeman *et al.* (1973) demonstrated dimethyl polysiloxane acts to stabilise the frying oil by providing a monomolecular layer at the oil surface.

The presence of natural antioxidants is also found to extend the frying life of a vegetable oil; therefore, refining processes are being modified to maximise the residual levels of tocopherols in vegetable oils and in some cases losses of tocopherol as a result of refining are made up by additions to the refined edible oil of natural or synthetic tocopherol.

Gertz *et al.* (2000), using a newly developed test to give the OSET index (oxidative stability at elevated temperature), have shown the influence of sterols, sesame oil, rosemary extract and other naturally occurring substances on the stability of frying oil at high temperatures. In the same study, experiments with other natural and synthetic antioxidants demonstrated that ascorbyl palmitate increased the oxidative stability of the oil.

It is now possible for fast-food-service restaurants to use a very wide range of frying oils, and those selected are governed by economic and nutritional issues. The higher the polyunsaturation in an oil or blend of oils, particularly linolenic acid, has been shown to give a shorter frying than oils with lower levels of polyunsaturation. The introduction of opaque liquid shortening followed from this consideration. Brush hydrogenated soybean oil and rapeseed oil, where the iodine value is reduced by, say, 20 units, shows a reduction in both linolenic and linoleic acids, with only a small increase in the amount of saturated and trans fatty acids (see figure 9.1). A clear, more stable, oil can be produced by 'winterisation'. Alternatively, the addition of a small proportion of fully hydrogenated oil followed by chilling produces an opaque pourable slurry. In products of this type, with the addition of dimethyl polysiloxane, an acceptable frying life can be obtained even in heavy-duty applications.

9.3.3.2   Industrial snack-food frying

In this case, products such as potato crisps (chips), tortilla chips and puffed snacks are fried and packaged for consumption, with a shelf-life of several months after manufacture. The frying oil used is crucial to the quality of the fried snack; hence the process is highly controlled.

The frying oil becomes a major component of the snack after the hot oil has first dehydrated the particle to concentrate the flavours. Fried snacks have a very high proportion of absorbed fat, in the order of 30–45%, so that the fat has an influence not only on the turnover of the fryer but also on the finish of the snack. A snack will have an oilier or brighter surface with a frying oil that is liquid at room temperature compared with one where the oil is solid at room temperature, giving a duller and greyer product.

In modern production plants, liquid oils or blends are used under tight control to minimise degradation in the frying oil in order to ensure an adequate shelf-life for the product. In some cases, to ensure a good shelf-life, phenolic antioxidants such as BHA, BHT and TBHQ (tert-butyl hydroquinone) are introduced, giving some carry-through protection for the snack food (Buck, 1981).

Oils that are popular in this application are cottonseed oil, palm olein, rapeseed oil and groundnut oil. These oils are often used as blends. Palm oil is used but is usually blended with liquid vegetable oil in order to ensure the oil is liquid at room temperature.

9.3.3.3   Bakery frying fat application

The major fried bakery product is the doughnut, although there are also fried pies and pastries. The fact that these products often have a coating as a finish, (e.g. sugar or icing) places a constraint on the frying medium that can be used; thus the frying fat as well as being a heat-transfer medium becomes a major ingredient of the product and must act as a binder for any added coating.

Frying fats that are solid at room temperature are preferred in this application. Hydrogenated soybean oil with a slip melting point of between 33°C and 40°C have been used. Palm oil and hydrogenated lard have also been used.

In selecting the particular melting characteristics a balance has to be struck between achieving a good appearance (i.e. no visible layer of fat, as such a layer could give a waxy mouth feel) and enough solid fat to ensure good sugar pickup. Finally, the solid fat content at body temperature must not be so high as to impair the eating quality. A profile of bakery frying fats is given in table 9.8.

Table 9.8 Typical solid fat index (SFI) profiles of bakery frying shortening (O'Brien, 1998)

Solid fat index			
Temperature (°C)	Solids (%)	Characteristic affected	Solid–liquid relationship effect
10.0	33–38	Shelf-life	The correct solids content will create a moisture barrier, keeping moisture inside the doughnut
	21–26		
21.1	21–26	Appearance	Too high an SFI value leaves a visible layer of fat on the doughnut, which provides a waxy mouth feel and can promote flaking of the sugar coating; too low an SFI content will leave the crust oily, which can promote oil soakage of the sugar
26.7	19–21	Sugar pickup	The correct ratio of hard to soft fractions in the frying shortening composition will help ensure proper sugar pickup; too high an SFI results in decreased sugar adherence; too low an SFI results in increased sugar disappearance
33.3	12–17		
40.0	7–12	Eating quality	SFI content of doughnut frying shortening at temperatures above body temperature have a direct effect upon eating quality; SFI values above 12% at 40°C may cause a waxy, unpleasant, mouth feel

9.4   Oils for roasting nuts

Many varieties of nuts are now available as snack foods, the major example being peanuts. When nuts are roasted in oil the surface is dehydrated and there is browning and the nut texture and appearance is changed. The nuts have a low moisture (about 5%) and high oil content and thus very little frying medium is absorbed, giving an extremely low turnover. Hence an oil with good oxidative stability is required. The preferred oils in this application are coconut oil and hydrogenated palm kernal oil, which have little unsaturation and which, because of the high content of short-chain and medium-chain fatty acids, have a low melting point.

9.5   Ghee

Ghee is clarified crystallised butter from buffalo's or cow's milk. It is the most common form in which butterfat is used in India and other countries of the Far East. Ghee manufacture in India is still a home industry (Ganguli and Jain, 1973). However, increasing industrialisation and growth of the urban population has led to the establishment of factories using both batch and continuous processes. Anhydrous butterfat is also widely sold as ghee.

Ghee is traditionally made from the boiled and cooled milk of the cow or buffalo by allowing it to set overnight after addition of a starter culture. The curd that is formed is then churned to butter prior to clarification at high temperature (100–120°C) to remove the water. This is the *desi* method and it is claimed that it gives the best flavour characteristic because during the moisture removal there is interaction between the fat and the fermented residue of the nonfat solids.

In factory manufacture the process is simplified, with the use of cream or sweet cream butter. Heat is applied to either the butter or the cream in open-jacketed vessels to remove the moisture through agitation. A prestratification method can also be used where after initial heating of the butter the bottom layer is discarded before the remainder is heated to the desired temperature.

The continuous method is based on pumping cream or butter through a steam-heated scraped-surface heat exchanger. The superheated cream or butter is passed into a flash evaporator, where the moisture is separated from the liquid fat. This process goes through multiple stages to remove the moisture completely.

The processes are selected on the basis of yield and energy efficiency as well as achieving the desired flavour, colour and shelf-life. The quality of the ghee depends on the quality of the milk, cream, the curd or butter and the temperature of clarification. The lower temperature of 110°C gives a mild flavour, whereas a temperature of 120°C gives a stronger and more 'cooked' flavour.

9.5.1 Ghee attributes and quality

Ghee has a shelf-life of up to eight months in tropical temperatures. This high stability is attributed to the low moisture and high phospholipid content. It is thought that amino acids in the fat phase from the phospholipid–protein complex form during culturing and cooking.

The highly characteristic flavour of ghee is generated mainly during the boiling-down process, where there is an interaction between protein and lactose. A range of ketones, alcohols, hydrocarbons and lactones have been identified as contributing to the flavour (Achaya, 1997).

The flavour, texture and colour have long served to characterise ghee, so they can also be used as indicators of its quality. Where the manufacture of ghee is still a cottage industry, using the *desi* method, flavour and texture are the only criteria used to judge quality. A nutty, lightly cooked aroma and flavour are generally prized. The texture can vary from granular to smooth and, in some cases, even show a tendency to separate out a liquid portion. These textural differences are fixed by local tradition; for example, a granular texture with no separation is preferred in India, whereas in Pakistan the product is favoured when it has larger, softer granules dispersed in a supernatant liquid.

In the industrial manufacture of ghee considerable effort has gone into providing a product of greater uniformity and extended shelf-life. The quality of ghee depends heavily on the milk, which in turn depends on the animal feed, the season and the health of the animals. For example, winter ghee has been shown to have a high acidity, melting point and grain size, and milk from animals fed on cottonseed gives ghee that increases in acidity less quickly, but that also leads to a product of lower melting point. Where the ghee preparation is by way of cream *dahi* or butter, the quality of these products also influence the quality of the ghee. Further, the method of preparation and temperature of clarification have an influence. These factors determine the physiochemical features of ghee. Typical analytical characteristics of cow and buffalo ghee are shown in table 9.9.

The quality of ghee (Sharma, 1981) is generally measured analytically by parameters such as acid value, peroxide value, flavour and shelf-life. Bacteriological quality is assured by reducing the moisture content to less than 0.3%.

The free fatty acid content of ghee varies with the method of preparation. Thus in ghee prepared from ripened cream or butter the free fatty acid is in the range of 0.34–0.40% whereas in unripened cream or butter it is 0.23–0.28%. The peroxide value is a less valuable indicator of quality than either free fatty acid content or flavour, and although in the fresh product it should be low in the

Table 9.9 Typical analytical characteristics of cow and buffalo ghee (Sharma, 1981)

	Buffalo	Cow
% Solid fat °C		
10	51·9	53·4
15	37·5	38·6
20	23·1	22·6
25	16·3	15·7
30	10·8	7·9
35	4·0	3·2
40	NIL	NIL
Slip point (°C)	29·9	33·4
IV	28·4	34·9
Lovibond colour (5 · 25 in cell)	2 · 5R 24 · 0Y	4 · 3R 44 · 0Y
% Moisture	0 · 3 max.	0 · 3 max.
% Free fatty acid		
Ripened milk	0 · 34–0 · 40	
Unripened milk	0 · 23–0 · 28	
Unsap. (mg/100)	390	450

stored product it varies considerably, particularly near the point where rancidity is detectable organoleptically.

The flavour of fresh ghee is greatly influenced by the temperature of clarification. Ghee prepared at 120°C has a distinctive 'cooked' flavour, whereas that prepared at 140°C has a 'burnt' flavour. The temperature of clarification also influences off-flavour development; product clarified at 110°C retains its flavour longer than that clarified at 120°C. The flavour of ghee is also dependent on the method of preparation. Ripening of the cream or butter is always considered to give an improved flavour, and ghee from *desi* butter is believed to be the best, having a 'nutty' flavour with a 'cooked' or 'caramelised' aroma.

It has been found that the keeping quality of *desi* ghee is better than that of direct cream or creamery butter ghee; longer heating has been found to improve the oxidative stability of ghee because of the liberation of phospholipids into the fatty matter. As well as organoleptic changes, ghee undergoes textural changes in storage. When filled into containers crystallisation occurs with the formation of solid, semisolid and liquid layers. Below 20°C ghee has a small-grained and compact texture. Storage temperatures of 28–29°C lead to a well-defined granular texture, though cooling to this too rapidly can give rise to a granular settled portion, a liquid oil portion and a floating hard flake, each layer having different chemical characteristics. At higher temperatures the texture becomes looser and a liquid oil layer then becomes evident. At 34°C the product is fully liquid.

The consumer's perception of quality is based on flavour, texture and colour, as these three indicate quality and purity. A uniform granular structure, with a white or off-white colour is expected. The flavour should be characteristic, with *desi* ghee the most popular.

Anhydrous milk fat, in spite of its more bland flavour, is now being supplied as ghee. The anhydrous milk fat is prepared at temperatures of 80°C, so it has a greater moisture content, but less protein.

9.5.2 Uses of ghee

Traditionally, ghee was the major culinary fat in the Middle and Far East and was used in a wide range of foods. It was used in the shallow frying of vegetables. In India it was used in curries, paratas and dosais. Certain foods such as puris and samosas were deep fried in ghee. Basting chicken and pilau with ghee imparts a distinctive and characteristic flavour. Certain recipes demand that the ghee be a solid. Finally, ghee is used as a spread in molten or semimolten form for chappatis or partially malted and mixed rice. A small proportion of the ghee produced is used in confectionery and to cook sweetmeats based on cereals, milk solids and fruit.

Changing culinary practices combined with the increasing popularity of vanaspati and liquid vegetable oils has meant that ghee consumption is declining. For example, less than 8% of households in India consume ghee (GCMMF, 1993).

Ghee is now used more in high-quality cooking. For example, pulses are flavoured with ghee, which is the equivalent of cooking them in butter. Ghee is also used in combination with vanaspati in order to enhance the flavour of the vanaspati. Cooking in ghee has a certain amount of status attached to it, rather like cooking in butter. However, the vast majority of recipes for dishes from the Indian subcontinent and Middle East call for the use of liquid vegetable oil or vanaspati.

9.6 Vanaspati

Vanaspati or vegetable ghee is a substitute for natural ghee and is based on hydrogenated vegetable oils. In India the economic situation caused the demand for animal fats to exceed production, resulting in a price increase so that the product was put beyond the reach of the general population.

Vanaspati was first imported into India from the Netherlands after the First World War as a substitute for ghee for the bulk users such as restaurants and sweetmeat manufacturers.

In 1930 production started in India, and 10 years after this imports ceased. The volumes produced have increased since that time, with a significant jump during the Second World War, when vanaspati was approved as a cooking fat for the armed forces.

The development of vanaspati in the marketplace has been controlled by the Vegetable Oil Products Control Order (1942) which regulates manufacture and control standards. Some of the important parameters are:

- moisture content, maximum 0.25%
- slipmelting point, 31–41°C
- free fatty acid (asoleic), maximum 0.25%
- unsaponifiable matter, maximum 2.0%
- nickel content, maximum 1.5 ppm

Colour and flavour are not permitted. The addition of refined sesame oil is mandatory. Vitamin A is a required additive at 25 IU; vitamin D addition is optional.

The oils permitted for use in vanaspati are under regular review and those currently permitted include soybean oil, cottonseed oil, rapeseed oil, sunflower oil, maize oil, palm oil, palm olein, rice bran oil, mahua fat, nigerseed oil, watermelon seed oil, sal fat (to maximum 10%) and sesame oil (used mainly as a marker). Recently included have been solvent-extracted expeller cake oils such as mustard–rapeseed, groundnut and sesame oil. Groundnut oil, on which the early vanaspati was developed, is no longer permitted to be used because of the demand for its use as a direct edible oil.

As vanaspati is made from hydrogenated oils a considerable amount of work has been done on the hydrogenation conditions to be applied and how they are varied for the individual oils. Selective hydrogenation conditions are applied, which lead to steep melting curves and the generation of high levels of trans fatty acids. The high level of trans fatty acids is felt to assist in the generation of the desired granular texture associated with ghee. It has been shown that as the hardened oil cools to 50°C the triacylglycerols containing 'trans' fatty acids in the liquid phase nucleate to initiate crystallisation with heat evolved, which is controlled by external cooling. Photomicrographs of vanaspati show the presence of large granules made up of radially arranged needle-like crystals. This crystal network, though containing large saturated triacyl glycerol crystals causing the granularity, can trap the liquid phase to prevent separation at ambient temperature.

Vanaspati is expected to be similar in texture and colour to natural ghee and achieve at least the same shelf-life. It is obviously not possible for the product to have the same flavour as natural ghee; however, there is now a growing acceptance of the product and in some countries flavours may be added to make a vanaspati even more similar to ghee. The culinary uses of vegetable ghee are equivalent to those of ghee, it being used in a range of basted and fried dishes. The improved oxidative stability of vanaspati compared with ghee and its higher melting point give it a wider application in confectionery.

In those countries where vanaspati has become the major culinary fat there is now a call for a fat product for commercial baking to make products such as puff pastry and patties. Shortening products are made as described elsewhere in this book (see chapter 2 of this volume) and typically use the same oils as used in vanaspati, though they are hydrogenated under processing conditions different from those used for vanaspati to give different melting curves and melting points in order to give the plastic characteristics desired for the given shortening. The shortenings are designed for use in a wide range of products such as crisp cookies, wafers and cream biscuits.

9.7 Speciality oils

Speciality oils are usually vegetable derived. The key reason for their purchase is their flavour. The oils are sourced from minor crops and so are available only in small quantities and are of high value. The market for speciality oils is growing as they are perceived as being nutritionally healthy. Also, there is a growing interest in the consumption of exotic foods, that require the use of these characteristically flavoured oils. Many of the speciality oils are not refined, in order to ensure that the distinctive flavour and colour characteristics are maintained. These requirements can cause significant problems, especially with consistency in quality with such minor crops.

Speciality oils are extracted mainly by cold pressing, and the use of solvents is avoided. Some speciality nuts and seeds are roasted prior to extraction in order to develop the desired flavours (e.g. toasted sesame oil).

There is an enormous number of nuts and seeds that can provide oils for food, pharmaceutical or cosmetic purposes. Some examples are as follows:

- apricot kernel oil
- blackcurrant oil
- cherry kernel oil
- macadamia nut oil
- meadowfoam oil
- borage oil
- passion fruit oil
- pistachio nut oil
- safflower seed oil

The major oils currently found on supermarket shelves are:

- groundnut oil
- grapeseed oil
- sesame seed oil

- walnut oil
- almond oil
- hazelnut oil

Olive oil can also be considered to be a speciality oil and constitutes a major member of this group of oils. The use of olive oil in cooking in countries with cooler climates has expanded significantly in recent years because of the highly characteristic flavour and the claimed nutritional benefits.

The grades and applications of olive oil are discussed in section 9.2 as a salad and cooking oil; however, the variations in colour and flavour are such it is now possible to purchase olive oils that are both specific to a variety of the fruit and to the growing area—making the oil highly specialist.

Table 9.10 shows the fatty-acid composition of some of the speciality oils. It can be seen that they generally contain high levels of unsaturated fatty acids, which in some cases leads to poor oxidative stability.

The speciality oils that are used in food applications are used almost exclusively for their flavour. Those of major interest are discussed below.

9.7.1 Almond oil

This is an oil with a sweet aromatic odour and flavour. It is popular in fish cookery and for use in some bakery products.

9.7.2 Groundnut oil

Though this oil at one period in Europe had the status of a commodity oil and in some parts of the world is still a major culinary oil its use has declined as it is a premium oil. In its unrefined state it is used as a speciality oil, giving a distinctive 'nutty' flavour to food. The absence of linolenic acid in groundnut oil provides it with greater oxidative stability compared with other speciality vegetable oils.

Groundnut oil is used extensively as both a shallow-frying and deep-frying medium. The fact that at low temperatures the higher melting triglycerides deposit in a gelatinous form limits its use in sauces and mayonnaise, as at low temperatures these separate.

9.7.3 Hazelnut oil

Though hazelnuts are grown mainly in Turkey, Italy and Spain, the major producer of the oil is France, where it is used as a salad oil. Hazelnut oil is sold both in a refined and an unrefined state, with the more strongly flavoured unrefined oil being more popular. The near equivalence of this oil to olive oil in terms of monounsaturated fatty acid content means that hazelnut oil has some of the same nutritional attractions.

Table 9.10 Data on speciality oils

Oil	Plant name	Iodine value	C16:0	C16:1	C18:0	C18:1	C18:2	C18:3a	C18:3g	C20:0	C20:1	C22:0	C22:1	C24:0	Other
Almond	Prunus amygdalus	98–105	4–9	0.8a	0–3	60–80	17–30	1.0a	–	–	–	–	–	–	–
Apricot kernel	Prunus armenica	98–112	3–8	1.0a	0–2	56–70	21–33	0–2	–	–	–	–	–	–	–
Avocado	Persia gratissima	80–95	10–20	3–10.5	0–1.5	56–75	8–16	0.4a	–	0–1	2–6	–	1–3.5	–	0.5–3.5b
Borage	Borago officinalis	140–155	9–13	0.6a	3–5	10–20	34–42	0.4a	18–25	2a	0.6a	0.5a	–	–	1–13c
Cherry kernel	Prunus avium	110–130	5.5–10	1.0a	1.5–3	23–39	40–48	1.0a	–	–	–	–	–	–	–
Corn (maize)	Zea mays	103–131	9–14	0.5a	0.5–4	24–42	34–62	2.0a	–	–	–	–	0.2a	–	–
Evening prmrose	Oenthera blennis	145–165	5.5–7	–	1–3	7–18	60–75	0.3a	7.5–11	–	1.0a	–	0.2a	–	–
Gold of pleasure	Camelina sativa	145–165	3–8	–	2–5	12–26	15–24	30–40	–	0–2	9–17	–	0–4	–	–
Grapeseed	Vitis vinefera	125–145	5–11	–	3–6	12–28	58–81	1.0a	–	–	–	–	–	–	–
Hazelnut	Corylus americana	87–102	4–10	–	1–4	70–84	9–19	1.5a	–	1.0a	–	2.0a	–	–	–
High oleic sunflower	Helianthus annus	80–90	3–5	–	3–5	77–84	4–15	1.0a	–	–	–	1.0a	–	–	0–1d
Macadamia	Macadamia ternifolia	–	7–10	16–24	2–45	54–65	1–3.5	–	–	1.5–3	1.5–3	1.0a	2–4e	–	15–23f
Meadowfoam	Limnanthes alba	90–102	1a	–	0.5a	4a	4a	–	–	–	60–65	0.2a	0.1a	0.2a	–
Olive oil	Olea europaea	80–88	7–20	3.5a	0.5–5	55–83	3.5–21	0.9a	–	0.5a	0.2a	0.2a	0.1a	0.2a	–
Passion flower	Passiflora Incarnata	132–145	8–12	0.3a	1.5–3	12–18	65–75	1.0a	–	0.5a	–	–	–	–	–
Peach kernel	Prunus persica	98–115	2–8	1.0a	0.5–2.5	54–67	23–35	0.8a	–	0.5a	–	–	–	–	–
Pistachio	Pistachia minor	90–120	9–20	0–2	1–3	40–60	28–38	2.0a	–	–	–	–	–	–	–
Pumpkin seed	Curcubita peppo	110–130	6–13	–	5–8	20–41	44–57	2.0a	–	–	–	–	–	–	–
Safflower	Carthamuc tinctorius	138–150	2–10	–	1–10	7–42	55–81	1.0a	–	–	–	–	–	–	–
Sesame	Sesamum indicum	103–118	7–12	–	3.5–6	35–50	35–50	1.0a	–	–	–	–	–	–	–
Toasted sesame	Sesamum indicum	103–118	7–12	–	3.5–6	35–50	35–50	1.0a	–	1.0a	–	–	–	–	–
Walnut	Junglans spp.	145–155	5–10	–	2–6	15–36	40–65	0.5–15	–	1.0a	2.0a	–	–	–	–
Wheatgerm	Triticum vulgare	115–140	11–21	0.5a	0.5–4	15–26	49–60	2–10	–	0.2a	2.0a	–	–	–	–

Fatty acid (% wt)

–Zero.
aMaximum.
bC24:1.
cC18:3 isomers.
dC14:0.
eDelta-5 isomer.
fC22:2.

9.7.4   Sesame seed oil

Before extraction of the oil the seeds are usually toasted to intensify the flavour of the oil. The extracted toasted oil is dark brown and has a very strong bitter 'nutty' flavour and odour. Unrefined sesame oil is used extensively in Oriental foods and is frequently added to stir-fry dishes in Chinese cookery to give a characteristic flavour. The unrefined oil is generally not used for cooking but for flavouring.

The refined oil, which is straw yellow in colour with a mild 'nutty' taste and exceptional oxidative stability, is commonly used for frying, roasting, stewing meat, fish and vegetables. Foods fried in sesame oil (e.g. snack foods) have been found to have an excellent shelf-life (Maiti *et al.*, 1988).

Sesame oil and groundnut oil are used interchangeably in Asian countries as frying oils, and sesame oil is often used as a replacement for olive oil. Sesame oil is a prized oil and so attracts a high price, which has led to the development of blends with oils such as groundnut, cottonseed and rapeseed oils; these are cheaper and the blend retains some of the attributes of the sesame oil.

The great advantage of sesame oil as a frying oil is its great oxidative stability, which is greater than would be expected from the tocopherol content. The unsaponifiable matter (see table 9.11) includes sesamol and phytosterols not found in other vegetable oils. The remarkable oxidative stability of the crude oil is now attributed to the presence of the endogenous phenolic antioxidants, sesamin, sesamolin and sesamol. A review by Deshpande *et al.* (1996) suggests sesamol is generated by various processes from sesamolin to the active antioxidant. Refining and deodorising of sesame oil reduces any released sesamol, lowering the oxidation stability of the refined and deodorised oil (Kikugawa *et al.*, 1983).

The fatty-acid composition of sesame oil is predominantly oleic and linoleic acid present in nearly equal amounts, making it an attractive oil from a nutritional and health viewpoint.

The presence of sesamolin in sesame oil has meant that it has long been used in margarine and vanaspati in order to detect their use as an adulterant in butter or ghee. The sesamolin has been found to react with furfural (or sucrose) in strong hydrochlonic acid to give a strong red coloration (Baudouin test).

9.7.5   Safflower oil

Safflower oil as an edible oil has one of the highest levels of polyunsaturation, which is virtually all linoleic acid. This is a strength and a weakness in that the oil is of value nutritionally; however, its stability to oxidation is low. The oil is considered to be a semidrying oil, though the low free-fatty-acid content in the

Table 9.11 Codex standards of the Food and Agricultural Organisations and the World Health Organisation for fatty acid composition and characteristics of sesame oil (Codex Standard 26-1981. Supplement 1, 1983)

	Range
Fatty acid (%):	
C< 14	< 0.1
C14:0	< 0.5
C16:0	7.0–12.0
C16:1	< 0.5
C18:0	3.5–6.0
C18:1	35.0–50.0
C18:2	35.0–50.0
C18:3	< 1.0
C20:0	< 1.0
C20:1	< 0.5
C22:0	< 0.5
Characteristic:	
iodine value	104–120
saponification value	187–195
unsaponifiables (%)	2.0[a]
acid value (%)	
virgin oil	4.0[a]
nonvirgin oil	0.6[a]
Peroxide value (meq kg^{-1})	10.0[a]

[a]Maximum.

crude oil and the absence of gums makes the oil easy to refine; however, exposure to air throughout the process must be avoided and nitrogen blanketing of the deodorised oil is necessary. Fresh safflower oil after refining and deodorisation is bland; however, it deteriorates very rapidly.

The oil has gained some popularity in spreads claiming to be 'high in polyunsaturates'. Liquid safflower can be used in shallow frying applications, though 'skinning' of utensils is a hazard. Mayonnaise and frozen salad dressings incorporating this oil have been found to exhibit excellent appearance, flavour and odour as well as good freeze–thaw characteristics.

9.7.6 Grapeseed oil

This very pale coloured oil is supplied for cooking purposes only in the refined and deodorised form and so has a nearly bland flavour. The oil is used in those hot and cold culinary applications where oils such as sunflower oil and groundnut oil can be used. The oil is popular in wine-producing countries but has gained status in other countries as a speciality oil, largely because of its perceived nutritional qualities. The oil is low in saturated fatty acid and high in unsaturated fatty acids, particularly linoleic acid. The oil shows good oxidative stability, possibly because of the fact the oil contains virtually no linolenic acid.

9.7.7 Walnut oil

This oil is usually supplied as a cold pressed virgin oil. The oil has a characteristic 'nutty' flavour and a golden colour. Walnut oil has the advantage of being very low in saturated fatty acids but high in unsaturated fatty acids. A major disadvantage is that the oil usually contains about 10% linolenic acid, causing it to have a very poor oxidative stability, leading to the rapid development of a flat, rather unpleasant, taste. This poor oxidative stability means that the oil is rarely used in hot applications and so is restricted to being used in salad dressing, where it provides a 'round' and 'nutty' flavour.

9.7.8 Rice bran oil

Rice bran oil is a culinary oil used extensively in Japan and other rice-producing countries. Production is likely to grow in the rest of the world for the benefits of an alternative oil source with good nutritional and performance characteristics. As a general-purpose frying oil, rice bran oil has been found in model trial situations to be equivalent to groundnut oil in performance (Ortheofer, 1995). Rice bran oil, when winterised, is a good salad oil and can be used in the manufacture of mayonnaise and salad dressing.

The fatty-acid composition of the rice bran oil shows it be to similar to groundnut oil in terms of the ratio of saturated to unsaturated fatty acids, except that groundnut oil contains long-chain fatty acids. Rice bran oil has a unusually high content of unsaponifiable matter, of which stereols represent the major part. Oryzganol, which is a group of compounds containing ferulate (4-hydroxy-3-methoxy cinnamic acid) esters of plant sterols and triterpene alcohols, are also found in the unsaponifiable matter.

Oryzganol intake has been associated with decreased cholesterol absorption, decrease in plasma cholesterol and decreased platelet aggregation (Nicolosi et al., 1992). The ferulic acid esters also have antioxidant properties, which, combined with the tocopherol content of the oil, contributes to the good stability of the oil.

One of the drawbacks in the use of rice bran oil is the very high losses of oil that can occur in neutralisation. The high losses have been assumed to be due to the presence of hydroxylated compounds (Hartman and Dos Reis, 1976). The high wax content (2–5%) is also a drawback in the production of the edible oil in that it contributes to there losses. Physical refining methods have been introduced now that not only reduce total losses but also ensure retention of up to 66% of the oryzganol content.

9.8 Concluding remarks

Frying is likely to continue to be a major method of food preparation, and a widening range of foods will be manufactured so that they can be prepared for the consumer by being deep fried.

Strategies will continue to be developed to improve the efficiency of deep frying systems and to extend the frying life of the oil. This in turn will allow the use of oils with high levels of monounsaturation and polyunsaturation so that the finished food product is nutritionally more acceptable. This development is exemplified by the improved frying life of high oleic sunflower oil containing sesame and rice bran oils (Kochhar, 2000). The use of solid and hydrogenated fats will continue to decline—again on nutritional grounds and the greater ease of handling pourable frying media.

Specialist frying fats produced by plant breeding, or even genetic modification, are likely to be developed to provide for the specific requirements of the fryer. The potential of zero-calorie products such as 'Olean' (Procter and Gamble) made from Olestra (generic name for sucrose polyesters) is still somewhat in doubt. Although the product was test marketed in 1996 and ultimately launched in the USA there has been no expansion of its use internationally (Yankah and Akoh, 2001).

The evaluation of frying media as they degrade during their use is under constant consideration. Laboratory analytical techniques have now advanced to the point where the breakdown products present can be estimated accurately. At present, reliable and simple techniques for evaluating residual frying life are not yet readily available with the degree of accuracy, reproducibility or ease of use that is desired. However, the search goes on!

References

Achaya, K.T. (1997) Ghee, Vanaspati and special fats in India. In *Lipid Technologies and Applications* (eds. F.D. Gunstone and F.B. Padley), Marcel Dekker, New York, pp. 369-390.

AOCS. (1981) *AOCS C11-53*, Analytical method in the official and tentative methods of the American Oil Chemists' Society. 3rd edn. AOCS Press, Champaign, IL.

Banks, D. (1996) Food service frying. In *Deep Frying* (eds. E.G. Perkins and M.D. Erickson), AOCS Press, Champaign, IL, pp. 246-247.

Blumenthal, M.M. and Stiers, R.F. (1991) Optimisation of deep frying operations. *Trends in Food Science and Technology*, June, 144-147.

Brinkmann, B. (2000) Quality criteria of industrial frying oils and fats. *Eur. J. Lipid Sci. Technol*, **102**, 539-541.

Buck, D. (1981) Antioxidants in soyabean oil. *J. Am. Oil Chem. Soc.*, **58**(3), 277-278.

Deshpande, S.S., Deshpande, U.S. and Salunke (1996) Sesame Oil, in *Bailey's Industrial Oil and Fat Products*, John Wiley & Sons, New York, 5th edn, volume 2, pp. 466-488.

EC Regulation 2568/91 (1991) *Official Journal of the European Parliament.*

Freeman, I.P., Padley, F.B. and Shepard, W.L. (1973) Use of silicones in frying oil. *J. Am. Oil Chem. Soc.*, **50**(4), 101-103.

Ganguli, N.C. and Jain, N.K. (1973) Ghee: its chemistry, processing and technology. *J. Dairy Sci.*, **56**(1), 19-25.

GCMMF (1993) *Ghee in India: Production, Statewise Consumption and Product Profile*, Gujarat Cooperative Milk Marketing Federation, Anand 388001, India.

Gertz, C. (2000) The future of frying. *Eur. J. Lipid Sci. Technol.*, **102**, 527.

Gertz, C., Klostermann, S. and Kochhar, S.P. (2000) Testing and comparing oxidative stability of fats at frying temperature. *Eur. J. Lipid Sci. Technol.*, **102**, 543-551.

Hartman, L. and Dos Reis, M.J.J. (1976) A study of rice bran oil refining. *J. Am. Oil Chem. Soc.*, **53**(4), 149-151.

Hom, K. (1990) *The Taste of China*, Pavilion, London.

Kikugawa, K., Arai, M. and Kurechi, T. (1983) Participation of sesamol in stability of sesame oil. *J. Am. Oil Chem. Soc.*, **60**, 1528.

Kochhar, S.P. (2000) Stable and healthful frying oil for the 21st century. *Inform*, **11** (June), 642-647.

Maiti, S., Hegele, M.R. and Chattopadhyay, Sesame (1988) In *Handbook of Annual Oilseed Crops*, Oxford University Press and IBH, New Delhi, pp. 109-137.

Nicolosi, R.J., Ausman, L. and Hegsted, M. (1992) Rice bran oil lowers serum total and low density lipoprotein cholesterol and ago B levels in non human primates. *Atherosclerosis*, **88**, 133-142.

O'Brien, R.D. (ed.) (1998) Shortening types. *Fats and Oils-Formulating and Processing for Applications* (ed. R.D. O'Brien), Technomic Publishers, Lancaster, PA, pp. 397-398.

Orthoefer, F.T. (1995) Rice bran oil-healthy source of lipids. *Food Technology*, **50**(12), 62-64.

Perkins, E.G. and Van Akkeren, L.A. (1965) Heated fats IV: chemical changes in fats subjected to deep fat frying processes: cottonseed oil. *J. Am. Oil Chem. Soc.*, **42**, 782.

Sharma, R.S. (1981) Ghee: a resume of recent researches. *J. Food Sci. Technol.*, **18** (March–April), 70-77.

Vegetable Oil Products Control Order (1942) Vanaspati Manufacturers Association of India, New Delhi, Statistical data 1994.

Warner, K. (1998) Chemistry of frying fats, In *Food Lipids* (eds. C.C. Akoh and D.B. Min), Marcel Dekker, New York, pp. 167-180.

Yankah, V.V. and Akoh, C.C. (2001) Zero energy fat-like substances: olestra. In *Structured and Modified Lipids* (ed. F.D. Gunstone), Marcel Dekker, New York, pp. 511-536.

Appendix   Nomenclature for fatty acids and triglycerides

Abbreviation	Name	Carbon number[a]
Fatty acids		
L	Lauric	12:0
M	Myristic	14:0
P	Palmitic	19:0
S	Stearic	18:0
A	Arachidic	20:0
B	Behenic	22:0
O	Oleic	18:1cis
E	Elaidic	18:1trans
Lin	Linoleic	18:2
Triglycerides (*XYZ*)[b]		
X	Fatty acid at the 1 position of the glycerol backbone[c]	
Y	Fatty acid at the 2 position of the glycerol backbone	
Z	Fatty acid at the 3 position of the glycerol backbone[c]	

[a] The first number refers to the number of carbon atoms in the chain; the second number indicates the number of double bonds, followed by a description of the isomeric arrangement around the double bond.

[b] *X*, *Y* and *Z* = *L*, *M*, *P*, *S*, ...

[c] For most purposes we do not distinguish between the 1 and 3 positions.

Index

α-tending emulsifier 23, 42, 238, 242, 244,
   262, 266
α polymorph 2, 4, 5, 7, 23, 33, 34, 208
acetic acid 48, 107
acetone insolubles 251
acetylated monoglyceride (AcMG) 42, 242-
   44, 265, 266
acetylated starch 258
acid
   gelation 282
   hydrolysis 257
   moiety 2, 15
   value 85
acidic food product 105
acidifier 270
acidity 111, 207
acrylonitrile butadiene styrene 202
active oxygen method 51
acyl chain 2
added-value product 76
additive 23, 114, 230
adsorbed protein 85, 118
adulteration 178
aeration 45, 49, 86, 116, 215, 231, 246
agglomeration 36
aggregation 6, 13, 282
aging process 72, 89, 102, 110
agitation 53, 129, 137
agitator 137
agricultural practices 31
air
   bubble 42, 43, 77, 85, 89, 114, 116, 117,
      231, 233, 237, 238, 239, 260, 265
   cell 43, 44
   distribution 49, 50
   incorporation 53, 54, 244, 261, 266, 267
   phase 239
air–liquid foam 228
air–oil interface 42
air–serum interface 116
air–solid interface 235
air–water interface 12, 116, 234, 235, 239,
   261, 262

albumin 39, 70, 234
alcohol 81, 240
alcoholic product 107
alginate 83, 202
aliphatic chain 264
alkyl ester 300
all-in method 42
almond oil 353
alternative fat 166, 168, 171, 173, 174, 178,
   180, 182, 185
aluminium foil 112, 202
ambient temperature 47
amino acid side-chain 234
Ammix process 217
ammonia 199
amphiphile 229
amphoteric 230
   lecithin 229
amylopectin 258, 270
amylose 50, 51, 258
analogue cheese product (ACP) 291, 295
anhydrous milk fat (AMF) 152, 211, 320, 350
anionic 230
   emulsifier 51, 249
   surfactant 237, 241, 249
anisidine value 51
annatto carotenoids 202
annular space 36, 37, 219
antibiotic dispersion 282
antibloom 17, 20, 187
antifoaming 84, 117, 207, 228, 344
antioxidant 32, 107, 187, 202, 207, 334, 344
antispattering agent 241, 259
antistaling properties 50, 264
appearance 9
apricot kernal oil 352
aqueous
   ingredients 43
   phase 43, 53, 69, 107, 202, 231, 237, 239
arabic 257
arachidic acid 249
aroma 214, 275, 277, 300, 338
aseptic
   conditions 94

filling 95, 112
  packing 97
asymmetric molecules 8
atmospheric pressure 129
autoclave 130, 133, 136-38, 141
automation 37, 45

Bacillus cereus 99
Bacillus stearothermophillus 92
bacteria 53, 94, 98, 109, 210, 287, 297
bacterial
  cell 69, 101, 102
  polymeric product 256
bacteriological standard 53
bactofugation 101, 102
bakery
  applications 59
  fats 33, 43, 54, 66
  frying 346
  industry 37, 42
  ingredient 31, 37, 39-42, 45, 59
  product 30, 40, 41, 44, 47, 66, 152
baking 43
batch method 87, 90, 220
batter 41, 43
  aeration 38, 41, 54
  density 266, 267
  preparation 41, 43
  stability 53
benzoic acid 212
berage oil 352
binary mixture 6, 7, 17
Bingham's concept of plasticity 208, 269, 270, 271
biosynthesis 213
biscuit 39, 40, 61, 187
  coating 61, 62
  digestive 62
  dough 62-64
  filling 64, 65, 183
  ginger-snap 62
  industry 186
  hard sweet 63
  laminated 64
  manufacture 38, 61
  sandwich-type 183
  semisweet 63
  shortbread 40, 62
  Shrewsbury 39, 40
  Viennese 39, 40
  wine 40
bixin 202

blackcurrant oil 352
blade 36, 147, 219
bleaching 138, 145, 250
blending 10, 11, 30, 33, 37, 123
bloom 9-11, 16, 19, 20, 41, 61, 63, 66, 182, 183, 186, 187
  formation 17
  inhibition 19, 175, 187 *see also* antibloom
  stability 15
bodying agent 228, 257
boiled paste 38, 40
boxed fats 61
brassidic acid 133
bread 193, 241
  dough 38
  improver 30, 60
  manufacture 54
  staling 50, 264
  type 72
  volume 30, 265
brine 64, 199
brittleness 43, 46, 48, 50, 52, 54, 62, 198, 209
Brownian movement 240
BSE crisis 196, 200
Buchner filter 147
bulk fats 8
bulk flow 43
bulking agent 228
BUSS loop reactor 139
butter 30, 37, 38, 41-45, 47, 48, 61, 70, 73, 82, 160, 167, 184, 192, 193, 200-202, 211, 216, 217, 220, 222, 259-61, 267, 333, 338, 347, 348
  blend 42
  churn 220
  consistency 215
  emulsion 215
  fat 34, 147, 152
  fractionated 47
  hard 60, 193
  making 213, 220
  milk 80, 82, 214, 222, 269
  mountain 152
  oil 42-44, 47, 79, 82, 152
  powder 59
  refrigerated 47
  sweet cream 215, 347
butter-like product 73, 76, 201
butylated hydroxyanisole (BHA) 334, 345
butylated hydroxytoluene (BHT) 334, 345

INDEX

CaCO₃ 237
cake 30, 38, 41, 42, 44, 45, 49, 56, 193, 241, 265, 266
    batter 49, 50, 231, 244, 256, 266
    emulsifier 264
    formulation 244
    margarine 43, 44
    madeira 41
    mix 30, 60, 245, 265
    texture 43, 50, 51
    volume 44, 50, 51
calcium 111-14
    alginate 257
    caseinate 318, 319
    chelation 292
    salt 51, 70, 83, 250
    stearate 250
    sulfate 260
calcium-to-casein ratio 277
Canadian Food and Drug Regulations 272
canola oil 106
caramel 160, 179
carbohydrate 70, 332
carbon dioxide 43, 51, 231
carbon steel tube 219
carboxylic acid group 257
carboxymethyl cellulose (CMC) 83
cardboard 202
carotene 224
carotenoid 71, 202
carrageenan gum 83, 257, 269, 271
casein 70, 72, 111, 113, 117, 212, 282-84, 290, 293
    aggregation 302
    concentration 307
    dehydration 287
    glycomacropeptide 303
    hydration 318
    micelle 282, 287
caseinate 222
casein-to-fat ratio 287
catalyst 120, 123-27, 134, 137, 138, 245, 247
    filtration 138
    slurry dosing system 138
cation 237
cationic cetyl trimethyl ammonium bromide 230
cell lamella 2
cellulose 259
    fibre 237, 265
    gum 269
centrifugal

filtration 151
force 74, 219
separation 101, 213
centrifugation 322
centripetal pump 76
cetyl trimethyl ammonium bromide (CTAB) 230
chain length 3, 16, 240
chain–chain interaction 5, 6, 15
Channel Island cattle 72
cheese 275-78, 281-331
    blue 300
    Cheddar 277, 278, 285, 288, 290, 292, 297, 298, 300, 302, 305, 319, 321
    composition 278, 282, 310
    French soft 300
    heated 295, 310, 313, 314, 319
    low-fat 277, 293, 297, 299, 305, 315
    making 277, 302
    manufacture 290, 295
    microstructure 287, 291, 292, 295
    milk 281, 295, 302, 321, 323
    mozzarella 277, 278, 287, 290, 293, 302, 309, 310, 311, 317, 321
    Parmesan 300, 306
    powder 275
    processed 275, 291, 306
    quality 282
    ready-to-use grated 275, 306
    rennet-curd 287, 293
    ripening 290, 297
    texture 304, 310
    yield 301-303
chemical refining 32
cherry kernal oil 352
chewiness 313
chilling drum 48, 64
Chinese cooking 338
chloride 83
chocolate 8, 9, 11, 15-17, 19, 60-62, 66, 152, 159, 160, 163, 165, 168, 170, 175, 177, 179-83, 187, 188, 234, 248
    coating 235, 248
    spread 192
cholesterol 67, 71, 169, 194, 196
choux paste 40
churning 152, 197, 212, 215, 220, 259, 260
cinnamon oil 257
cis isomer 124, 131
citric acid 48, 107, 261
citrus oil 257
clarification 74

cleaning-in-place (CIP) 94, 151
climate 31
cling 259
Clostridium botulinum 92
cloud point 144
clouding 271
clumping *see* flocculation
coagulation 12, 43
coalescence 13, 43, 53, 233, 235-38, 240,
    260, 261, 290, 293, 297, 302, 318, 323
coating 60, 235, 267, 268
cocoa butter 2, 3, 8-11, 15-19, 25, 152, 153,
    159-62, 165, 166, 168, 173, 180, 182,
    187, 235
    equivalent (CBE) 15, 16, 151, 153, 166,
        168, 178, 185
    replacer (CBR) 15, 16, 151
    substitute (CBS) 170
cocoa
    liquor 179
    mass 9
    powder 86
cocoa-based coating 234
coconut oil 34, 56, 159, 170, 171, 175, 200,
    223, 262, 333
Code of Food Regulations 273
Codex standards 207
coffee 111, 114, 115
    cream 11, 60, 80, 111-13, 263
    whitener 80, 246, 263, 264
cold pressing 352
cold-water paste 40
collected phase 236
colloid mill 102, 107
colloidal
    phase 70
    solution 69, 70
colony-forming unit 96
colour 86, 117, 194, 198, 199, 200, 207, 338
colour test 200
colouring 105, 201, 202
comminution 307
complector 48, 199
compound formulation 6, 52
concentric tube 36
conching 10, 180
condensation 240
condensed phase 230, 231
condiments *see* sauces
cone penetronomy 209
confectionery
    fat 3, 17, 61, 151-53

industry 186
market 17
product 66, 159, 160, 175, 179, 181, 183,
    184, 188, 267
confocal laser scanning microscopy
    (CLSM) 288, 294, 296
conical sieve centrifuge 150, 151, 153
consistency 9, 13, 38, 87, 103, 107, 123, 209,
    228
consumer 1, 349
contact angle 234, 235
contamination 96, 188
continuous
    column chromatography 154
    flow method 96
    metering system 53
    phase 9, 13, 230, 231, 233, 235, 236, 240,
        267
    plate pasteurisation system 91
control valve 76
convection current 43
convenience food 60
cooking
    ingredient 271
    oil 333
    performance 312
cooling 12, 314
    curve 161
    rate 4, 12, 20, 36, 144
corn oil 106, 125, 143, 250, 333, 338
Cornish pasty 40
cottonseed oil 14, 32, 34, 106, 123, 141, 200,
    250, 333, 338
cow 72, 73, 220
cows' milk, 72, 82, 96, 347
crackers 64
cream 43, 64, 65, 69, 70, 73, 74, 76, 77, 81,
    82, 86, 90, 94, 97, 98, 100-105, 109,
    111, 113, 114, 152, 160, 193, 211,
    212, 214, 217, 259, 310, 347, 348
    categories 78
    cheese 192
    cultured 214
    dairy 73
    dessert 79
    fillings 62, 64, 183
    high-fat 110
    liqueur 81, 112
    low-fat 111
    nondairy 79, 80
    plug 110
    preparation 115

pumping 115
recombined 79
ripening tank 214
single 78, 100-12
whipped 11, 12, 69, 79, 111, 114-16, 234, 260, 261
yield of 77
creaming 12, 39, 40, 42, 43, 71, 110, 235-38, 240
creep compliance 305
critical micelle concentration (CMC) 232
cross-linked starch 258, 270
crude oil 32
crumb
    firming 241, 264
    grain 265
    structure 54
    texture 41, 42
crust browning 53
crystal 1, 8, 10, 12, 13, 25, 36, 38, 42, 66, 70, 234, 333
    aggregate 42
    formation 89
    growth 10, 12, 22, 37, 143, 256
    growth rate 22
    inhibitor 107, 268, 269, 271
    modifier 263
    morphology 1, 3, 6, 12, 147, 186
    network 8, 12, 13, 25, 36
    nuclei 43, 59
    orientation 12
    polymorph modifier 245
    size distribution 8, 12
    structure 19, 31, 33, 36, 37
crystalline 2, 238, 265
    fat 38
    gel structure 49
    monoglyceride 49
    properties 30, 43
crystallisation 2, 6, 11-14, 16, 17, 20, 21, 25, 31, 35-37, 48, 52, 72, 84, 107, 142, 144, 145, 180, 209
    acceleration 60
    column 154
    kinetics 1
    performance 34
    rate 20, 21, 142, 143
    temperature 4, 20, 21, 37
crystallised
    butter 347
    starch 258
crystalliser 147

β crystals 55
β' crystals 55
cubic structure 49
curd 282, 290, 293
    granule junction 287
cutting method 209
cyclone 58

dairy product 73, 269
decaglycerol monostearate 254, 267
de-emulsification 9, 12, 260
deep-fat frying 152, 332, 337
degumming 145, 250
delta lactone 201
demolding 10
densitometer 76, 77
density 71, 73, 236
    measurement 77
    transmitter 77
deodorisation 32, 125, 160-62, 197, 213, 333
deodorised oil 52
desi method 347, 349
detergent 228
dewaxing 333
dextrin 58
diacetyl 201
    tartaric acid 48
    tartaric ester 51, 55
    tataric ester of monoglyceride (DATEM) 242-45, 249, 265
diacylglycerol 14, 20, 25, 211
diene 124, 136
diet 1, 72, 200, 276
dietary fibre 66, 234
differential scanning colorimetry (DSC) 18, 20, 21, 25, 161
diffusible-calcium-reduced (DCR) 112
digestion 47
diglyceride 44, 85, 201, 241, 243, 264
dilactylic dimer 249
dilatometry 52
dilution 19
dimethyl polysiloxane 344
1,2-dipalmitoyl-3-acylglycerols (PPn) 12
1,3-dipalmitoyl-sn-2-oleoyl-glyerol (POP) 14
1,3-dibehenoyl-sn-2-oleoyl-glyerol (BOB) 11, 177, 187
disappearance, rate of 128
discontinuous phase 230, 235
dispersed (solid) phase 70, 119, 235, 236
dispersion 80, 104, 228, 233-37
displacement 311

distilled monoglyceride 79, 211, 241, 242,
      245, 264
divalent ion 237
divert valve 91
docosahexaenoic acid (DHA) 225
dodecanol 240
dosing 56
double
      bond 4, 124
      chain-length structure (DCL) 2, 3, 7, 14,
        16, 18
      homogenisation 104
dough 39, 40, 46-48
      conditioning 48, 54
      strengthener 55, 242, 249, 264, 265
draining 239
dried gum 238
drop, rate of 140
droplet
      cross-section 233
      size 20, 57, 232, 233, 236, 238, 270
droplet–droplet interaction 20
drum 216
dry
      foam 239
      ingredients 56
      mix 234
drying chamber 58
DSC melting peak 17

eating habits 30
eating quality 38, 40, 48, 52, 83
eccentric shaft 219
edible
      gum 105, 107
      protein 207
egg 105, 265, 270
      albumin 234, 270
      content 45
      protein 43
      white 234, 238, 239, 249, 250, 271
      yolk 86, 106, 201, 212, 228, 250, 252,
        270, 271, 335
eicosapentaenoic acid (EPA) 225
elaidic acid 5, 173, 242
elastic
      deformation 304
      shear modulus 306
elasticity 208
electrical
      conductance 209
      double layer 237

heating 92
potential 237
repulsion 237
elongation 290
emulsification 12, 21, 38, 58, 197, 233, 267,
      281, 319
emulsified fat globule (EFG) 282
emulsifier 11, 12, 14, 20, 42, 44, 45, 48-51,
      52, 55, 56, 60, 79, 80, 84-87, 89, 201,
      207, 211, 212, 215, 228, 230, 232-35,
      238, 240, 241, 247, 248, 252-55, 257,
      261, 267, 272, 335, 338
      molecule 232
      salt 228
      system 30, 44, 54, 60, 71
emulsifier–stabiliser system 66, 87
emulsifying salt 292
emulsion 8, 12, 13, 20-22, 53, 54, 59, 66, 69,
      70, 82, 107, 109, 110, 116, 209, 233,
      235-38, 240, 244, 270, 271
      droplet 58, 59
      instability 13
      preparation 213
      stability 12, 109, 116, 210, 238, 239
      structure 208
      technology 30
endospore 96
energy 230
English method 46, 48
enriched sponge 45
enrobing 11, 267
entrained air 77
entrapped air 235
enzyme 71
enzyme-modified cheese (EMC) 275
equilibrium 230
erucic acid 133
ESL process 101
esterification 21, 48, 245, 248
ethoxylated monoglyceride (EMG) 242, 243,
      244, 246, 247
EU Chocolate Directive 175
EU Yellow Fat Spreads regulation 205
European Parliament and Council
      Directive 273
eutectics 52, 181
      behaviour 36
      effect 17
      interaction 184
      phase 6, 7, 16, 18, 19
      point 16
exothermic 136

extracellular proteases 94
extrusion 89, 209

fast food 343, 345
fat
    blend 9, 65, 213
    breakdown 300
    content 275, 276, 285, 297, 298, 315
    crystal network 2, 11
    distribution 49, 295, 297
    formulation 57
    globule 38, 44, 72, 81, 85, 102, 109, 113,
        115, 116, 118, 260, 282, 287, 288, 302
    globule membrane 38, 71, 78, 110, 212,
        263
    layer 47
    mixture 6
    powder 30, 56, 57, 60
    recovery 302
    replacer 80
fat-blend recipe 38
fat–casein interface 287, 288
fat-free
    dressing 257, 259
    sponge cake 49
fat-globule
    diameter 282
    membrane 282
fat–protein
    interaction 118
    interface 59
fatty-acid 16, 17, 34, 48, 50, 84, 128, 153,
        241
    composition 12, 134, 242
    content 51
    methyl ester 52
    moiety 5, 12, 25
    salt 254
fatty alcohol 240
fat–water emulsion 53
feathering 111, 114, 264
fermentation 48, 51, 214
filler 83, 86
film 237
    breakage 240
    cake 143, 144, 147
filter cloth 148
filterability 144
filtering system 74
filtration 236
fine emulsion 210

finished goods 37
firmness 33, 36, 37, 54, 277, 308
fish oil 38, 124, 200, 216, 225
flaked fat 56, 57, 60
flavour 30, 32, 38, 40, 44, 45, 47, 53, 86, 97,
        105, 107, 117, 160, 180, 182, 184,
        201, 207, 209-11, 213, 214, 224, 257,
        277, 283, 299-301, 332, 336, 338, 348,
        349
    taint 214
flocculation 235-38
floccule 110
Florentine filter 145, 147
flotation 73
flour 39, 41, 55, 241
flour–water dough 39
flow
    properties 57, 248, 311
    rate 37
    resistance 313
    shear effect 233
    velocity 8
flowability 235, 313, 315, 317, 320, 321
fluid 38
    batter 43
    shortening 55, 56, 66
foam
    destabilisation 239
    stability 239, 240
foaming 42, 69, 70, 79, 82, 84, 114-17, 234,
        235, 237-39, 260, 261, 266
folded carton 202
folic acid 83
fondant 60
Food and Drug Administration (FDA) 207,
        228
food
    emulsifiers 9, 240
    poisoning 53
    safety 89
    service industry 275
    standards code 175
    system 228
    technology 229
form
    IV 10, 163
    V 10, 11, 163, 180, 186, 187
    VI 163, 186, 187
formula 84
Fourier transform Near Infra Red
        Analysis 141

fractionation 17, 31, 33, 65, 123-25, 141, 142,
      144, 145, 147, 151-53, 166, 169, 188,
      333
fracture
      strain 308
      stress 308
fracture-related properties 307
frame press 148
free
      amino acid (FAA) 283
      fatty acid (FFA) 78, 79, 238, 348
      oil 318
free-fat content 110, 114
free-flowing powder 59, 238
freeze–thaw stability 257, 356
freezing 87
freezing point 83
French dressing 106, 119, 335 see also
      mayonnaise, salad dressing
French method 46
French rolls 197, 198
freshness 96
fried food 338
frozen product 256
frozen state 69
fruit
      filling 257
      product 39
frying
      media 55, 66, 339
      oil 339, 340-42, 344, 345
functional
      food 66
      group 253
      properties 37, 89, 310, 313, 314
functionality 30, 38, 54, 61, 66, 252, 310

gas
      chromatography 52, 178, 242
      phase 116, 230, 234
      volume fraction 238
gateaux base 45
gear pump 219
gel 50, 110, 115, 240, 258, 270
gelatine 83, 118, 202, 212, 319
gelatinisation 40, 43, 45, 258, 270
gelation 240, 302
gel-forming gum 257
gel-forming protein 282
generally recognised as safe (GRAS) 272
genetic modification 67, 196, 200
Geotrichum candidum 300

ghee 192, 216, 332, 333, 347-50
gliadin 39
globulin 39, 234
gloss 169, 246
glucose 259
      ester 269
gluten 39, 40, 46, 51, 62, 63, 265
gluten-forming protein 45
glutenin 39
glyceride 143, 208, 209, 224
glycerin 241
glycerol 48, 230, 241, 247
      carbon 4
      group conformation 15
      monostearate (GMS) 79, 240, 247, 264
goats' milk, 72
gossypol 33
grading 211
graining 16
Gram-negative psychotroph 94
granular crystal 13, 14
grapeseed oil 352, 356
gravitational
      constant 236
      effect 237
gravity 237, 239
grinding 10
groundnut oil 34, 200, 333, 338, 352, 353
gum 228, 236, 238, 240, 250, 256, 257, 267,
      335
gumminess 54
gum-stabilised emulsion 238

halvarine products 194, 196, 202
hardness 8, 13, 43, 50, 111, 124, 184, 197,
      308
hazelnut oil 353
health 169, 276, 339
heart disease 33, 66, 276
heat
      exchanger 36, 55, 87, 90-92, 100, 137,
            145, 199, 218, 219
      treatment 81, 89, 91, 92, 98, 101
heating cycle 90
heat-resistant organism 96
heat-transfer process 142, 339
Henry's law 140
heptane 232
hermetic centrifuge 101
heterogeneous interaction 10, 22
n-hexadecane 25
hexagonal structure 49

high rotation speed 219
high-melting
    fat 1, 25, 168
    fraction 20, 107, 123
    triglyceride 260, 271
high-melting-point 13
    glyceride 71
high-molecular-weight
    emulsifying agent 236
    polysaccharide 256
high-pressure pump 102
high-ratio
    cake 44, 45, 50, 54, 241, 265
    shortening 49
high-shear mixer 102, 107, 115
high-temperature short-time (HTST) 90, 91
high-trans fat 159, 168, 173, 174, 179, 182, 183
high-viscosity gum 240, 257
Holstein milk 72
homogenisation 12, 37, 79-81, 87, 102-104, 111-15, 236, 263, 281, 282, 295, 307, 309-11, 317, 320, 321, 322
homogeniser 102-104, 112, 115
hopper 48
Horiuti–Polanyi mechanism 134
horizontal isosolid line 52, 173
hydrated monoglyceride 241, 250
hydration 39, 49
hydraulic pressure 160, 322
hydrocarbon 34, 49, 229, 253
hydrocolloid 83
hydrogen 125, 128-31, 133, 134, 137-40
    bond 55, 236
    dissolution 133, 138
    peroxide 250
hydrogenated
    fat 240
    sunflower oil 223
hydrogenation 33, 55, 57, 61, 123-26, 131, 133, 137, 138, 140, 153, 334
hydrolysis 78, 269, 276, 340
hydrophilic 234
    base 59
    chain 234, 239
    head 71
    layer 85
    part 229, 230, 238
    particles 11, 72
    portion 252
hydrophilic–lipophilic balance (HLB) 84, 85, 211, 240, 245, 247, 248, 252-55, 267

hydrophobic
    emulsifier 20, 25
    fatty acid 239
    flavour oil 257
    interaction 237
    part 229, 234
    polyglycerin ester 25
    segment 262
    side-chain 234, 239
    tail 71
hydroxypropyl methyl cellulose 259
hygiene conditions 74, 89, 90
hypothiocyanite group 99

ice cream 11, 12, 60, 69, 70, 82, 83, 85-87, 89-91, 102-105, 109, 114, 116-18, 152, 175, 257, 260, 262, 263
ice crystals 82, 118
icing 241, 246, 267, 346
immiscible fluid 70
impervious layer 38
impurity 32
Indian cooking 337
indigenous lipase enzyme 78
industrialisation 31
instant desserts 60
interesterification 9, 15, 33, 66, 123, 124, 153, 197, 223, 245
interface 228, 232
interfacial
    film 238, 261
    force 233
    free energy see surface free energy
    region 229
    tension see surface tension
intermolecular interaction 116
interstitial cavity 150, 208
inulin 66
iodine value 52, 124, 242, 345
ion exchange 112
ionic
    bond 236
    repulsion 236
    strength 230, 237-39
    surfactant 237
isomer 241
isomeric formulation 6
isomerisation 124, 133, 136
isosolid 36
    diagram 52, 173

jelly 257

jellying agent 228
juice drink 109

kinetics 96, 125, 133
knife rotor 219
Krafft point 49

lactalbumin 212
lactation 72
lactic
    acid 48, 214, 249
    acid ester 50, 245, 261
    acid monoglyceride 55
    butter 214
    cream 215
    culture 300
β-lactoglobulin 116
lactoperoxidase system (LPS) 99
lactose 53, 117
lactulose value 101
lactylated monoglyceride (LacMG) 242, 261,
    266
lactylated monostearin 262
lactylic
    acid 249
    tetramer 249
    trimer 249
lamellae 118
    mesophase 264
    structure 49
laminating fat 46, 48
lamination 47, 63
lard 34, 40, 61, 66, 123, 242, 333, 338, 346
large strain deformation 306
lauric
    acid 20, 67, 170, 225
    fat 16, 60, 64, 159, 168, 171, 173, 179,
        181-83, 200, 262
lauryl sulfate 229-31
layering fat 47, 48, 63
leavening 267
lecithin 72, 79, 86, 201, 212, 235, 250-52,
    259
legislation 202
lemon juice 106, 107
lethality value 92
lifestyle 30
light
    microscopy 295
    reflectance 80
lightness 116

linoleic acid 5, 126-128, 134, 136, 165, 225,
    242, 300, 335
linolenic acid 55, 124, 127, 225
lipase 94, 301
lipid 229
    bilayer 49
lipid–starch form 264
lipolysis 276, 300
lipolytic bacteria 53
lipophilic 84, 229, 230, 234, 252
lipoprotein lipase 290
liquefaction 297
liquid 235, 237
    ammonia 64
    crystalline mesophase 49
    emulsion 36, 238
    fat 38
    food 228
    nitrogen 141
    shortening 54, 244
liquid–liquid emulsion 228
liquid–solid phase change 208
long-chain
    fatty acid 15, 17, 72, 230
    polyunsaturated fatty acid (LCPUFA) 225
long-life
    cream 91
    milk 104
low coloured edible oil 31
low-calorie product 194
low-fat
    content 299, 312, 315
    dressing 257, 259
    spread 202, 205, 208, 210, 211, 222
low-iodine-value fat 214, 215
low-melting
    fat 168
    fraction 123
low-melting-point 13, 17
low-trans-fatty-acid margarines 223
low-viscosity gum 257, 267
lubrication 84
lubricity 38
lypolytic action 71

macadamia nut oil 352
machine wrapping 202
macromolecule 87
macrostructure 8
magnesium 83, 112
Maillard reaction 184
maize oil 200

maltodextrin 262

margarine 8, 13, 30, 31, 36-39, 41-44, 46, 48,
49, 51-53, 56, 123-25, 153, 192-94,
200-202, 205, 207, 211, 225, 241, 259,
338
commercial 44
emulsion 48, 215
manufacturing 241
vegetable 47

marine oil 38

mass
flow 208
transfer 125, 133

mass-transfer equation 140

mastication 275, 299

maturation 290, 297

maxima point 36

mayonnaise 69, 105, 107, 108, 119, 120, 212,
226, 250, 269, 271 *see also* French
dressing, salad dressing

meadowfoam oil 352

meat
filling 40
product 39, 40, 193

mechanical
agitation 37
energy 230, 232
force 260
stirring 232
stress 46
working 209, 217

medium-chain-length triglyceride 67

medium-melting-point 13

medium-viscosity gum 257

melange 194, 202

melt resistance 311

melting 2, 9, 11, 12, 15, 35, 123, 208
curve 63, 167
enthalpy 17
point 3, 5, 6, 13, 14, 16, 17, 33, 35, 36,
48, 56, 224, 238
properties 36, 83, 117, 170, 313
temperature 20

membrane
filter 101, 153
plate 148
press 147, 148, 150

membrane–membrane interaction 260

meringue 257

meta-stable form 2, 10

methyl
end stacking 15

ketone 276, 300

methylcellulose 259

methylsilicone 107

micelle 49, 111

microbial
cell 101
contamination 53
enzyme lipase 78

microbiological quality 53, 69, 79, 81, 94, 96,
115, 117, 212

microcomponent 70

microcrystal 269

microcrystalline cellulose 259, 269, 271

microencapsulation 58, 60

microfiltration 101

microfluidation 281, 282

microorganisms 96, 109, 213

microscopy 209

microstructure 8, 9, 78, 116, 287, 291, 292,
295

middle-melting fraction 17

migration 66, 186

milk 16, 60, 69, 70, 72, 74, 76, 82, 94, 97, 99,
100, 102, 109, 115, 212, 259, 307,
310, 319
condensed 82, 83
cultured 201, 202
evaporated 82
fat 6, 15-20, 44, 59, 71, 79, 80, 152, 159,
160, 165, 205, 260, 300
gel 283, 287
poor quality 78
powder 9, 11, 58, 59, 82, 202
product 207
production 96
protein 60, 83, 184, 212
raw 79, 94, 96, 236
ripening 202
salts 83
semi-skimmed 76
separation 74, 75, 89, 109, 115
silo 109
skimmed 53, 74, 76, 78, 79, 90, 101, 102,
109, 114, 115
solid 53, 79, 80, 117, 338

milk-fat globule diameter 213

milking equipment 96

milk-protein-stabilised cream emulsion 81

mimetics 277

minarine 207

minima point 36

minor lipids 1, 19, 20

miscible phase 6, 7, 16
mixed-acid triacylglycerols 4, 5, 6, 21
mixing 10
    energy 233
    high-speed 42
    phase 7
    single-stage 42
modification 30, 36
β' modification 43
moisture 265, 318
    content 53, 61
    ingress 39
moisture-to-casein ratio 302
moisture-to-protein ratio 277, 303, 318
molecular
    compound crystals 7
    distillation 241, 245
    packing 34, 35
    size 35
    weight 253
molecule 228, 229
molten fat 59
monoacid
    triacylglycerols 4, 5
    triglyceride 34
monoacylgycerol 211
monodiglyceride 50, 52, 55, 59
monoene 124, 133-36
monoene-to-diene ratio 131, 133
monoester 246
monoglyceride 44, 49, 50, 55, 85, 201, 228, 230, 240-42, 244, 245, 254, 259, 264, 265, 267
monolayer 12
monooleate 245, 246, 261
monostearate 261
    derivative 246
monostearin 241
monounsaturated fatty acid 5, 30, 66, 126, 131, 201
monovalent ion 237
morphology 2, 4, 8
mould 53, 89, 109, 300
mouth feel 8, 37, 41, 45, 47, 63, 102, 123, 152, 170, 180, 235, 258, 263, 267, 299
multiphase system 43
multiplex roller 197
multi-triacylglycerols 2
Mycobacterium tuberculosis 96
myristic acid 67, 170, 249

nanionic 1-monoglyceride 229

native milk fat globule membrane (NMFGM) 290
natural fat 25
neat liquid 4
needle-like crystals 4, 10
negative charge 230, 237
network 1, 3
neutralisation 145
Newton 208, 304
nickel 138
noncalorific fat substitute 248
nondairy 80
nonfat
    dry solids 60
    milk solids 117
    solids 79, 80, 83
nonionic 230
nonpolar 229, 234
nonstarter
    bacteria 298
    lactic acid bacteria (NSLAB) 297, 299
nonsterile environment 94
nontempering 168
nonvolatile degradation product 341
Novagel™ 293
novel oil 30
nozzle separator 150
nuclear magnetic resonance (NMR) 52, 56, 141
nucleation 4, 10, 12, 20, 22, 24, 25, 36, 89
nut oil 347, 352, 353
nutrient 69
nutrition 30, 31, 86, 339

obesity 276
odour *see* aroma
off-flavour *see* rancidity
oil
    blend 33, 43, 48
    droplet 107, 230-34, 236-38, 270, 271
    exudation 209
    phase 236
    solubility 238, 250
oil–egg emulsion 270
oil–fat separation 13
oiling off 110
oil-in-water (O/W) emulsion 9, 11-13, 20-25, 43, 69, 70, 80, 84, 85, 87, 105, 106, 110, 111, 113, 212, 226, 233, 236-38, 252, 260
oilseed 31
oil-soluble flavour 210

oil–water interface 42, 116, 230, 238, 244, 257, 259
oil–water mixture 232
oleic acid 5, 20, 165, 173, 225, 242, 335
    moiety 15
olestra 248, 358
oligosaccharide 257
olive oil 106, 125, 143, 200, 216, 333, 335-37, 353
omega-3-rich nonhydrogenated fish oil 216
organic
    solvent phase 232
    spread 196
organoleptic characteristics 83, 86, 94, 97, 100, 114
orzyganol 333, 357
OSET index 344
OSO mixture 7
overrun 79, 102, 114, 115, 117, 261
oxidation 64, 81, 99, 107, 120, 187, 276, 300, 340
    resistance 61
    breakdown 344
    stability 55, 61, 224, 333, 344

Pacillac process 217
packaging 117, 202, 203
packing machine 199
palatability 31, 33, 83, 117, 209
pallet 202
palm
    fraction 33
    kernal oil 25, 34, 56, 64, 151, 170, 171, 183, 185, 200, 223, 262, 333
    mid-fraction (PMF) 2, 20, 22-24, 147, 151, 153, 167-69
    mid-fraction (PMF)/water emulsion 23, 24
    oil 2, 6, 13, 14, 20-23, 31, 34, 37, 38, 64, 66, 143, 145, 152, 166, 167, 200, 201, 223, 333, 338, 346
    olein 21, 22, 64, 131, 132, 141, 147, 151, 153, 169, 188, 223
    stearin 21, 22, 33, 38, 141, 147, 151, 223
palmitic acid 20, 22, 67, 165, 225, 242, 249
pan frying *see* shallow frying
para-casein matrix 277, 287, 288, 290, 292, 293, 295, 302, 304, 311, 318
parenteral emulsion 282
particle
    diameter 236
    distribution 12

size 12, 41, 42, 70, 103, 120
particle–particle interaction 13
particle-size reduction 102
passion fruit oil 352
pasteurisation 12, 87, 89-91, 94, 96-98, 101, 102, 109, 213, 214
    temperature 277
pasteurised processed-cheese product (PCP) 275, 277, 291, 295
pastry 193
    Danish 48
    flaky 30, 38, 46
    manufacture 46
    puff 30, 38, 46-48, 63
    savoury 39-41
    short 30, 38-40, 44, 56, 62
    sweet 40
    unsweetened 40
pasty 39
pathogen 69, 96
peanut butter 60
Pearson's rectangle 76
penetration method 19, 46
Penicillium camemberti 300
penicillium lipase 300
Penicillium roqueforti 300
percentage solids 57
performance 36, 44, 52
permeability 290
peroxide value 51, 348
pesticide 160
pH sensitivity 257
pH value 53, 69, 213, 230, 254, 258, 269, 270, 277, 278
phase
    behaviour 2, 6, 35, 52
    diagram 36, 49
    inversion process 215, 217, 259
    separation 3, 7
phosphate 83
phosphate ester 258
phosphatidyl choline 230
phospholipid 20, 71, 72, 160, 212, 250, 251
physical
    chemistry 229
    properties 1, 4, 6-9, 15, 31, 35, 83, 123
    refining 32
phytosterol 32, 66, 200
pistachio nut oil 352
plastic 38, 43
    behaviour 36, 47
    container 112, 193, 202

properties 37, 41, 42, 46, 48
    range 44
plasticisation 290
plasticised fat 42, 43, 54, 344
plasticity 9, 13, 48, 208, 210
plate press 148
Plateau's border 239
platelet structure 8
poison 138
polar 229
polar lipids 20
polyacrylamide gel electrophoresis
        (PAGE) 283, 284
polyene 133
polyethylene 202
    container 107
polyglycerol 254
        ester 55, 107, 247, 269
        polyricinoleate (PGPR) 248
polyhydric alcohol 84
polymer 228
polymerisation 240, 340
β polymorph 2-5, 10, 13, 14, 33-36, 40, 41,
        173, 187, 208
β' polymorph 2-5, 7, 10, 13-15, 33-36, 41, 42,
        168, 174, 208
β_1 polymorph 208
β_2 polymorph 11, 208
polymorphic
    change 186
    form 35, 208
    properties 1, 3, 5-7, 13, 14, 19, 23
polymorphism 2, 4, 6, 8, 10-12, 14, 20, 25,
        31, 33, 163, 165, 168
polyoxyethylene chain 265
polyoxyethylene derivative 238, 246
polypropylene 202
polysaccharide 87, 256, 257
polysorbate 234, 246, 247, 254, 261, 262,
        264, 266, 267
polyunsaturated fatty acid 30, 63, 66, 131,
        194, 200, 201
polyvalent alcohol 48
polyvinyl chloride 202
POO 15
POP 7, 8, 14, 21, 159, 163, 179
POS 159, 163, 179
positional
    isomer 244
    selectivity 126, 128
positive charge 230
positive displacement pump 102

postcrystallisation 4, 48, 197
post-production handling 96
potassium 83, 254
potato crisps 345
pourable
    product 79, 257, 268
    shortening 55
powder 102
powdered
    fat 30, 56, 57, 60
    product 60
PPO 7, 14, 15
PPP 4, 7
pre-emulsification 12
pregastric esterase 301
preservative 207, 210, 212
pressure 231
    differential 91, 231
primary hydroxyl (3-hydroxyl) of
        monoglyceride 242
propylene glycol 48, 59, 245
    alginate 257, 269
    ester 50, 257
    monoester (PGME) 42, 242-44, 266
    monostearate (PGMS) 55, 232, 238, 247,
        262
protein 12, 39, 42, 70, 71, 84, 89, 116, 212,
        228, 234, 236, 237, 239, 332
    chain 234
    desorption 260, 262
    matrix 292
    membrane 259, 260, 282
    molecule 239
    stabiliser 239
protein–protein interaction 116
protein-stabilised foam 244
proteolysis 278, 283, 284, 286, 290, 318
pseudo-plastic material 208
psychotrophic spore 96
pure emulsion 24

quality indicator 341, 342, 349
quiche Lorraine 39, 40

raffinose 67
rancidity 61, 82, 120, 187, 188, 212
Rancimat test 52, 125
random fluctuation 240
rapeseed oil 34, 38, 55, 64, 106, 125, 133,
        138, 200, 220, 333, 334, 338, 345
ready-prepared meal 275
recipe 45, 47, 49

reconstituted low-heat skim milk powder (RSMP) 320
recording method 91
recrystallisation 66, 264
rectangular needle morphology 13
redispersion 236
refining 32, 179 *see also* chemical refining, physical refining
refrigerant 36, 54
regulation 272
    EC 336
    national 82, 90
rehomogenisation 12
rehydration enhancer 228
rennet 282, 301
retardation 19, 20
retrogradation 258, 264
reverse osmosis 82
reversed hexagonal micelle 25
rheological properties 1, 2, 8-12, 15, 20, 37, 208, 268, 275, 304, 307, 309, 311
rice bran oil 143, 333, 357, 358
rich paste 39
ricinoleic acid 5
rotor speed 220
'rubbing in' method 39

saccharide 257
safflower oil 14, 352, 355, 356
salad
    cream 70
    dressing 69, 105, 107, 109, 119, 120, 124, 125, 143, 236, 254, 257, 258, 268-70, 333, 337 *see also* French dressing, mayonnaise
salt 52, 70, 83, 87, 198, 199, 207, 212, 230, 303
    concentration 238
    water-soluble 107
saponification number 85
saturated
    emulsifier 262
    fatty acid 4, 5, 15, 17, 41, 49, 66, 124, 201, 240, 241
    mixed acid triacylglycerols 6, 7
    mono-acid triacylglycerols 7
    monoglyceride 50
saturation 136
sauces 193, 257, 259, 271, 275, 335
savoury product 39-41, 64
scanning electron microscopy (SEM) 119, 288, 289, 292

scoopability 116
Scotch method 46
scraped-surface
    cooling 216, 217
    heat exchanger *see* heat exchanger
scraper blade 36, 37, 87
scroll 150
seasonality effect 220
seasoning 106
seaweed extract 256
secondary (2 position) hydroxyl 242
secondary oxidation product 51
sedimentation 74, 236, 240
    velocity 73
seed crystals 10
seeding 72, 142
seepage 297
selectivity 126, 128-31, 133, 135, 136, 139, 141
semisolids 13, 70, 333
sensory characteristics 275
separation 114
    efficiency 77, 147, 148, 151
    temperature 90
separator 74, 76-78, 90, 115
serum
    drain 79
    leakage 114, 117
    oil 200, 338, 345, 352, 355, 358
setting agent 228
shaft 36
    rotation speed 37, 87
shallow frying 201, 337, 338
shear 8, 114, 209, 258
    action 78
    characteristics 53
    force 10, 233, 269
    stress 11, 311
shelf life 30, 32, 45, 48, 61, 94, 96, 97, 100, 109, 111, 114, 119, 210, 265, 348, 349
shock chilling 48
short chain 15, 17
    fatty acid 12, 170, 300, 301
shortening 30, 33, 36-39, 41-44, 50-52, 60, 123, 124, 241, 244, 265, 267, 346, 352
shrink wrapping 202
shrinkage 10, 118
Shukoff method 161
shunt reaction 128
single phase 173
size distribution 11
sizing 209

skim fraction 101
skimming 73, 109
slow crystallisation 4
sludge 74
slurry 55
small-angle spectrum 14
smoothness 8
sn-2 position 15
snack food 66, 345
snap 8-10
soap 250
    solution 240
sodium 51, 83, 254
    alginate 257
    caseinate 81, 113, 262
    chloride 237
    dodecyl sulfate (SDS) 249, 250
    laurate 240
    lauryl sulfate 235, 249, 250
    stearate 233, 250
    steroyl fumarate 249
    steroyl lactylate (SSL) 54, 55, 241, 247,
        249, 250, 264, 265
    trimetaphosphate 270
    tripolyphosphate 258
softening 17, 19, 165, 182, 183, 208, 209,
    221
solid 38, 70
    fat 159, 235, 344
        content (SFC) 8, 9, 11, 13, 16, 19, 33,
            43, 52, 56, 63, 123, 128, 141, 161,
            224, 260
        index (SFI) 346
    film 244
    oil 333
    phase 267
    solution 6, 35, 36, 163
    surface 235, 237
solidification 1, 2, 9, 12, 15, 25, 64, 208
solid–liquid
    dispersion 228
    phase change 208
solid-to-liquid ratio 33, 36, 41, 43, 47, 48, 52
solid–water interface 235
solubilising agent 228
solubility 39, 116, 117, 130, 133, 140, 142
solution phase 4
solvent 144, 145, 151, 153, 166, 169
    extraction 160, 182
sorbic acid 212
sorbitan 240, 254
    ester 245

monolawate polyglycol ether 20
monostearate 240, 245, 246, 254
    ring 245
    tristearate 56, 269
sorbitol 48, 67, 240, 245
SOS 7, 8, 15, 159, 163, 166-68, 178-80, 185,
    187
soup 59, 70, 275
soya
    powder 80
    slurry 80
soybean 201, 212, 220
    oil 14, 34, 38, 55, 64, 106, 123-25, 131,
        137, 139, 141, 153, 200, 240, 241,
        250, 262, 333, 334, 338, 345, 346
spattering 338 *see also* antispattering agent
speciality oil 352, 354
spectra 14
spice 107, 236, 258, 269, 337
spoilage 53, 97
sponge goods 45
spoonable dressing 269, 270
spore 101, 102
    count 98-100
spray
    chilling 57
    cooling 57
    drying 57, 58
    fats 64, 66
spray-dried flavour 238
spreadability 8, 9, 13, 209, 234, 260, 269
spreadable fat 1, 8, 13, 152, 192-94, 212, 216,
    220
spreading coefficient 234
SSS 7
stabilisation 11, 42, 48, 49, 235, 238
stabiliser 79, 80, 83, 87, 102, 105, 116, 117,
    212, 228, 236, 250, 256
stabiliser–emulsifier system 79
stabilising agent 60
stability 8, 9, 20, 32, 105, 111, 124, 125, 142,
    246
    test 51
stable
    foam 45, 114, 240
    polymorph 7
stachyose 67
staling retardation 241
starch 45, 48, 50, 55, 58, 86, 105, 202, 212,
    213, 240, 258, 264, 269, 270
starch-complexing
    ability 84

agent 264
starch–water–protein matrix 55
starter
  bacteria 298
  culture 277, 347
  lactococci 298
stearate 242
stearic acid 20, 22, 126, 128, 129, 165, 240, 245, 264
stearoyl 2-lactylate 51
stearoyl lactylic acid 266, 267
sterene 178
stereo-specific number 5
steric hindrance 238, 239
sterile product 94
sterilisation 104, 112
sterol 196, 333, 344
sterol degradation product 178
stiffness 79, 261
stir-frying 338
Stokes's law 73, 236
storage 14, 56, 94, 114, 115, 222
  condition 94, 222
straight-chain
  aliphatic compound 264
  hydrocarbon molecule 50
stress, applied 47
stretchability 311, 313, 315
structure 30, 38
  lipid 67
submicelle 282
succinic anhydride 244
succinyl monoglyceride (SMG) 242-44, 249
succinylated monoglyceride 51
sucrose 48, 248
  ester 254, 266, 269
  fatty-acid oligoester (SOE) 21-25
  polyester 66, 358
sugar 9, 11, 81, 83, 207, 235, 241, 253, 265, 267, 346
  alcohol sorbitol 245
  batter 41
sulfate 230
  ester 257
sulfhydryl group 99
summer temperature 11
sunflower oil 34, 106, 125, 131, 135, 141, 143, 200, 262, 333, 338, 358
Supercoating 180, 181, 186
supercooling 36, 37, 147, 214
superglycerination 44
surface

appearance 313
charge 237
electron microscopy 116
energy 231
free energy 230, 231, 233
gloss 89
negative charge 237
spray 62
tension 103, 211, 228, 230-33, 235, 259, 266
surface-active
  agent 228
  material 236, 251, 270
surfactant 228-30, 232, 234, 235, 241, 264
  concentration 232
suspension see emulsion
sweet paste 38-40
sweetener 83
Swift's test 51
symmetric triacylglycerols 8
symmetrical-type fat see SOS fat
synchrotron radiation X-ray beam 7, 25
syneresis 299, 302

talcum powder 260
tallow 34, 35, 38, 40, 123, 152, 199, 200, 240, 333, 338
  mid-fraction 152
  olein 152, 192, 200
  stearin 152
taste 82, 194, 210, 275
temperature 9, 16, 20, 111, 134, 213, 214, 258, 260, 290, 295, 302, 338
  differential 37
  variation 8, 11
tempering 10, 177, 187
  machine 10
β-tending 14
β'-tending 23
ternary mixture 6
tert-butyl hydroquinone (TBHQ) 345
textural properties 1, 2, 13, 33, 36-41, 86, 276, 277, 309
texture 13, 48, 103, 107, 117, 180, 194, 209, 269, 275, 283, 332
Theobroma cacao 160
thermal
  degradation 340
  measurement 8
  stability 339
  treatment 2
thermoduric count 98-100

thermodynamic
    activity 232
    free energy 229
thermopeaks 18
thickener *see* stabiliser
thickness 105
thin-film transmission electron microscopy (TEM) 119
thixotropy 259, 269
three-bladed shaft 219
three-dimensional crystal network 8, 13
time-lapse photography 231
Tirtiaux plant 145
tocopherol 32, 125, 344
toffee 160, 184
tomato 107
toppings 60
tragacanth 257
trans
    fatty acid 1, 30, 33, 66, 124, 125, 131, 135, 169, 200, 351
    selectivity 128, 131
trans-fatty-acid-free oil blend 222
transformation 1-3, 6, 8, 10, 11, 14, 25
β'-β transformation 3, 14
trans-isomer 131, 133, 134, 138, 139, 152, 153, 242
transition metal 120
transmission electron microscopy (TEM) 288
transport 56
trehalose 67
triacylglycerol (TAG) 1-8, 12, 14-16, 19, 32, 351
trieleaidin 173
triene 124
triester 246
triglyceride 33-36, 39-41, 43, 72, 78, 119, 123, 154, 159, 166, 168, 245
    selectivity 126, 127
    liquid 52
    solid 52
tripalmitoyl glycerol 4
triple chain-length structure (TCL) 2, 3, 7, 14-17
triplex piston pump 219
tristearate 245, 246, 261
tristearin 34, 35
trivalent salt 237
troglycerol tristearate 254
tropical oil 31
truffle 185
tubular action 217

turbidity 8
turbulence 36
two-bladed shaft 219

ultra high temperature (UHT) treatment 91, 94, 95, 111, 112, 114, 115
ultrafiltration 82
ultrasonic
    velocity measurement 20, 21, 25
    velocity value 21-23
    wave 102
ultrasound irradiation 11
uniform distribution 102
unpasteurised milk 302
unsaturated
    emulsifier 262
    fatty acids 4, 5, 15, 17, 49, 107, 120, 124, 242, 276
    mixed acid triacylglycerol 6, 7
unsaturation, degree of 224
US Department of Agriculture (USDA) 207

vacreator 214
vacuum filtration 147, 151
valve design 103
van der Waal's forces 237
van Slyke formula 301
vanaspati 216, 332, 333, 350
vanilla extract 267
vapour 235
    pressure 259
vegetable 236, 269
    fat 71
    oil 43, 46, 64, 66, 79, 80, 106, 221, 236, 344
Vegetable Oil Products Control Order 351
Venturi tube 139
very-low-yellow fat emulsions 1
vinegar 106, 107, 257, 258, 270
virgin olive oil *see* olive oil
viscoelastic
    deformation 304
    properties 116, 304
viscosity 9, 11, 37, 50, 57, 84, 86, 107, 112, 113, 120, 144, 180, 208, 235, 236, 239, 240, 251, 256-58, 269, 313
viscous
    deformation 304
    gel 49
    solution 228
vitamins 66, 71, 83, 107, 194, 201, 202, 207, 224

deficiency 194
volatile degradation product 341
vol-au-vent 54
votator 201

walnut oil 353, 357
warehouse 11
water
  continuous emulsion 81, 89, 109
  droplet 13, 53, 210, 233
  exudation 257
  migration 256
  phase 11, 12, 42, 236, 239
  solubility 39, 84, 250
  treatment plant 236
  vapour 43
water-binding capacity 256, 257
water-in-oil (W/O) emulsion 9, 13, 43, 52,
    69, 84, 199, 233, 252, 259, 260, 271
water–oil interface 259
water-soluble material 241
wax 333
  crystals 144
  removal 141, 143
weight 36
wet foam 239, 240
wetting agent 228, 234
wheat
  bread 50
  gluten 48
  protein 39
  starch 50
whey 82, 202
  protein 87, 116, 117, 282, 302, 303
  solids 53
whipped topping 60, 246, 257, 261, 262
whipping 221, 234
  agent 228, 249
  air 45, 87, 89
  properties 9, 12, 114, 115
  rate 114
white powder 65
whitening 80, 81
wine making 244
winter temperature 11
winterisation 55, 107, 119, 141-43, 271, 333,
    345
wooden box 202
worker unit 37, 114
working temperature 43
wrapping 193, 202

xanthan gum 257, 269
x-ray
  analysis 17, 34
  diffraction 14, 18, 23, 25

yeast 48, 51, 53, 109, 241
yield
  characteristics 297
  strength 209
  stress 208
  value 208, 268, 269, 275

zero-calorie product 358
zero-fat spread 196
zinc
  distearate 233
  sulfate 237